HANDBOOK OF
SINC NUMERICAL
METHODS

CHAPMAN & HALL/CRC
Numerical Analysis and Scientific Computing

Aims and scope:
Scientific computing and numerical analysis provide invaluable tools for the sciences and engineering. This series aims to capture new developments and summarize state-of-the-art methods over the whole spectrum of these fields. It will include a broad range of textbooks, monographs, and handbooks. Volumes in theory, including discretisation techniques, numerical algorithms, multiscale techniques, parallel and distributed algorithms, as well as applications of these methods in multi-disciplinary fields, are welcome. The inclusion of concrete real-world examples is highly encouraged. This series is meant to appeal to students and researchers in mathematics, engineering, and computational science.

Proposals for the series should be submitted to one of the series editors above or directly to:
CRC Press, Taylor & Francis Group
4th, Floor, Albert House
1-4 Singer Street
London EC2A 4BQ
UK

Published Titles

Classical and Modern Numerical Analysis: Theory, Methods and Practice
Azmy S. Ackleh, Edward James Allen, Ralph Baker Kearfott,
and Padmanabhan Seshaiyer

A Concise Introduction to Image Processing using C++
Meiqing Wang and Choi-Hong Lai

Decomposition Methods for Differential Equations:
Theory and Applications
Juergen Geiser

Grid Resource Management: Toward Virtual and Services Compliant Grid
Computing
Frédéric Magoulès, Thi-Mai-Huong Nguyen, and Lei Yu

Fundamentals of Grid Computing: Theory, Algorithms and Technologies
Frédéric Magoulès

Handbook of Sinc Numerical Methods
Frank Stenger

Introduction to Grid Computing
Frédéric Magoulès, Jie Pan, Kiat-An Tan, and Abhinit Kumar

Mathematical Objects in C++: Computational Tools in a Unified Object-
Oriented Approach
Yair Shapira

Numerical Linear Approximation in C
Nabih N. Abdelmalek and William A. Malek

Numerical Techniques for Direct and Large-Eddy Simulations
Xi Jiang and Choi-Hong Lai

Parallel Algorithms
Henri Casanova, Arnaud Legrand, and Yves Robert

Parallel Iterative Algorithms: From Sequential to Grid Computing
Jacques M. Bahi, Sylvain Contassot-Vivier, and Raphael Couturier

HANDBOOK OF
SINC NUMERICAL METHODS

Frank Stenger
University of Utah
Salt Lake City, USA

CRC Press
Taylor & Francis Group
Boca Raton London New York

CRC Press is an imprint of the
Taylor & Francis Group, an **informa** business

CRC Press
Taylor & Francis Group
6000 Broken Sound Parkway NW, Suite 300
Boca Raton, FL 33487-2742

First issued in paperback 2017

© 2011 by Taylor and Francis Group, LLC
CRC Press is an imprint of Taylor & Francis Group, an Informa business

No claim to original U.S. Government works

ISBN-13: 978-1-4398-2158-9 (hbk)
ISBN-13: 978-1-138-11617-7 (pbk)

To Carol

Contents

Preface   **xv**

1 One Dimensional Sinc Theory   **1**
  1.1 Introduction and Summary   1
     1.1.1 Some Introductory Remarks   2
     1.1.2 Uses and Misuses of Sinc   5
  1.2 Sampling over the Real Line   7
     Problems for Section 1.2 11
  1.3 More General Sinc Approximation on $\mathbb{R}$ 18
     1.3.1 Infinite Term Sinc Approximation on $\mathbb{R}$ 18
     1.3.2 Finite Term Sinc Approximation on $\mathbb{R}$ 25
     Problems for Section 1.3 31
  1.4 Sinc, Wavelets, Trigonometric and Algebraic Polynomials and Quadratures 32
     1.4.1 A General Theorem 35
     1.4.2 Explicit Special Cases on $[0, 2\pi]$ 36
     1.4.3 Wavelets and Trigonometric Polynomials . . . 40
     1.4.4 Error of Approximation 42
     1.4.5 Algebraic Interpolation and Quadrature 48
     1.4.6 Wavelet Differentiation 60
     1.4.7 Wavelet Indefinite Integration 62
     1.4.8 Hilbert Transforms 63
     1.4.9 Discrete Fourier Transform 65
     Problems for Section 1.4 68
  1.5 Sinc Methods on Γ 70
     1.5.1 Sinc Approximation on a Finite Interval 70
     1.5.2 Sinc Spaces for Intervals and Arcs 72
     1.5.3 Important Explicit Transformations 79
     1.5.4 Interpolation on Γ 82

1.5.5   Sinc Approximation of Derivatives 87
1.5.6   Sinc Collocation 89
1.5.7   Sinc Quadrature 90
1.5.8   Sinc Indefinite Integration 92
1.5.9   Sinc Indefinite Convolution 93
1.5.10 Laplace Transform Inversion 100
1.5.11 More General $1 - d$ Convolutions 101
1.5.12 Hilbert and Cauchy Transforms 105
1.5.13 Analytic Continuation 113
1.5.14 Initial Value Problems 116
1.5.15 Wiener–Hopf Equations 118
        Problems for Section 1.5 120
1.6   Rational Approximation at Sinc Points 125
1.6.1   Rational Approximation in $\mathbf{M}_{\alpha,\beta,d}(\varphi)$ 126
1.6.2   Thiele–Like Algorithms 127
        Problems for Section 1.6 128
1.7   Polynomial Methods at Sinc Points 129
1.7.1   Sinc Polynomial Approximation on $(0, 1)$ . . . 130
1.7.2   Polynomial Approximation on Γ 133
1.7.3   Approximation of the Derivative on Γ 134
        Problems for Section 1.7 137

2 Sinc Convolution–BIE Methods for PDE & IE   139
2.1   Introduction and Summary 139
2.2   Some Properties of Green's Functions 141
2.2.1   Directional Derivatives 141
2.2.2   Integrals along Arcs 142
2.2.3   Surface Integrals 142
2.2.4   Some Green's Identities 143
        Problems for Section 2.2 150
2.3   Free–Space Green's Functions for PDE 150
2.3.1   Heat Problems 151
2.3.2   Wave Problems 151
2.3.3   Helmholtz Equations 152
2.3.4   Biharmonic Green's Functions 154
2.4   Laplace Transforms of Green's Functions 155
2.4.1   Transforms for Poisson Problems 158
2.4.2   Transforms for Helmholtz Equations 163
2.4.3   Transforms for Hyperbolic Problems 168
2.4.4   Wave Equation in $\mathbb{R}^3 \times (0, T)$ 169

2.4.5 Transforms for Parabolic Problems 172
2.4.6 Navier–Stokes Equations 173
2.4.7 Transforms for Biharmonic Green's Functions . 180
Problems for Section 2.4 184
2.5 Multi–Dimensional Convolution Based on Sinc 187
2.5.1 Rectangular Region in $2 - d$ 187
2.5.2 Rectangular Region in $3 - d$ 191
2.5.3 Curvilinear Region in $2 - d$ 192
2.5.4 Curvilinear Region in $3 - d$ 199
2.5.5 Boundary Integral Convolutions 207
Problems for Section 2.5 209
2.6 Theory of Separation of Variables 209
2.6.1 Regions and Function Spaces 210
2.6.2 Analyticity and Separation of Variables 222
Problems for Section 2.6 242

3 Explicit 1–d Program Solutions via Sinc–Pack   243
3.1 Introduction and Summary 243
3.2 Sinc Interpolation 245
3.2.1 Sinc Points Programs 246
3.2.2 Sinc Basis Programs 248
3.2.3 Interpolation and Approximation 251
3.2.4 Singular, Unbounded Functions 257
Problems for Section 3.2 257
3.3 Approximation of Derivatives 258
Problems for Section 3.3 261
3.4 Sinc Quadrature 262
Problems for Section 3.4 265
3.5 Sinc Indefinite Integration 266
Problems for Section 3.5 268
3.6 Sinc Indefinite Convolution 270
Problems for Section 3.6 274
3.7 Laplace Transform Inversion 275
Problems for Section 3.7 278
3.8 Hilbert and Cauchy Transforms 280
Problems for Section 3.8 283
3.9 Sinc Solution of ODE 284
3.9.1 Nonlinear ODE–IVP on $(0, T)$ via Picard . . . 285
3.9.2 Linear ODE–IVP on $(0, T)$ via Picard 286
3.9.3 Linear ODE–IVP on $(0, T)$ via Direct Solution 289

3.9.4   Second-Order Equations 292

3.9.5   Wiener–Hopf Equations 298

3.10 Wavelet Examples 300

    3.10.1 Wavelet Approximations 302

    3.10.2 Wavelet Sol'n of a Nonlinear ODE via Picard . 309

4 Explicit Program Solutions of PDE via Sinc–Pack   315

4.1   Introduction and Summary 315

4.2   Elliptic PDE 315

    4.2.1   Harmonic Sinc Approximation 315

    4.2.2   A Poisson-Dirichlet Problem over $\mathbb{R}^2$ 320

    4.2.3   A Poisson–Dirichlet Problem over a Square . . 323

    4.2.4   Neumann to a Dirichlet Problem on
            Lemniscate 332

    4.2.5   A Poisson Problem over a Curvilinear Region
            in $\mathbb{R}^2$ 338

    4.2.6   A Poisson Problem over $\mathbb{R}^3$ 350

4.3   Hyperbolic PDE 356

    4.3.1   Solving a Wave Equation Over $\mathbb{R}^3 \times (0,T)$ . . 356

    4.3.2   Solving Helmholtz Equation 364

4.4   Parabolic PDE 365

    4.4.1   A Nonlinear Population Density Problem . . . 365

    4.4.2   Navier–Stokes Example 383

4.5   Performance Comparisons 404

    4.5.1   The Problems 404

    4.5.2   The Comparisons 405

5 Directory of Programs   409

5.1   Wavelet Formulas 409

5.2   One Dimensional Sinc Programs 410

    5.2.1   Standard Sinc Transformations 410

    5.2.2   Sinc Points and Weights 411

    5.2.3   Interpolation at Sinc Points 412

    5.2.4   Derivative Matrices 413

    5.2.5   Quadrature 414

    5.2.6   Indefinite Integration 416

    5.2.7   Indefinite Convolution 416

    5.2.8   Laplace Transform Inversion 418

    5.2.9   Hilbert Transform Programs 418

    5.2.10 Cauchy Transform Programs 419

  5.2.11 Analytic Continuation 420

  5.2.12 Cauchy Transforms 421

  5.2.13 Initial Value Problems 421

  5.2.14 Wiener–Hopf Equations 422

5.3 Multi–Dimensional Laplace Transform Programs . . . 422

  5.3.1 Q – Function Program 422

  5.3.2 Transf. for Poisson Green's Functions 422

  5.3.3 Transforms of Helmholtz Green's Function . . . 423

  5.3.4 Transforms of Hyperbolic Green's Functions . . 423

  5.3.5 Transforms of Parabolic Green's Functions . . 424

  5.3.6 Transf. of Navier–Stokes Green's Functions . . 425

  5.3.7 Transforms of Biharmonic Green's Functions . 425

  5.3.8 Example Programs for PDE Solutions 426

Bibliography   **429**

Index   **461**

Preface

This handbook provides a package of methods based on Sinc approximation. It contains about 450 MATLAB® programs for approximating almost every type of operation stemming from calculus, as well as new and powerful methods for solving ordinary differential equations (ODE), partial differential equations (PDE), and integral equations (IE). This work reflects advances with Sinc the author has made since 1995.

The mathematics of Sinc methods is substantially different from the mathematics of classical numerical analysis. One of the aims for creating this handbook is to make Sinc methods accessible to users who may choose to bypass the complete theory behind why the methods work so well. This handbook nevertheless provides sufficient theoretical detail for those who do want a full working understanding of this exciting area of numerical analysis. The MATLAB® programs, presented in Chapters 3 and 4, apply to problems which range from elementary to complex ones having actual real world interest. The ability to apply Sinc methods to new engineering problems will depend to a large extent on a careful reading of these chapters.

Sinc methods are particularly adept at solving one dimensional problems: interpolation, indefinite and definite integration, definite and indefinite convolution, approximation of derivatives, polynomial and wavelet bases for carrying out Sinc approximation, approximation of Fourier and Laplace transforms and their inverse transforms, approximation of Hilbert transforms, analytic continuation, linear and nonlinear ODE initial value problems using both Sinc and wavelet collocation, and Wiener–Hopf problems.

In more than one dimension, this handbook derives and illustrates methods for solving linear and nonlinear PDE problems with emphasis on elliptic, hyperbolic and parabolic PDE, including Navier–Stokes equations, over bounded, unbounded and curvilinear regions. Underlying these methods and the basis for the so–called Sinc method of *separation of variables* is the ability to transform a multi-dimensional problem into a small number of one dimensional possibly nonlinear indefinite convolutions, which in turn can be efficiently computed through simple operations on small sized matrices.

We believe that beyond providing approximation and solution algorithms, Sinc represents an "ideal road map" for handling numeric problems in general, much like calculus represents a *way of life* for scientists in general. Using the mathematical tools developed in Chapters 1 and 2, Chapters 3 and 4 provide details which support this Sinc perspective.

Let me comment on the layout of the handbook.

Chapter 1 of the handbook gives a theoretical presentation of one dimensional Sinc methods. This differs somewhat from [S1], in that part of the present version is developed using Fourier transforms in place of analytic function methods. This is an easier approach for beginning students. Ultimately, however, analyticity first becomes necessary for the derivation of Sinc indefinite convolution in §1.5.9, and then, as Chapters 1 and 2 unfold, takes over and becomes the most mathematically expedient setting for developing our results. Indeed, this handbook supports the principle that the mathematics of numerical analysis is best carried out in the complex plane.

New procedures derived in Chapter 1 which do not appear in [S1] are: derivations of functional relations between Sinc series trigonometric and algebraic polynomials, thus establishing connections between known and new finite term quadratures; more explicit development of polynomial–like interpolation for approximating derivatives [S13]; formulas for Laplace transform inversion; formulas for evaluating Hilbert transforms; approximating nonlinear convolutions, such as

$$r(x) = \int_a^x k(x, x - t, t)\, dt$$

and advances in analytic continuation. We also derive easy-to-use periodic Sinc–wavelet formulas and expressions based on the connection between Sinc series and Fourier polynomials [SBCHP]. The final section of this chapter derives a family of composite polynomials that interpolate at Sinc points. The derivative of these interpolating polynomials enables accurate approximation of the derivative of the Sinc interpolated function.

Sinc methods in one dimension developed historically as an interplay between finding Sinc approximation algorithms for functions defined on finite and infinite intervals and identifying spaces $\mathbf{M}_{\alpha,\beta,d}(\varphi)$ best suited to carry out these algorithms. This proved to be a powerful, fruitful process. Previously open questions in approximation theory were tackled and new results achieving near optimal rates of convergence were obtained [S1]. When Sinc was then applied to solving PDE and IE problems, these same spaces, as we shall see, again proved to be the proper setting in which to carry out Sinc convolution integral algorithms, and as an ideal repository for solution to these problems.

Chapter 2 provides a new theory that combines *Sinc convolution* with the boundary integral equation approach for achieving accurate approximate solutions of a variety of elliptic, parabolic, and hyperbolic PDE through IE methods. Fundamental to this are derivations of closed form expressions of Laplace transforms for virtually all standard free–space Green's functions in $\mathbb{R}^d \times (0,\infty)$, $d = 1, 2, 3$. These multidimensional Laplace transforms enable accurate approximate evaluation of the convolution integrals of the IE, transforming each integral into a finite sum of repeated one dimensional integrals. We also derive strong intrinsic analyticity properties of solutions of PDE needed to apply the Sinc separation of variables methods over rectangular and curvilinear regions.

In general, we can achieve separation of variables whenever a PDE can be written in the form

$$\mathbf{L}\,u = f(\bar{r}\,,t\,,u\,,\nabla u)$$

with $\bar{r}$ in some region of $\mathbb{R}^d$, $d = 1, 2, 3$ and time t in $(0\,,\infty)\,,$ where $\mathbf{L}$ is one of the linear differential operators

$$\mathbf{L}\,u = \begin{cases} \Delta\,u\,, \\[2mm] \dfrac{\partial u}{\partial t} - \Delta\,u\,, \text{ or} \\[3mm] \dfrac{\partial^2 u}{\partial t^2} - \Delta\,u. \end{cases} \tag{1}$$

Analyticity, a natural result of calculus–like derivations, then allows us to solve (1) relying solely on one dimensional Sinc methods, thereby eliminating dealing with the large sized matrices typically associated with finite difference and finite element methods. We shall see how elementary assumptions on the way PDE are derived allow us to break away from restrictions postulated in Morse and Feshbach [MrFe], who assert 3–d Laplace and Helmholtz equations can be solved by separation of variables only in thirteen coordinate systems.

Chapter 3 provides examples and MATLAB® programs for the full repertoire of one dimensional problems.

Chapter 4 provides MATLAB® programs which illustrate Sinc solution of Poisson, wave and heat problems over rectangular and curvilinear regions, as well as solutions formed as analytic continuations.

Chapter 5 contains a complete list of of the MATLAB® programs of this handbook.

The advent of modern computers has strongly influenced how differential equations are modeled, understood, and numerically solved. In the past sixty years, finite difference, finite element and spectral methods are the most prominent procedures to have emerged. Sinc, having its origins as a spectral method, first established itself as a means to strengthen and unify classical and modern techniques for approximating functions. Since 1995 a deeper understanding has been developing how Sinc can perform superbly as a means for numerically solving partial differential equations. Popular numeric solvers, i.e., finite difference and finite elements, have severe limitations with regard to speed and accuracy, as well as intrinsic restrictions making them ill fit for realistic, less simplified engineering problems. Spectral methods also have substantial difficulties regarding complexity of implementation and accuracy of solution in the presence of singularities.

Research in Sinc methods is ongoing and vibrant. Hence this handbook is far from a complete story. All of the programs have been tested. It will be possible over time to make many of them more efficient, e.g. multidimensional algorithms are particularly amenable to parallel processing.

I shall be grateful to be informed of errors or misprints the reader may discover.

Acknowledgments. Knowledgeable readers of this handbook will be able to discern the influence of mathematicians who guided me throughout this work:

My friend, Philip Davis, whose encyclopedic command of mathematics has been such a strong influence;

My mentor, John McNamee, who introduced me to applications replete with beautiful mathematics;

My friend, Frank Olver, for his expertise in special functions, asymptotics, and use of complex variables;

My friend, Richard Varga, who has inspired me with numerous publications full of beautiful pure and applied mathematics;

My Ph.D. adviser, Eoin Whitney, who imbued me with his contagious love of the classical geometric theory of functions; and

My students, who demonstrated Sinc's theoretical and practical vitality.

I would like to acknowledge, in addition, that this handbook was made possible in part by the USAF contract # F33615–00–C–5004, and in part through the support of Richard Hall, who envisioned the use of these Sinc methods for the efficient solution of aircraft stress–strain problems.

Earlier programs based on this work were written in FORTRAN by Jim Schwing, Kris Sikorski, Michael O'Reilly, and Stenger. An ODE package [SGKOrPa] was written in *FORTRAN* by Sven Gustafson, Brian Keyes, Michael O'Reilly, Ken Parker, and Stenger, which solved ODE-initial value problems by means of a combination of Sinc and Newton's method. This *FORTRAN* version can be downloaded from *Netlib*. The present handbook uses either successive approximation, or direct solution of the collocated system of equations to solve such problems. Some of the solutions to PDE problems were originally written in MATLAB® by Ahmad Naghsh-Nilchi, Jenny Niebsch, and Ronny Ramlau. These have been changed in the present handbook for sake of consistency and simplicity of the programs. Excellent programming contributions have also been made by Michael Hohn and Ross Schmidtlein, and more recently by Baker Kearfott and Michael O'Reilly, although nearly all of the programs of this package were written by me.

Finally, my thanks go to Michael Freedman for his editing assistance.

For MATLAB® and Simulink® product information, please contact:

The MathWorks, Inc.
3 Apple Hill Drive
Natick, MA 01760-2098 USA
Tel: 508 647 7000
Fax: 508-647-7000
E-mail: info@mathworks.com
Web: www.mathworks.com

Frank Stenger
November 8, 2010

1

One Dimensional Sinc Theory

ABSTRACT   This chapter presents the theory of one dimensional Sinc methods. Our presentation differs somewhat from that in [S1], in that Fourier and Laplace transformations are used in place of complex variables to derive many of the identities. Whenever possible, references are given to programs of this handbook following each derived formula.

At the outset we present identities of band limited classes of functions, which can be obtained via use of Fourier series and Fourier integrals. In the next section we show that these identities are, in fact, accurate approximations of certain functions which are not band limited. We also introduce Fourier wavelets, and polynomials, which are a subset of Sinc series.

We then introduce spaces of functions in which the finite sums of these approximations converge rapidly, enabling accurate approximations with a relatively small number of evaluation points. These results are then extended to an arbitrary arc Γ using a special class of transformations, and then to various procedures of numerical approximation.

In the final section of Chapter 1, we derive composition polynomials for approximating data given at Sinc points.

1.1   Introduction and Summary

This chapter contains derivations of one dimensional Sinc approximations. Whereas previous derivations of these results were done using complex variables, the majority of the derivations in this section are done using Fourier transforms. We thus present derivations of Sinc approximation, over a finite interval (a, b), over a semi–infinite interval $(0, \infty)$, over the whole real line $(-\infty, \infty)$, and more generally over arcs in the complex plane. Sinc approximations are provided for the following types of operations:

   Interpolation;
   Approximation of derivatives;
   Quadrature;
   Indefinite integration;
   Indefinite convolution;

Fourier transforms;
Hilbert transforms;
Cauchy transforms;
Analytic continuation;
Laplace transforms, and their inversion;
Solutions of Volterra integral equations;
Solutions of Wiener Hopf integral equations; and
Solutions of ordinary differential equation initial value problems.

1.1.1 SOME INTRODUCTORY REMARKS

Let me make a few general remarks about Sinc methods.

The *B–spline*,

$$B_N(x) = \prod_{k=1}^{N} \left(1 - \frac{x^2}{k^2} \right) \tag{1.1.1}$$

becomes the "sinc function" sinc(x) in the limit as $N \to \infty$, where

$$\mathrm{sinc}(x) \equiv \frac{\sin(\pi x)}{\pi x} = \lim_{N \to \infty} \prod_{k=1}^{N} \left(1 - \frac{x^2}{k^2} \right). \tag{1.1.2}$$

Figure 1.1 is a graph of sinc($x - 1$).

For k an element of $\mathbb{Z}$, the set of integers, and h a positive number, it is notationally convenient to define the *Sinc* function $S(k, h)$ as follows,

$$S(k, h)(x) = \mathrm{sinc}\left(\frac{x}{h} - k \right). \tag{1.1.3}$$

Sinc functions do well as approximating functions. Let f be defined for all x on the real line, and form the sum

$$F_h(x) = \sum_{k \in \mathbb{Z}} f(k\,h)\, S(k, h)(x). \tag{1.1.4}$$

If convergent, this series is known as the "Whittaker Cardinal Function". My mentor, J.J. McNamee, called this function "a function of royal blood, whose distinguished properties separate it from its bourgeois brethren". It is replete with many identities, and it

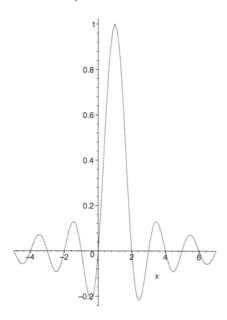

FIGURE 1.1.   sinc(x-1)

enables highly accurate approximation of smooth functions f defined on $\mathbb{R}$ such as $f(x) = 1/(1 + x^2)$, or $f(x) = 1/\cosh(x)$, and $f(x) = \exp(-x^2)$. Indeed, if the maximum difference for all $x \in \mathbb{R}$ between $f(x)$ and $F_h(x)$ is ε, then the maximum difference between $f(x)$ and $F_{h/2}(x)$ is less than ε^2. Thus replacement of h by $h/2$ leaves every second *sinc interpolation point* $k\,h$ unchanged and enables automatic checking of the accuracy of approximation. These desirable features of sinc approximation gave impetus for sinc wavelet approximation, the most used basis for wavelet approximation on $\mathbb{R}$.

The Fourier transform of $F_h(x)$ is just the well known discrete Fourier transform (DFT),

$$\widehat{F_h}(t) = \begin{cases} \displaystyle\sum_{n=-\infty}^{\infty} f(k\,h)\,e^{i\,k\,h\,t} & \text{if } |x| < \pi/h \\[2mm] 0 & \text{if } |x| > \pi/h. \end{cases} \qquad (1.1.5)$$

This enables the *fast Fourier transform* (FFT) approximation of the Fourier transform of f (see §1.4.10).

The above infinite expansion, $F_h(x)$, while accurate on $\mathbb{R}$, is not nearly as accurate when it is used for approximation on a finite interval. Fortunately, there exist certain transformations that enable us to preserve the two features of Sinc approximation – identities which yield simple to apply approximations – and very rapid convergence to zero of the error of approximation, for finite, and semi–infinite intervals, and even for arcs. See the references at the end of this text. For example, the transformation w and its inverse x

$$w \;=\; \varphi(x) = \log\left(\frac{x}{1-x}\right)$$

$$x \;=\; \varphi^{-1}(w) = \frac{e^w}{1+e^w}$$

(1.1.6)

yield a highly accurate approximation of any function g defined on $(0,1)$, provided that g is analytic in a region containing the interval $(0,1)$, of class **Lip**$_\alpha$ on $[0,1]$, and which vanishes at the end–points of $(0,1)$. This approximation takes the form

$$g(x) \approx \sum_{k=-M}^{N} g\left(\varphi^{-1}(k\,h)\right) S(k\,,h) \circ \varphi(x). \qquad (1.1.7)$$

Approximations (1.1.7) are termed *Sinc approximations*.

For example Fig. 1.2 graphs the *mapped* Sinc function, $S(1,1)(\varphi(x))$, where $\varphi(x) = \log(x/(1-x))$.

In this case, the Sinc points $z_k = \varphi^{-1}(k\,h) = e^{k\,h}/\left(1+e^{k\,h}\right)$ are all on the interval $(0,1)$, and "bunch up" at the end–points 0 and 1 of this interval. Furthermore, the approximation in (1.1.7) is accurate to within a uniform error, that converges to 0 at the rate

$$\exp\left(-C\,N^{1/2}\right),$$

with C a positive constant, even for badly behaved functions, such as $g(x) = x^{1/3}\left(1 - x^{1/2}\right)$, and even if the exact nature of the singularity of a function at an end–point of the interval is not explicitly known. This accuracy is preserved under the calculus operation differentiation and definite and indefinite integration applied to functions g on $(0,1)$.

The function g in (1.1.7) must vanish at the end–points of $(0,1)$ in order for us to obtain uniform accuracy of approximation. It is also possible for g to be of class **Lip**$_\alpha$ in a neighborhood of the end–point

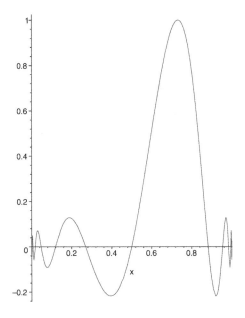

FIGURE 1.2. $\text{sinc}(\varphi(x) - 1)$

0 of $(0, 1)$, and of class $\mathbf{Lip}_\beta$ in a neighborhood of the end–point 1.
In §1.5.4 of this handbook we give an explicit simple modification of
the Sinc basis to the expression

$$g(x) \approx \sum_{k=-M}^{N} g(z_k)\,\omega_k(x)\,, \tag{1.1.8}$$

where the ω_k are defined in Definition 1.5.12. Expression (1.1.8) enables uniform accuracy of approximation on an interval or arc Γ for
all functions g that are bounded on the closure of Γ, analytic on the
open arc Γ, of class $\mathbf{Lip}_\alpha$ in a neighborhood of the end–point 0 of
Γ, and of class $\mathbf{Lip}_\beta$ in a neighborhood of the end–point 1.

1.1.2   Uses and Misuses of Sinc

Sinc methods require careful computations of function values near
singularities.

Example 1.1.1   – The accuracy of Sinc quadrature (i.e., numerical
integration) is illustrated in [S11], in the approximation of the
integral

$$I = \int_{-1}^{1} \frac{dx}{\sqrt{1 - x^2}}$$

by means of the Sinc quadrature formula

$$I \approx I_N = h \sum_{k=-N}^{N} \frac{w_k}{\sqrt{1 - x_k^2}},$$

where

$$h = \pi/\sqrt{N}, \quad w_k = \frac{2\,e^{kh}}{(1 + e^{kh})^2}, \quad x_k = \frac{e^{kh} - 1}{e^{kh} + 1}.$$

Points x_k and weights w_k were first evaluated and then substituted into the formula for I_N. This yielded only about 10 places of accuracy in double precision, the reason being that the x_k bunch up near the end-points of the interval. For example, near $x = 1$, x_k may have a decimal expansion of the form $x_k = .9999999978...$, leaving relatively few places of accuracy in the evaluation of $1 - x_k^2$. The problem can be circumvented by noting, in view of the above expression for x_k that $1 - x_k^2 = 4\,e^{kh}/(1 + e^{kh})^2$, an expression which can be computed in double precision to 16 places of accuracy, and which gives approximation of I to 16 places of accuracy.

– The series F_h of (1.1.4) enables uniform approximation to f on all of $\mathbb{R}$. Furthermore, the derivative of F_h yields a uniform approximation to f' on $\mathbb{R}$. Similarly, the series on the right of (1.1.7) yields a uniformly accurate approximation to g on $(0, 1)$. On the other hand the derivative of the series in (1.1.7) yields a uniformly accurate approximation to g' only on strict subintervals of $(0, 1)$. This is because when computing the derivative over an interval other than $\mathbb{R}$, the factor φ' appears, which becomes infinite at a finite end–point.

– Even when f has a bounded derivative on a closed interval with a finite end–point differentiation of the Sinc approximation does not yield a uniformly accurate approximation of the derivative in a neighborhood of the finite end–point. This problem is effectively dealt with in the §1.7.

– MATLAB® computes eigenvalues to within an absolute (not relative) error; this can result in larger relative errors associated
with the typically small eigenvalues of Sinc indefinite integration matrices in neighborhoods of finite end–points. These
eigenvalue errors yield known limited accuracy in conjunction
with Sinc–convolution methods of this handbook.

– The programs of this handbook are written for approximation
with the Sinc bases functions $\{\omega_k\}_{k=-M}^{N}$, derived in §1.5.4.
When $M = N$ and functions are analytic and of class $\mathbf{Lip}_\alpha$ we
achieve uniform convergence at the rate $O\left(\exp\left(-c'N^{1/2}\right)\right)$,
with c' a positive constant independent of N, this rate being
numerically achievable without having to determine α.

1.2 Sampling over the Real Line

Let $\mathbb{R}$ denote the real line $(-\infty, \infty)$, and let $\mathbb{C}$ denote the complex
plane, $\{(x, y) : x + iy : x \in \mathbb{R}, y \in \mathbb{R}\}$, with $i = \sqrt{-1}$.

Well known in engineering literature is the *sinc function* defined for
all real and complex z by

$$\text{sinc}(z) = \frac{\sin(\pi z)}{\pi z}. \tag{1.2.1}$$

For positive number h and an arbitrary integer k, we define *Sinc
function*

$$S(k, h)(x) = \text{sinc}\left(\frac{x}{h} - k\right). \tag{1.2.2}$$

Let $\mathbf{L}^2(a, b)$ be the family of all complex valued functions f defined
on the interval (a, b) such that $\|f\|^2 = \int_a^b |f(x)|^2 \, dx < \infty$. It is well
known that any function $G \in \mathbf{L}^2(-\pi/h, \pi/h)$ may be represented on
$(-\pi/h, \pi/h)$ by its *Fourier series*,

$$G(x) = \sum_{k \in \mathbb{Z}} c_k \, e^{ikhx} \tag{1.2.3}$$

with

$$c_k = \frac{h}{2\pi} \int_{-\pi/h}^{\pi/h} G(x)\, e^{-i\,k\,h\,x} dx\,, \qquad (1.2.4)$$

where the convergence is in the $\mathbf{L}^2$ norm $\|\cdot\|$.

It turns out that the Fourier series of the complex exponential is its Sinc expansion over $\mathbb{R}$.

Theorem 1.2.1 *Let $\zeta \in \mathbb{C}$, and let $\mathbb{Z}$ denote the set of all integers. Then*

$$e^{i\zeta x} = \sum_{n \in \mathbb{Z}} S(n,h)(\zeta)\, e^{i\,n\,h\,x} \qquad -\pi/h < x < \pi/h. \qquad (1.2.5)$$

Moreover, for any $m \in \mathbb{Z}$,

$$\sum_{n=-\infty}^{\infty} S(n,h)(\zeta)\, e^{i\,n\,h\,x}$$

$$= \begin{cases} e^{i\zeta(x-2\,m\,\pi/h)} & \text{if} \quad (2\,m-1)\,\pi/h < x < (2\,m+1)\,\pi/h \\ \cos(\pi\,\zeta/h) & \text{if} \quad x = (2\,m \pm 1)\,\pi/h. \end{cases}$$

$$(1.2.6)$$

Proof. Considered as a function of x, the function $G(x) = e^{i\zeta x}$ clearly belongs to $\mathbf{L}^2(-\pi/h, \pi/h)$, and therefore has a Fourier series expansion of the form (1.2.3) on the interval $(-\pi/h, \pi/h)$, with the c_k given by (1.2.4). Thus

$$c_k = \frac{h}{2\,\pi} \int_{-\pi/h}^{\pi/h} e^{ix(\zeta - k\,h)}\, dx = S(k,h)(\zeta). \qquad (1.2.7)$$

The identity (1.2.5) therefore follows.

However, the series sum on the right hand side of (1.2.5) represents a periodic function of x on the real line $\mathbb{R}$, with period $2\,\pi/h$. Thus we deduce the top line on the right hand side of (1.2.6). It should perhaps be emphasized that the Fourier series of the function $G(x) = e^{i\zeta x}$, which is identical to this function on $(-\pi/h, \pi/h)$, defines a new function $H(x)$, which is a periodic extension of G to all of $\mathbb{R}$. In fact, $H(x) = G(x)$ if $-\pi/h < x < \pi/h$, while if $|x| > \pi/h$, and

$x = \xi + 2\,m\,\pi/h$, with $-\pi/h < \xi < \pi/h$, then H is defined by $H(x) = H(\xi + 2\,m\,\pi/h) = H(\xi)$.

The bottom line of (1.2.5) describes the fact that the function H is discontinuous at $x = (2\,m+1)\,\pi/h$, and at such points, the Fourier series of h takes on its average value, i.e.,

$$H((2\,m+1)\,\pi/h)$$

$$= \lim_{t \to 0^+} \frac{G((2\,m+1)\,\pi/h - t) + G((2\,m+1)\,\pi/h + t)}{2}$$

$$= \frac{G(-\pi/h) + G(\pi/h)}{2}$$

$$= \cos(\pi\,\zeta\,/h).$$

∎

Analogous to (1.2.3), given function $f \in \mathbf{L}^2(\mathbb{R})$, the *Fourier transform*, $\tilde{f}$ of f, is defined by

$$\tilde{f}(x) = \int_{\mathbb{R}} f(t)\,e^{ixt}\,dt. \tag{1.2.8}$$

The function $\tilde{f}$ also belongs to $\mathbf{L}^2(\mathbb{R})$.

Given $\tilde{f}$ we can recover f via the *inverse transform* of $\tilde{f}$,

$$f(t) = \frac{1}{2\pi} \int_{\mathbb{R}} e^{-ixt}\,\tilde{f}(x)\,dx. \tag{1.2.9}$$

Starting with a function $F \in L^2(-\pi/h, \pi/h)$, let us define a new function $\tilde{f}$ on $\mathbb{R}$,

$$\tilde{f}(x) = \begin{cases} F(x) & \text{if } \ x \in (-\pi/h, \pi/h) \\[2mm] 0 & \text{if } \ x \in \mathbb{R},\ x \notin (-\pi/h, \pi/h). \end{cases} \tag{1.2.10}$$

(Here and henceforth we will not include values of Fourier series or Fourier transforms at end–points of an interval.)

Calculating the inverse Fourier transform of this function $\tilde{f}$, gives back

$$
\begin{aligned}
f(t) &= \frac{1}{2\pi} \int_{\mathbb{R}} e^{-ixt}\, \tilde{f}(x)\, dx \\[2ex]
&= \frac{1}{2\pi} \int_{-\pi/h}^{\pi/h} e^{-ixt}\, F(x)\, dx.
\end{aligned}
\tag{1.2.11}
$$

Every function f defined in this way is said to belong to the Wiener space $\mathbf{W}(\pi/h)$, *or equivalently, is said to be band limited.*

In particular, it follows from (1.2.10) and (1.2.11), that

$$
f(nh) = \frac{1}{2\pi} \int_{-\pi/h}^{\pi/h} e^{-inhx}\, \tilde{f}(x)\, dx.
\tag{1.2.12}
$$

By Theorem 1.2.1 we then get the *Cardinal function* representation of f,

$$
\begin{aligned}
f(t) &= \frac{1}{2\pi} \int_{-\pi/h}^{\pi/h} \tilde{f}(x) \sum_{n\in\mathbb{Z}} S(n,h)(t) e^{-inhx}\, dx \\[2ex]
&= \sum_{n\in\mathbb{Z}} f(nh)\, S(n,h)(t).
\end{aligned}
\tag{1.2.13}
$$

The *Cardinal function* representation of f is denoted by $C(f,h)$, and given for $t \in \mathbb{R}$ (and generally for any complex t) by

$$
C(f,h)(t) = \sum_{n\in\mathbb{Z}} f(nh)\, S(n,h)(t).
\tag{1.2.14}
$$

Sinc methods are based on the use of the Cardinal function. In this handbook, t in $S(k,h)(t)$ is frequently replaced by a transformation, $\varphi(t)$. This allows Cardinal function approximation over intervals other than the real line.

It turns out that the set of all functions f which satisfy $f = C(f,h)$ is precisely the set of band limited functions $\mathbf{W}(\pi/h)$.

The identities of the following exercises are important, and will be used in the sequel. These can be derived applying standard calculus operations or Fourier series and Fourier transform formulas to the Cardinal function. See [Zy, RaSe, Sh, S1] for help with these problems.

PROBLEMS FOR SECTION 1.2

1.2.1 Deduce, using the expansion (1.2.5), that for all $x \in \mathbb{C}$,

$$\frac{1}{2\pi} \int_{-\pi/h}^{\pi/h} h\, e^{ikht - ixt}\, dt = S(k,h)(x)\,;$$

1.2.2 Use *Parseval's theorem*, which states that if $\tilde{f}$ and $\tilde{g}$ are Fourier transforms of f and g (resp.), then

$$\int_{\mathbb{R}} f(t)\, \overline{g(t)}\, dt = \frac{1}{2\pi} \int_{\mathbb{R}} \tilde{f}(x)\, \overline{\tilde{g}(x)}\, dx\,,$$

where e.g., $\overline{g(t)}$ denotes the complex conjugate of $g(t)$, to deduce that the functions in the sequence of Sinc functions $\{S(k,h)\}_{-\infty}^{\infty}$ are "continuous" orthogonal, i.e., $\forall k$ and $\ell \in \mathbb{Z}$,

$$\int_{\mathbb{R}} S(k,h)(x) S(\ell,h)(x)\, dx = \frac{h^2}{2\pi} \int_{-\pi/h}^{\pi/h} e^{i(k-\ell)ht}\, dt = h\, \delta_{k-\ell}\,,$$

where $\delta_{k-\ell}$ is defined in Problem 1.2.6 (i) below.

1.2.3 Deduce that the functions of the sequence $\{S(k,h)\}_{-\infty}^{\infty}$ are "discrete" orthogonal, i.e., $\forall k$ and $\ell \in \mathbb{Z}$,

$$S(k,h)(\ell h) = \begin{cases} 1 & \text{if } k = \ell \\ 0 & \text{if } k \neq \ell \end{cases}\,;$$

1.2.4 Prove that if $\Im z > 0$, then

$$\frac{1}{2\pi i} \int_{\mathbb{R}} \frac{S(k,h)(t)}{t - z}\, dt = \frac{e^{ix(z-kh)/h} - 1}{2i\pi(z - kh)/h}\,, \quad \Im z > 0\,,$$

$$\rightarrow S(k,h)(x) \quad \text{as } z = x + iy \rightarrow x \in \mathbb{R}\,.$$

1.2.5 Parseval's Theorem states that if $\tilde{f}$ denotes the Fourier transform of f, and if $\tilde{g}$ denotes the Fourier transform of g, then

$$\int_{\mathbb{R}} f(x)\,\overline{g(x)}\,dx = \frac{1}{2\pi}\int_{\mathbb{R}} \tilde{f}(x)\,\overline{\tilde{g}(x)}\,dx.$$

Deduce using Parseval's theorem that for $z = x + iy$, with $x \in \mathbb{R}$ and $y \in \mathbb{R}$,

$$\Im\frac{1}{\pi i}\int_{\mathbb{R}} \frac{S(k,h)(t)}{t - x - iy}\,dt = \Im\left\{\frac{e^{i\pi(x-kh)/h} - 1}{i\pi(z - kh)/h}\right\}$$

$$\rightarrow \frac{P.V.}{\pi}\int_{\mathbb{R}} \frac{S(k,h)(t)}{t - x}\,dt = \frac{\cos[\pi(x - kh)/h] - 1}{\pi(x - kh)/h}$$

where "P.V." denotes the *principal value* of the integral, i.e., using the above notation, and letting $\mathcal{S}\,f$ denote the Hilbert transform, we have

$$(\mathcal{S}\,\tilde{f})(x) = \frac{P.V.}{\pi i}\int_{\mathbb{R}} \frac{\tilde{f}(t)}{t - x}\,dt = \int_{\mathbb{R}} e^{i\,x\,t}\,\mathrm{sgn}(t)\,f(t)\,dt.$$

with

$$\mathrm{sgn}(t) = \left\{\begin{array}{rl} 1 & \text{if } t > 0 \\ 0 & \text{if } t = 0 \\ -1 & \text{if } t < 0 \end{array}\right.$$

1.2.6 Let $f \in \mathbf{W}(\pi/h)$. Prove that:

(a) For all $x \in \mathbb{R}$,

$$\tilde{f}(x) = \int_{\mathbb{R}} f(t)\,e^{ixt}\,dt$$

$$= \left\{\begin{array}{ll} h\displaystyle\sum_{k=-\infty}^{\infty} f(kh)\,e^{ikhx}, & \text{if } |x| < \pi/h; \\ 0, & \text{if } |x| > \pi/h. \end{array}\right.$$

(b) f may be "reproduced" for any $t \in \mathbb{C}$ by the formula

$$f(t) = \frac{1}{h} \int_{\mathbb{R}} f(x) \frac{\sin\left(\frac{\pi}{h}(t-x)\right)}{\frac{\pi}{h}(t-x)} \, dx.$$

(c) f satisfies the relation

$$\int_{\mathbb{R}} |f(t)|^2 \, dt = h \sum_{n=-\infty}^{\infty} |f(nh)|^2.$$

(d) Let $u_{xx} + u_{yy} = 0$ in $\Omega_+ = \{(x,y) : x \in \mathbb{R}, y > 0\}$, and let $u(x, 0^+) = f(x)$. Then, with $z = x + iy$,

$$u(x,y) = \sum_{k \in \mathbb{Z}} f(kh) \, \Re \left\{ \frac{e^{i\pi(z-kh)/h} - 1}{i\pi(z-kh)/h} \right\}, \quad \Im z > 0.$$

$$(1.2.15)$$

(e) Let $u_{xx} + u_{yy} = 0$ for $y = \Im z > 0$, and let $u(x, 0^+) = \mathcal{S}f(x)$, with $\mathcal{S}f$ denoting the Hilbert transform,

$$\mathcal{S}f(x) = \frac{P.V.}{\pi i} \int_{\mathbb{R}} \frac{f(t)}{t - x} \, dt, \quad x \in \mathbb{R}.$$

Then, $\forall \, \Im z > 0$,

$$u(x,y) = \sum_{k=-\infty}^{\infty} f(kh) \, \Im \left\{ \frac{e^{i\pi(z-kh)/h} - 1}{i\pi(z-kh)/h} \right\}.$$

(f) Set $\{\mathcal{S}\}S(k, h) = T(k, h)$, and prove that

$$T(k,h)(x) = \frac{\cos[\pi(x - jh)/h] - 1}{\pi i(x - jh)/h}.$$

(g) Let $T(k, h)$ be defined as in (f) above. Prove that

$$\sup_{x \in \mathbb{R}} |T(k,h)(x)| = 1 \quad \text{and} \quad \sum_{k \in \mathbb{Z}} |T(k,h)(x)|^2 = 1.$$

(h) Prove that if $f \in \mathbf{W}(\pi/h)$, and if $\mathcal{S}f$ denotes the Hilbert transform of f as defined in part (e), then

$$\mathcal{S}(f)(x) = \sum_{k \in \mathbb{Z}} f(kh) \, T(k,h)(x), \quad x \in \mathbb{R}.$$

(i) If $f \in \mathbf{W}(\pi/h)$, then $f' \in \mathbf{W}(\pi/h)$. Hence, if $m \in \mathbb{Z}$, $n \in \mathbb{Z}$, and if $\delta_{m-n}^{(k)}$ is defined by

$$\delta_{m-n}^{(k)} = \left(\frac{d}{dx}\right)^k S(n,1)(x)|_{x=m}, \quad k = 0,1,2,\cdots, \quad (1.2.16)$$

then

$$f^{(k)}(nh) = h^{-k} \sum_{n=-\infty}^{\infty} \delta_{m-n}^{(k)} f(n\,h).$$

Notice that this definition differs from that in [S1], inasmuch as the present definition is more convenient for our purposes. This difference should not matter appreciably, since the method of solution of PDE in this handbook differs from that used in [S1].

In particular,

$$\delta_{m-n}^{(0)} = \begin{cases} 1 & \text{if } m = n \\ 0 & \text{if } m \neq n, \end{cases}$$

$$\delta_{m-n}^{(1)} = \begin{cases} 0 & \text{if } m = n \\ \dfrac{(-1)^{m-n}}{m-n} & \text{if } m \neq n, \end{cases}$$

$$\delta_{n-m}^{(2)} = \begin{cases} -\pi^2/3 & \text{if } m = n \\ \dfrac{-2(-1)^{m-n}}{(m-n)^2} & \text{if } m \neq n. \end{cases}$$

This handbook contains programs which construct square matrices that have $\delta_{m-n}^{(k)}$ as their $(m,n)^{th}$ entry.

The constants $\delta_{m-n}^{(k)}$ enable us to write

$$f^{(k)}(x) = h^{-k} \sum_{n=-\infty}^{\infty} \left\{ \sum_{m=-\infty}^{\infty} f(mh)\, \delta_{m-n}^{(k)} \right\} S(n,h)(x),$$

an identity if $f \in \mathbf{W}(\pi/h)$.

(j) If $g(x) = \int_{-\infty}^{x} f(t)\, dt$ exists and is uniformly bounded for all $x \in \mathbb{R}$, then for all $k = 0, \pm 1, \pm 2, \cdots$,

$$g(k\,h) = \int_{-\infty}^{kh} f(t)\, dt = h \sum_{\ell=-\infty}^{\infty} \delta_{k-\ell}^{(-1)} f(\ell h)$$

with

$$\delta_k^{(-1)} = \frac{1}{2} + \sigma_k,$$

$$\sigma_k = \int_0^k \frac{\sin(\pi t)}{\pi t}\, dt.$$

Moreover, if $g \in \mathbf{L}^2(\mathbb{R})$, then for all $x \in \mathbb{C}$,

$$\int_{-\infty}^{x} f(t)\, dt = h \sum_{k=-\infty}^{\infty} \left\{ \sum_{\ell=-\infty}^{\infty} \sigma_{k-\ell} f(\ell h) \right\} S(k, h)(x).$$

1.2.7* Verify the validity of the identity

$$\sum_{n=0}^{\infty} \left\{ \binom{n+1-x/h}{2n+1} \delta^{2n} g(0) + \binom{n+x/h}{2n+1} \delta^{2n} g(h) \right\}$$

$$= \sum_{n=-\infty}^{\infty} g(nh) S(n, h)(x)$$

in the case

$$|g(0)| + |g(h)| + |g(-h)| + \sum_{n=2}^{\infty} [|g(nh)| + |g(-nh)|] \cdot \frac{\log n}{n} < \infty.$$

Here

$$\delta g(x) = g(x + h/2) - g(x - h/2)$$

$$\delta^2 g(x) = \delta\{\delta g(x)\}, \quad \text{etc.}$$

1.2.8 Prove that if $a \in \mathbb{C}$, $z \in \mathbb{C}$, then for $h > 0$,

$$\sum_{n=-\infty}^{\infty} \frac{1}{a^2 + n^2 h^2} S(n, h)(z) = \frac{1}{a^2 + z^2} \left[1 + \frac{z \sin(\pi x/h)}{a \sinh(\pi a/h)} \right].$$

1.2.9 Let Γ denote the Gamma function, which is defined for $\Re z > 0$ by $\Gamma(z) = \int_0^\infty t^{z-1} e^{-t}\, dt$, and which may be continued analytically to all of $\mathbb{C}$ via the formula $\Gamma(z + 1) = z\Gamma(z)$. Use the identity $1/\Gamma(z) = \Gamma(1 - z) \sin(\pi z)/\pi$ to deduce the formula

$$(*) \qquad \frac{1}{\Gamma(z)} = \sum_{n=1}^{\infty} \frac{S(n, 1)(z)}{(n-1)!} + \frac{\sin(\pi z)}{\pi} \int_1^\infty e^{-t} t^{-z}\, dt,$$

which is valid for all $z \in \mathbb{C}$. Discuss the error in approximating $1/\Gamma(z)$ by the sum ($\sum_{n=1}^{\infty}$) on the right–hand side of $(*)$.

1.2.10 Show that if $\theta \in \mathbb{R}$, $0 < \alpha < 2\pi$, then

$$\sum_{n=-\infty}^{\infty} \left(\frac{\sin(n\,\alpha + \theta)}{n\,\alpha + \theta} \right)^2 = \frac{\pi}{\alpha}.$$

1.2.11 (a) Deduce, using Theorem 1.2.1, that if $x \in \mathbb{R}$, then

$$\sup_{t \in \mathbb{R}} |S(\ell, h)(t)| = \sum_{k \in \mathbb{Z}} (S(k, h)(x))^2 = 1\,;$$

(b) Prove that for any positive integer N,

$$(*) \qquad \|C_N\| \equiv \sup_{x \in \mathbb{R}} \sum_{k=-N}^{N} |S(k, h)(x)| \leq \frac{2}{\pi} [3/2 + \gamma + \log(N + 1)]$$

with γ denoting Euler's constant,

$$\gamma = \lim_{N \to \infty} 1 + \frac{1}{2} + \frac{1}{3} + \cdots + \frac{1}{N-1} - \log(N) = .577 \ldots.$$

[Hint: Prove first that the sum $(*)$ is even in x; next, in the case of $0 \leq x \leq h$, prove that

$$\max_{0 \le x \le h} \frac{\sin(\pi x/h)}{\pi} \left\{ \frac{1}{x/h} + \frac{1}{1 - x/h} \right\} = \frac{4}{\pi}.$$

On the interval $0 \le x \le h$, prove that for the remaining sum,

$$b_N(x) \equiv$$

$$\frac{\sin(\pi x/h)}{\pi} \left\{ \left[\frac{1}{1 + x/h} + \frac{1}{2 - x/h} \right] \right.$$
$$+ \left[\frac{1}{2 + x/h} + \frac{1}{3 - x/h} \right] + \cdots$$
$$+ \left. \left[\frac{1}{N - 1 + x/h} + \frac{1}{N - x/h} \right] + \frac{1}{N + x/h} \right\},$$

each of the terms in square brackets has the same maximum value at $x = 0$ and at $x = h$, and hence

$$|b_N(x)| \le b_N \equiv \frac{2}{\pi} \left\{ -\frac{1}{2} + \sum_{k=1}^{N} \frac{1}{k} \right\}$$

$$\le \frac{2}{\pi} \{ \gamma + \log(N + 1) - \frac{1}{2} \}.$$

Finally, show that if $k > 0$, and if $k\,h \le x \le (k+1)\,h$, then the sum in $(*)$ is less than it is for $0 \le x \le h$.]

1.2.12 Let $\zeta \in \mathbb{C}, \quad x \in \mathbb{R}$

(a) Prove that

$$e^{i x \zeta} = \sum_{k=-\infty}^{\infty} [1 + (i\,x)\,(\zeta - n\,h)]\, S^2(n, h)(\zeta)\, e^{i n h x}$$

$$(1.2.17)$$

(b) Prove that if for some $G \in \mathbf{L}^2(-2\,\pi/h\,, 2\,\pi/h)$ we have

$$f(\zeta) = \frac{1}{2\,\pi} \int_{-2\pi/h}^{2\pi/h} e^{-i x \zeta}\, G(x)\, dx, \qquad (1.2.18)$$

then

$$f(\zeta) = \sum_{n \in \mathbb{Z}} \left[f(n\,h) + (\zeta - n\,h)\, f'(n\,h) \right] S^2(n,h)(\zeta).$$

$$(1.2.19)$$

1.3   More General Sinc Approximation on $\mathbb{R}$

When no longer being an exact representation for all functions, $C(f, h)$ provides a highly accurate approximation on $\mathbb{R}$ for functions f whose Fourier transforms decay rapidly. We consider in this section the approximation of functions f that are not band limited. At the outset we consider approximation of those functions f whose Fourier transform $\tilde{f}$ have a decay rate on $\mathbb{R}$, of the form $\tilde{f}(y) = \mathcal{O}(\exp(-d|y|))$, for d a positive number. Later we consider the use of a truncated form of $C(f, h)$ to achieve an accurate approximation of functions for which $\tilde{f}(y) = \mathcal{O}(\exp(-d|y|))$ and for which $f(x) = \mathcal{O}(\exp(-\alpha|x|))$, with α a positive number.

1.3.1   Infinite Term Sinc Approximation on $\mathbb{R}$

An important result relating Fourier transforms and Fourier series is *Poisson's Summation formula.*

Theorem 1.3.1 Poisson Summation Formula. *Let $f \in \mathbf{L}^2(\mathbb{R})$, and let f and its Fourier transform $\tilde{f}$ for t and x on $\mathbb{R}$, satisfy the conditions*

$$
\begin{aligned}
f(x) &= \lim_{t \to 0^+} \frac{f(x - t) + f(x + t)}{2} \\[2mm]
\tilde{f}(x) &= \lim_{t \to 0^+} \frac{\tilde{f}(x - t) + \tilde{f}(x + t)}{2}.
\end{aligned}
$$

$$(1.3.1)$$

Then, for all $h > 0$,

$$h \sum_{m \in \mathbb{Z}} f(m\,h)\, e^{i\,m\,h\,x} = \sum_{n \in \mathbb{Z}} \tilde{f}\left(\frac{2\,n\,\pi}{h} + x \right).$$

$$(1.3.2)$$

Proof.   Set

$$F(x) = \sum_{n \in \mathbb{Z}} \tilde{f}\left(\frac{2n\pi}{h} + x\right). \tag{1.3.3}$$

Then F is evidently periodic, with period $2\pi/h$, so that

$$F(x) = \sum_{m \in \mathbb{Z}} c_m\, e^{imhx}, \quad x \in \mathbb{R}, \tag{1.3.4}$$

with

$$
\begin{aligned}
c_m &= \frac{h}{2\pi} \int_{-\pi/h}^{\pi/h} F(x)\, e^{-imhx}\, dx \\[2mm]
&= \frac{h}{2\pi} \int_{-\pi/h}^{\pi/h} \sum_{n \in \mathbb{Z}} \tilde{f}\left(\frac{2n\pi}{h} + x\right) e^{-imhx}\, dx \\[2mm]
&= \frac{h}{2\pi} \sum_{n \in \mathbb{Z}} \int_{(2n-1)\pi/h}^{(2n+1)\pi/h} \tilde{f}(x)\, e^{-imhx}\, dx \\[2mm]
&= \frac{h}{2\pi} \int_{\mathbb{R}} \tilde{f}(x)\, e^{-imhx}\, dx \\[2mm]
&= h\, f(mh),
\end{aligned}
\tag{1.3.5}
$$

which proves (1.3.2).

In view of (1.3.1), and since the Fourier series of a piecewise smooth function takes on its average value at a point of discontinuity, the relations in (1.3.5) hold even if f and $\tilde{f}$ have jump discontinuities. ∎

Let us use the notation $\|f\|$ to denote $\sup_{x \in \mathbb{R}} |f(x)|$.

Theorem 1.3.2 *Let f be defined on $\mathbb{R}$, and let Fourier transform $\tilde{f}$, be such that for some positive constant d,*

$$|\tilde{f}(y)| = \mathcal{O}\left(e^{-d|y|}\right) \quad y \to \pm\infty. \tag{1.3.6}$$

For $0 < d' < d$, and $x \in \mathbb{R}$, set

$$\tilde{C}(f,h)(x) = \begin{cases} h \sum_{k \in \mathbb{Z}} f(kh)\, e^{ikhx} & \text{if} \quad |x| < \pi/h \\[2mm] 0 & \text{if} \quad |x| > \pi/h. \end{cases} \tag{1.3.7}$$

Then, as $h \to 0$,

$$\left\| f - \sum_{k \in \mathbb{Z}} S(k,h)\, f(k\,h) \right\| = \mathcal{O}\left(e^{-\frac{\pi d}{h}}\right); \tag{1.3.8}$$

$$\int_{\mathbb{R}} f(t)\, dt - h \sum_{k \in \mathbb{Z}} f(kh) = \mathcal{O}\left(e^{-\frac{2\pi d}{h}}\right); \tag{1.3.9}$$

$$\left\| \tilde{f} - \tilde{C}(f,h) \right\| = \mathcal{O}\left(e^{-\frac{\pi d}{h}}\right); \tag{1.3.10}$$

$$h^j \left| f^{(j)}(\ell h) - h^{-j} \sum_{k \in \mathbb{Z}} \delta^{(j)}_{\ell-k}\, f(kh) \right| = \mathcal{O}\left(e^{-\frac{\pi d}{h}}\right); \tag{1.3.11}$$

$$\left| \int_{-\infty}^{\ell h} f(t)\, dt - h \sum_{k \in \mathbb{Z}} \delta^{(-1)}_{\ell-k}\, f(kh) \right| = \mathcal{O}\left(e^{-\frac{\pi d}{h}}\right); \tag{1.3.12}$$

$$\left\| \frac{P.V.}{\pi} \int_{\mathbb{R}} \frac{f(t)}{t-x}\, dt - \sum_{k \in \mathbb{Z}} \frac{\cos[\pi(x-kh)/h] - 1}{\pi(x-kh)/h}\, f(kh) \right\| = \mathcal{O}\left(e^{-\frac{\pi d}{h}}\right); \tag{1.3.13}$$

$$\frac{1}{h} \int_{\mathbb{R}} f(t)\, S(j,h)(t)\, dt - f(jh) = \mathcal{O}\left(e^{-\frac{\pi d}{h}}\right). \tag{1.3.14}$$

where in (1.3.11) and (1.3.12),

$$\begin{aligned} \delta^{(m)}_{\ell} &= \left. \left(\frac{d}{dx}\right)^m \operatorname{sinc}(x) \right|_{x=\ell} \\[2mm] \delta^{(-1)}_{\ell} &= \int_{-\infty}^{\ell} \operatorname{sinc}(x)\, dx. \end{aligned} \tag{1.3.15}$$

Remark. We may note, that if h is replaced by $h/2$ in these formulas, then they yield (roughly) twice as many significant figures of accuracy. Thus, when using these formulas, we can have the computer determine when a particular accuracy has been achieved. For, when the difference between a Sinc approximation with step size h and one with step size $h/2$ differs by $\varepsilon^{1/2}$, then the actual error in the $h/2$ approximation is of the order of ε. For example, if $I(f) = \int_{\mathbb{R}} f(t)\, dt$, and $T(f,h) = h \sum_{n \in \mathbb{Z}} f(nh)$, then we know from (1.3.9) that $I(f) - T(f,h) = \mathcal{O}\left(\varepsilon^{1/2}\right)$, and $I(f) - T(f,h/2) = \mathcal{O}(\varepsilon)$. Thus, if $T(f,h/2) - T(f,h) = (I(f) - T(f,h)) - (I(f) - T(f,h/2)) = \mathcal{O}\left(\varepsilon^{1/2}\right)$, then $I(f) - T(f,h/2) = \mathcal{O}(\varepsilon)$.

We also remark, that the rates of convergence appearing in Theorem 1.3.2 hold uniformly with respect to h, over finite intervals $(0, h_0)$, $h_0 > 0$. Therefore in each equation, "$= \mathcal{O}\left(e^{-\pi d/h}\right)$, $h \to 0$" can be replaced by an inequality "$\leq C\, e^{-\pi d/h}$, $h \in (0, h_0)$", for C some constant independent of h.

   Proof of (1.3.8): By Theorem 1.2.1 we have

$$f(t) - \sum_{n \in \mathbb{Z}} f(nh)\, S(n,h)(t)$$

$$= \frac{1}{2\pi} \int_{\mathbb{R}} \tilde{f}(x) \left[e^{-ixt} - \sum_{n \in \mathbb{Z}} e^{-inhx}\, S(n,h)(t) \right] dx$$

$$= \frac{1}{2\pi} \int_{|x| > \pi/h} \tilde{f}(x) \left[e^{-ixt} - \sum_{n \in \mathbb{Z}} e^{-inhx}\, S(n,h)(t) \right] dx.$$

$$(1.3.16)$$

But it follows from (1.2.6) that if $x \in \mathbb{R}$ and $|x| > \pi/h$ then

$$\sum_{n \in \mathbb{Z}} e^{inhx}\, S(n,h)(t) = e^{i\xi t}$$

for some $\xi \in \mathbb{R}$, so that this sum is bounded by 1 on $\mathbb{R}$. The first term e^{-ixt} inside the square brackets of the last equation in (1.3.16) is also bounded by 1 on $\mathbb{R}$. That is, the term in square brackets in the last equation of (1.3.16) is bounded by 2. Furthermore, under our assumption on $\tilde{f}$, there exists a constant C such that $|\tilde{f}(x)| \leq C \exp(-d\,|x|)$ for all $x \in \mathbb{R}$. Therefore, we have

$$\sup_{t \in \mathbb{R}} |f(t) - C(f, h)(t)| \le \frac{4C}{2\pi} \int_{\pi/h}^{\infty} e^{-dx} \, dx = \frac{4C}{2\pi d} e^{-\frac{\pi d}{h}}. \quad (1.3.17)$$

Proof of (1.3.9): Taking $x = 0$ in the Poisson Summation Formula expression (1.3.2), we have

$$\int_{\mathbb{R}} f(t) \, dt - h \sum_{m \in \mathbb{Z}} f(mh)$$

$$= \tilde{f}(0) - h \sum_{m \in \mathbb{Z}} f(mh) \qquad (1.3.18)$$

$$= - \sum_{n \in \mathbb{Z}, n \ne 0} \tilde{f}\left(\frac{2n\pi d}{h}\right).$$

That is, the difference in (1.3.9) is bounded as follows:

$$\left| \sum_{n \in \mathbb{Z}, n \ne 0} \tilde{f}\left(\frac{2n\pi}{h}\right) \right|$$

$$\le 2C \sum_{n=1}^{\infty} \exp\left(\frac{-2n\pi d}{h}\right) \qquad (1.3.19)$$

$$= \frac{2C e^{-\frac{2\pi d}{h}}}{1 - e^{-\frac{2\pi d}{h}}}$$

$$= \mathcal{O}\left(e^{-\frac{2\pi d}{h}}\right), \quad h \to 0.$$

Proof of (1.3.10): If $x \in (-\pi/h, \pi/h)$, then using the Poisson Summation Formula

$$\int_{\mathbb{R}} f(t) e^{ixt} \, dt - \tilde{C}(f, h)(x)$$

$$= \tilde{f}(x) - h \sum_{m \in \mathbb{Z}} f(mh) e^{ixt} \qquad (1.3.20)$$

$$= - \sum_{n \in \mathbb{Z}, n \ne 0} \tilde{f}\left(\frac{2n\pi}{h} + x\right).$$

That is, the difference (1.3.10) is bounded by

$$\sup_{|x|<\pi/h} \left| \sum_{n\in\mathbb{Z},n\neq 0} \tilde{f}\left(\frac{2n\pi}{h}+x\right) \right|$$

$$\leq 2C \sum_{n=1}^{\infty} \exp\left(-(2n-1)\frac{\pi d}{h}\right) \tag{1.3.21}$$

$$= \frac{2Ce^{-\pi d/h}}{1-e^{-2\pi d/h}}$$

$$= \mathcal{O}\left(e^{-\pi d/h}\right), \quad h \to 0.$$

On the other hand, if $x \notin [-\pi/h, \pi/h]$, then $\tilde{C}(f,h)(x) = 0$, and the difference on the left hand side of (1.3.10) is bounded by $|\tilde{f}(x)| \leq Ce^{-d|x|} \leq Ce^{-\pi d/h}$.

Proof of (1.3.11): For $x \in (-\pi/h, \pi/h)$, we take the jth derivative of both sides of (1.2.5) with respect to ζ to get

$$(-ix)^j e^{-itx} = \sum_{n=-\infty}^{\infty} S^{(j)}(n,h)(t)e^{-inhx} \quad -\pi/h < x < \pi/h.$$

Then, since $h^{-j}\delta_{\ell-n}^{(j)} = h^{-j}\mathrm{sinc}^{(j)}(\ell-n) = S^{(j)}(n,h)(\ell h)$,

$$f^{(j)}(\ell h) - h^{-j} \sum_{n\in\mathbb{Z}} \delta_{\ell-n}^{(j)} f(nh)$$

$$= \frac{1}{2\pi} \int_{\mathbb{R}} \tilde{f}(x) \left\{ (-ix)^j e^{-i\ell hx} - \sum_{n=-\infty}^{\infty} S^{(j)}(n,h)(\ell h)e^{-inhx} \right\} dx$$

$$= \frac{1}{2\pi} \int_{|x|>\pi/h} \tilde{f}(x) \cdot$$

$$\cdot \left\{ (-ix)^j e^{-i\ell hx} - \sum_{n=-\infty}^{\infty} S^{(j)}(n,h)(\ell h)e^{-inhx} \right\} dx.$$

$$\tag{1.3.22}$$

The final infinite sum of (1.3.22) is periodic in x, with period $2\pi/h$; because this series is the Fourier series expansion of $(-ix)^j e^{-itx}$, it is bounded by $(\pi/h)^j$ on $\mathbb{R}$. Also, $|\tilde{f}(x)| \leq Ce^{-d|x|}$ on $\mathbb{R}$. Hence

$$\frac{1}{2\pi}\left|\int_{|x|>\pi/h}\tilde{f}(x)\sum_{n=-\infty}^{\infty}S^{(j)}(n,h)(\ell h)\,e^{-inhx}\,dx\right|$$

$$\leq\frac{C}{\pi}\left(\frac{\pi}{h}\right)^{j}\int_{\pi/h}^{\infty}\exp(-dx)\,dx=\frac{C}{\pi d}\left(\frac{\pi}{h}\right)^{j}\exp\left(-\frac{\pi d}{h}\right).$$

$$(1.3.23)$$

Integration by parts, we get

$$\frac{1}{2\pi}\left|\int_{|x|>\pi/h}\tilde{f}(x)\,(-ix)^{j}\,e^{-i\ell hx}\,dx\right|$$

$$\leq\frac{C}{\pi}\int_{\pi/h}^{\infty}e^{-dx}\,x^{j}\,dx=\frac{j!\,C}{\pi d}\left(\frac{\pi}{h}\right)^{j}\exp\left(-\frac{\pi d}{h}\right)S,$$

with

$$S=\sum_{k=0}^{j}\left(\frac{1}{(j-k)!}\left(\frac{h}{\pi d}\right)^{k}\right)$$

In the case when $h/(\pi d)\leq 1$, then

$$S\leq\sum_{k=0}^{j}\frac{1}{k!}\left(\frac{h}{\pi d}\right)^{k}<\exp\left(\frac{h}{\pi d}\right),$$

whereas if $h/(\pi d)>1$, then

$$S\leq\left(\frac{h}{\pi d}\right)^{j}e.$$

It thus follows that in either of these cases, we have

$$\frac{1}{2\pi}\left|\int_{|x|>\pi/h}\tilde{f}(x)\,(-ix)^{j}\,e^{-i\ell hx}\,dx\right|=\mathcal{O}\left(h^{-j}\exp\left(-\frac{\pi d}{h}\right)\right).$$

$$(1.3.24)$$

The inequality (1.3.11) thus follows from (1.3.23) and (1.3.24).

Proof of (1.3.12): A proof of this inequality may be found in [S1, §4.5].

Proof of (1.3.13): This result may be established using an argument similar to that of the proof of (1.3.8), by means of the identity (1.2.5). It is left to Problem 1.3.5.

Proof of (1.3.14): This result can also be readily verified using the identity (1.2.5). It is left to Problem 1.3.6. ■

1.3.2 FINITE TERM SINC APPROXIMATION ON $\mathbb{R}$

Even though Sinc approximation with infinite number of terms may be very accurate, as indicated by Theorem 1.3.2, there may still be a problem from the point of view of numerical computation. In particular, this can happen when the number of terms in the summations that we require to achieve this accuracy is large.

For example (see Problem 1.2.8) if $f(t) = (1 + t^2)^{-1}$, then $\tilde{f}(x) = \exp(-|x|)$, so that $d = 1$ by Theorem 1.3.2, and the difference between f and its infinite–term Cardinal series converges to zero rapidly as $h \to 0$. On the other hand, the function $f(t)$ decreases to zero relatively slowly as $t \to \pm\infty$, and replacing $\sum_{-\infty}^{\infty}\{\cdot\}$ in (1.3.8) by $\sum_{-N}^{N}\{\cdot\}$ yields a large error, unless N is taken to be very large. This situation is circumvented when not only the Fourier transform $\tilde{f}$ but also f decay rapidly to zero on $\mathbb{R}$ for large argument.

In what follows, "id" signifies the identity map.

Definition 1.3.3 (i) *Corresponding to positive numbers α and d, let $\mathbf{L}_{\alpha,d}(\mathrm{id})$ denote the family of all functions f, with Fourier transforms $\tilde{f}$, such that*

$$f(t) = \mathcal{O}\left(e^{-\alpha |t|}\right), \quad t \to \pm\infty,$$
$$\tilde{f}(x) = \mathcal{O}\left(e^{-d |x|}\right), \quad x \to \pm\infty. \tag{1.3.25}$$

It will be convenient in §1.5.9 and in Chapter 2 of this handbook, to replace the second equation in (1.3.25) with the assumption that f is analytic in the domain $\mathcal{D}_d = \{z \in \mathbb{C} : |\Im(z)| < d\}$.

(ii) *Let g defined on $\mathbb{R}$ be such that g has finite limits at $\pm\infty$. Suppose the above constants α and d in (i) are restricted so that $0 < \alpha \leq 1$ and $0 < d < \pi$. Define Lg by*

$$(Lg)(t) = \frac{g(-\infty) + e^t\, g(\infty)}{1 + e^t}. \tag{1.3.26}$$

$\mathbf{M}_{\alpha,d}(\mathrm{id})$ *will denote the family of all functions g such that* $g - Lg \in \mathbf{L}_{\alpha,d}(\mathrm{id})$.

Example 1.3.4 Consider the function $f(t) = 1/\cosh(\alpha t)$, where $\alpha \in (1/2, 1]$. The Fourier transform $\tilde{f}(x)$ is $(\pi/\alpha)/\cosh(\pi x/(2\alpha))$. Since $f(t) = \mathcal{O}\left(e^{-\alpha|t|}\right)$, $t \to \pm\infty$, and $\tilde{f}(x) = \mathcal{O}\left(e^{-d|x|}\right)$, $x \to \pm\infty$, it follows that $f \in \mathbf{L}_{\alpha,d}(\mathrm{id})$, with $d = \pi/(2\alpha) < \pi$. Also, $g(t) = e^{\alpha t}/\cosh(\alpha t)$ is $\mathcal{O}\left(e^{-2\alpha|t|}\right)$ as $t \to -\infty$, and $g(t) \to 1$ as $t \to \infty$. If we now set $f_1 = g - Lg$ since $Lg(t) = e^t/(1 + e^t)$, then, with $\gamma = \min(2\alpha, 1)$,

$$f_1(t) = \begin{cases} \mathcal{O}\left(e^{-\gamma|t|}\right), & t \to -\infty \\ \\ \mathcal{O}\left(e^{-\alpha t}\right) & t \to \infty. \end{cases}$$

Furthermore, it is readily shown that $\tilde{f}_1(x) = \mathcal{O}\left(e^{-d|x|}\right)$, $x \to \pm\infty$, with $d = \pi/(2\alpha) < \pi$, and thus $g \in \mathbf{M}_{\alpha,d}(\mathrm{id})$. Notice also, that g has singularities at the points $\pm i\,\pi/(2\alpha)$, which shows also that both g and also f_1 are analytic in the strip $\mathcal{D}_d$.

Remark. The introduction of the class of functions $\mathbf{M}_{\alpha,d}$ in the previous definition enables us to approximate using Sinc methods functions that do not vanish at $\pm\infty$. The reasons for restricting the constants α and d in the above definition will become evident. We use the notation of Theorem 1.3.2 in the following theorem.

Theorem 1.3.5 *Let $f \in \mathbf{L}_{\alpha,d}(\mathrm{id})$, and corresponding to a positive integer N, set*

$$h = \left(\frac{\pi d}{\alpha N}\right)^{1/2},$$

$$h^* = \left(\frac{2\pi d}{\alpha N}\right)^{1/2},$$

$$\varepsilon_N = N^{1/2} \exp\left\{-(\pi d\alpha N)^{1/2}\right\},$$

$$\varepsilon_N^* = \exp\left\{-(2\pi d\alpha N)^{1/2}\right\}. \tag{1.3.27}$$

Then, as $N \to \infty$,

$$\left\| f(t) - \sum_{k=-N}^{N} f(k\,h)\,S(k,h)(t) \right\| \;=\; \mathcal{O}\left(\varepsilon_N\right),$$

$$(1.3.28)$$

$$\left| \int_{\mathbb{R}} f(t)\,dt - h^* \sum_{k=-N}^{N} f(k\,h^*) \right| \;=\; \mathcal{O}\left(\varepsilon_N^*\right);$$

$$(1.3.29)$$

$$\left\| \int_{\mathbb{R}} e^{i\,x\,t} f(t)\,dt - h \sum_{k=-N}^{N} e^{i\,k\,h\,x}\, f(k\,h) \right\| \;=\; \mathcal{O}\left(\varepsilon_N\right);$$

$$(1.3.30)$$

$$h^j \left| f^{(j)}(\ell h) - h^{-j} \sum_{k=-N}^{N} \delta_{\ell-k}^{(j)}\, f(kh) \right| \;=\; \mathcal{O}\left(\varepsilon_N\right);$$

$$(1.3.31)$$

$$\left| \int_{-\infty}^{\ell h} f(x)\,dx - h \sum_{k=-N}^{N} \delta_{\ell-k}^{(-1)}\, f(kh) \right| \;=\; \mathcal{O}\left(\varepsilon_N\right);$$

$$(1.3.32)$$

$$\left\| \frac{P.V.}{\pi} \int_{\mathbb{R}} \frac{f(t)}{t-x}\,dt - \sum_{k=-N}^{N} \frac{\cos\left[\frac{\pi}{h}(x-kh)\right] - 1}{\frac{\pi}{h}(x-kh)}\, f(kh) \right\| \;=\; \mathcal{O}\left(\varepsilon_N\right);$$

$$(1.3.33)$$

$$\frac{1}{h} \int_{\mathbb{R}} f(t)\,S(j,h)(t)\,dt - f(jh) \;=\; \mathcal{O}\left(\varepsilon_N\right).$$

$$(1.3.34)$$

Remark. The order relations on the right hand sides of the equations in Theorem 1.3.5 actually hold uniformly with respect to N, for all $N > 1$. Thus, the right hand sides of these equations can be replaced by "$\leq C\,\varepsilon_N$, $N \in (1,\infty)$", and C a constant independent of N.

Proof. We prove only the first of these, (1.3.28), leaving the remaining proofs for problems at the end of this section.

We have,

$$\left| f(t) - \sum_{k=-N}^{N} f(k\,h)\, S(k,h)(t) \right|$$

$$\leq \left| f(t) - \sum_{k=-\infty}^{\infty} f(k\,h)\, S(k,h)(t) \right| + \left| \sum_{|k|>N} f(k\,h)\, S(k,h)(t) \right|$$

$$\leq E + \sum_{|k|>N} |f(k\,h)\, S(k,h)(t)|.$$

$$(1.3.35)$$

where by Theorem 1.3.2, $E = \mathcal{O}\left(\exp(-\pi\, d/h)\right)$. Under our assumption on f, there exists a constant C' such that for all $t \in \mathbb{R}$, we have $|f(t)| \leq C'\, e^{-\alpha|t|}$. Hence, since $|S(k,h)(t)| \leq 1$ on $\mathbb{R}$,

$$\left| \sum_{|k|>N} f(k\,h)\, S(k,h)(t) \right| \leq \sum_{|k|>N} |f(k,h)|$$

$$\leq 2\,C' \sum_{k>N} e^{-\alpha\,k\,h}$$

$$(1.3.36)$$

$$= 2\,C'e^{-\alpha\,N\,h}\, \frac{1}{e^{\alpha h} - 1}$$

$$\leq \frac{2\,C'}{\alpha\,h}\, e^{-\alpha\,N\,h}.$$

It therefore follows from (1.3.35) that

$$\left| f(t) - \sum_{k=-N}^{N} f(k\,h)\, S(k,h)(t) \right| \leq C_1\, e^{-\pi\,d/h} + \frac{C_2}{\alpha\,h}\, e^{-\alpha\,N\,h}, \quad (1.3.37)$$

where C_1 and C_2 are constants. Substituting the definition of h given in (1.3.27) into (1.3.37), we get (1.3.28). ∎

Remark. As already mentioned, the programs of this hand–book deal with approximation in $\mathbf{M}_{\alpha,\beta,d}(\varphi)$, with $0 < \alpha, \beta \leq 1$. In this case we would use approximating sums of the form $\sum_{-M}^{N}$ rather than

$\sum_{-N}^{N}$, and as a consequence, the error term (1.3.37) now takes the form

$$\left| f(t) - \sum_{k=-M}^{N} f(k\,h)\,S(k\,h)(t) \right|$$

$$\leq C_1\,e^{-\pi\,d/h} + \frac{C_2}{\alpha\,h}\,e^{-\alpha\,M\,h} + \frac{C_3}{\beta\,h}\,e^{-\beta\,N\,h}.$$

We would now set $M = [\beta\,N/\alpha]$, and we would then determine h such that $\beta\,N\,h = \pi\,d/h$.

With reference to the class of functions $\mathbf{L}_{\alpha,d}(\mathrm{id})$, it is usually easier to determine the constant α than it is to determine d. For example, the function $f(t) = 1/\cosh(\alpha\,t)$, decays to zero like $\exp(-\alpha\,|t|)$. On the other hand, in order to find d, we need its Fourier transform

$$\int_{\mathbb{R}} \frac{e^{i\,x\,t}}{\cosh(\alpha\,t)}\,dt = \frac{\pi}{\alpha\,\cosh\left(\frac{\pi\,x}{2\,\alpha}\right)}. \qquad (1.3.38)$$

This decays to zero like $\exp\left(-\frac{\pi\,|x|}{2\,\alpha}\right)$, showing that $d = \pi/(2\,\alpha)$. The following theorem may at times be useful for determining these constants. To apply Sinc methods and obtain exponential convergence $\mathcal{O}\left(e^{-c\,N^{1/2}}\right)$ it is not necessary to know α and d exactly, as shown in Corollary 1.3.7 below. However, to obtain optimal exponential convergence, it is desirable to know these constants. The following theorem may be useful in this regard. In this theorem α, d, α_i and d_i $(i = 1, 2)$ denote positive constants.

Theorem 1.3.6   (i) If $f \in \mathbf{L}_{\alpha,d}(\mathrm{id})$, then $\tilde{f} \in \mathbf{L}_{d,\alpha}(\mathrm{id})$;

(ii) If $\alpha_1 \leq \alpha_2$, and $d_1 \leq d_2$, then $\mathbf{L}_{\alpha_1,d_1}(\mathrm{id}) \supseteq \mathbf{L}_{\alpha_2,d_2}(\mathrm{id})$;

(iii) If $f_i \in \mathbf{L}_{\alpha_i,d_i}$, $i = 1, 2$, then $f_3 = f_1 \pm f_2 \in \mathbf{L}_{\alpha,d}(\mathrm{id})$ with $\alpha = \min(\alpha_1, \alpha_2)$ and $d = \min(d_1, d_2)$;

(iv) If $f_i \in \mathbf{L}_{\alpha_i,d_i}(\mathrm{id})$, $i = 1, 2$, and if $f_3 = f_1\,f_2$, then $f_3 \in \mathbf{L}_{\alpha,d}(\mathrm{id})$ with $\alpha = \alpha_1 + \alpha_2$; and

$$d = \begin{cases} \min(d_1, d_2) & \text{if } d_1 \neq d_2 \\ d_1 - \varepsilon & \text{if } d_1 = d_2 \,; \end{cases} \qquad (1.3.39)$$

(v) If $f_i \in \mathbf{L}_{\alpha_i, d_i}(\mathrm{id})$, $i = 1, 2$, and if f_3 is the convolution of f_1 and f_2, i.e., $f_3(t) = \int_{\mathbb{R}} f_1(\tau) f_2(t - \tau) \, d\tau$, then $f_3 \in \mathbf{L}_{\alpha, d}(\mathrm{id})$ with $d = d_1 + d_2$ and

$$\alpha = \begin{cases} \min(\alpha_1, \alpha_2) & \text{if } \alpha_1 \neq \alpha_2 \\ \alpha_1 - \varepsilon & \text{if } \alpha_1 = \alpha_2. \end{cases} \qquad (1.3.40)$$

In (iv) and (v), $\varepsilon > 0$ is otherwise arbitrary.

Proof. Parts (i), (ii) and (iii) of the theorem are straightforward, and we leave them to the problems at the end of this section.

The first part of (iv) i.e., that for $f_3 = f_1 f_2$ we have $\alpha = \alpha_1 + \alpha_2$ is also obvious from the definition of $\mathbf{L}_{\alpha, d}(\mathrm{id})$. To prove the second half of Part (iv), let $\tilde{f}_3 = \widetilde{f_1 f_2}$ denote the Fourier transform of $f_1 f_2$. Then we have, for $x > 0$, and some constant C,

$$\left| \tilde{f}_3(x) \right| \leq C \left(I_1 + I_2 + I_3 \right),$$

with

$$I_1 = \int_{-\infty}^{0} e^{d_1 t + d_2(t-x)} \, dt = \frac{e^{-d_2 x}}{d_1 + d_2},$$

$$I_2 = \int_{0}^{x} e^{-d_1 t + d_2(t-x)} \, dt = \begin{cases} \dfrac{e^{-d_2 x} - e^{-d_1 x}}{d_1 - d_2} & \text{if } d_1 \neq d_2, \\ x \, e^{-d_2 x} & \text{if } d_1 = d_2, \end{cases}$$

$$I_3 = \int_{x}^{\infty} e^{-d_1 t - d_2(t-x)} \, dt = \frac{e^{-d_1 x}}{d_1 + d_2}.$$

Thus the constant d for $\tilde{f}_3(x)$ satisfies $(1.3.39)$, which proves the statement of the theorem for the case of $x > 0$. The proof for $x < 0$ is similar.

Part (v) is a consequence of Parts (i) and (iv). ■

Remark. The next corollary states that we can always get exponential convergence in Sinc approximation without the positive constants α and d being explicitly known.

Corollary 1.3.7 *Let $f \in \mathbf{L}_{\alpha,d}(\mathrm{id})$, and for arbitrary $\gamma > 0$ set*

$$h = \frac{\gamma}{N^{1/2}}. \qquad (1.3.41)$$

Then ε_N on the right hand sides of (1.3.28)–(1.3.34) in Theorem 1.3.5 can be replaced by δ_N, where

$$\delta_N = N^{1/2} \, e^{-\beta N^{1/2}}, \qquad (1.3.42)$$

and where

$$\beta = \min\left\{ \frac{\pi d}{\gamma}, \alpha\gamma \right\}. \qquad (1.3.43)$$

Proof. The result follows directly for (1.3.28), upon substitution of $h = \gamma/N^{1/2}$ in (1.3.37), and in a like manner, for (1.3.29)–(1.3.34). ∎

This corollary tells us that for $f \in \mathbf{L}_{\alpha,d}(\mathrm{id})$, regardless of the values of α and d, the convergence can always be exponential. Note that the fastest convergence is achieved when β is a maximum, obtained approximately, by setting $\pi d/\gamma = \alpha\gamma$ and solving for γ.

Problems for Section 1.3

1.3.1 Let the Fourier transform $\tilde{f}$ exist and satisfy the relation $|\tilde{f}(x)| \leq C \exp(-d|x|)$ for all $x \in \mathbb{R}$.

   (a) Deduce that f is analytic in the strip

$$\mathcal{D}_d = \{ z \in \mathbb{C} : |\Im(z)| < d \}$$

   (b) Show that $\mathcal{S}f$, the Hilbert transform of f, is also analytic in the strip $\mathcal{D}_d$.

   (c) Show that the derivative of f as well as the indefinite integral of f are also analytic in the strip $\mathcal{D}_d$.

1.3.2 Does $f(t) = e^{-\alpha|t|}$ belong to the space $\mathbf{L}_{\alpha,d}(\mathrm{id})$? Why, or why not ? Compare with (1.3.38).

1.3.3 Set $g = f'$, let $\tilde{f}$ and $\tilde{g}$ denote the Fourier transforms of f and g respectively, and assume that for some positive constants C_1, C_2 and d, we have for $x \in \mathbb{R}$, that

$$|\tilde{f}(x)| \le C_1 \, e^{-d|x|}, \quad |\tilde{g}(x)| \le C_2 \, e^{-d|x|}.$$

Prove that there exists a constant C such that for all $x \in \mathbb{R}$,

$$\left| \frac{\tilde{g}(x)}{x} \right| \le C \, e^{-d|x|}.$$

1.3.4 Let $f \in \mathbf{L}_{\alpha,d}(\mathrm{id})$. Why is it no longer true that $\mathcal{S}f \in \mathbf{L}_{\alpha,d}(\mathrm{id})$, where $\mathcal{S}f$ denotes the Hilbert transform of f?

1.3.5 Give a detailed proof of (1.3.13). [Hint: Start with the representation

$$\frac{P.V.}{\pi} \int_{\mathbb{R}} \frac{f(t)}{t - x} \, dt = \frac{1}{2\pi} \int_{\mathbb{R}} \mathrm{sgn}(t) \, \tilde{f}(t) \, e^{-i\,x\,t} \, dt,$$

use (1.3.6) and then proceed as in the proof of (1.3.8).]

1.3.7 Give a detailed proof of (1.3.14). [Hint: Apply Parseval's theorem to the integral part of (1.3.14).]

1.4   Sinc, Wavelets, Trigonometric and Algebraic Polynomials and Quadratures

In the present section we shall study relationships between the Cardinal series trigonometric polynomials and wavelets, and algebraic polynomials.

Most wavelet applications are based on use of the Cardinal series (1.1.4), although only special cases of (1.1.4) satisfy all of the properties of wavelets [Gz]. Our aim in this section is not the development nor the illustration of wavelet algorithms, but rather to present new insights into the relationships between the cardinal series, trigonometric polynomials and algebraic polynomials, and also, the development of novel expressions which are, in fact, wavelets [Gz]. (We assume that users of wavelet methods have access to wavelet algorithms.)

Surprisingly, if f is periodic with period 2π on $\mathbb{R}$, then with suitable selection of h, (1.1.4) reduces to an explicit finite sum of Dirichlet kernels times function values, thus providing a trigonometric interpolation polynomial. Well–known identities then enable these trigonometric polynomials to be transformed into algebraic polynomials.

Algebraic polynomials are of course equivalent to trigonometric cosine polynomials (see p. 106 of [S1]), and so they are a subset of all Fourier polynomials. On the other hand, the results of this section show that Fourier polynomials (hence cosine and sine polynomials) are a subset of Cardinal series expansions. The hierarchy relation *Algebraic Polynomials* $\subset$ *Fourier polynomials* $\subset$ *Cardinal series expansions* thus follows. We shall show that trigonometric polynomial approximation of finite degree on $(0, \pi)$ is equivalent to approximation on $(-1, 1)$, with functions of the form $P(x) + \sqrt{1 - x^2}\, Q(x)$, with $P(x)$ and $Q(x)$ denoting polynomials in x.

Finally, the fact that all algebraic and trigonometric polynomials can be represented as Cardinal series enables us to derive novel error bounds for approximation of periodic analytic functions via Cardinal series, via trigonometric polynomials, and via algebraic polynomials, and for quadrature.

We shall use the following well known identities several times in this section. The second of these is, in fact a special case of the first.

Lemma 1.4.1 *If $\mathbb{C}$ denotes the complex plane, and if $z \in \mathbb{C}$, then*

$$\frac{\pi}{\tan(\pi z)} = \lim_{N \to \infty} \sum_{k=-N}^{N} \frac{1}{z - k}. \tag{1.4.1}$$

Also,

$$\frac{\pi}{\sin(\pi z)} = \sum_{k=-\infty}^{\infty} \frac{(-1)^k}{z - k}. \tag{1.4.2}$$

Throughout the remainder of this section, we shall use the Dirichlet kernel notations

$$D_e(N, x)$$

$$= \frac{1}{2N} \left(1 + \cos(Nx) + 2 \sum_{k=1}^{N-1} \cos(kx) \right) \qquad (1.4.3)$$

$$= \frac{\sin\{Nx\}}{2N \tan(x/2)},$$

and

$$D_o(N, x) \;=\; \frac{1}{2N+1} \left(1 + 2 \sum_{k=1}^{N} \cos(kx) \right)$$

$$\qquad (1.4.4)$$

$$\;=\; \frac{\sin\{(N+1/2)x\}}{(2N+1)\sin(x/2)},$$

with subscript e denoting "even", corresponding to $h = 2\pi/(2N) = \pi/N$, and with subscript o denoting "odd", corresponding to $h = 2\pi/(2N+1)$.

These even and odd kernels are closely related through

$$(2N+1)\,D_o(N, x) - 2N\,D_e(N, x) = \cos(Nx). \qquad (1.4.5)$$

Remark. The following identities are easily verified, for any given integer k,

$$D_e(N, x - (2N+k-b)h) \;=\; D_e(N, x - (k-b)h)$$

$$D_o(N, x - (2N+1+k-b)h) \;=\; D_o(N, x - (k-b)h).$$

$$\qquad (1.4.6)$$

These identities serve to simplify the expressions in the parts (ii) and (iii) of these Corollaries.

Denoting p and q to be integers with $p \le q$, we shall write

$$\sum_{j=p}^{q}{}' a_j = (1/2)\,a_p + a_{p+1} + \ldots + a_{q-1} + (1/2)\,a_q, \qquad (1.4.7)$$

i.e., the prime indicates that the first and last terms of the sum are to be halved.

1.4.1 A GENERAL THEOREM

The condition that $f(x+t) - 2f(x) + f(x-t) \to 0$ as $t \to 0$ in the theorem which follows just means that that f takes on its average value at points of discontinuity on $\mathbb{R}$.

Theorem 1.4.2 Let f, have period 2π on $\mathbb{R}$, let $f(x+t) - 2f(x) + f(x-t) \to 0$ as $t \to 0$, with $(x,t) \in \mathbb{R}^2$, and take $h > 0$. For arbitrary $b \in \mathbb{R}$, write

$$f_h(b,x) = \sum_{k \in \mathbb{Z}} f((k-b)h) \, \frac{\sin\left\{\frac{\pi}{h}(x-(k-b)h)\right\}}{\frac{\pi}{h}(x-(k-b)h)}, \qquad x \in \mathbb{R}. \quad (1.4.8)$$

(i) If $Nh = \pi$, where N is a positive integer, then

$$f_h(b,x) = \sum_{k=-N+1}^{N} f((k-b)h) \, D_e(N, x-(k-b)h). \quad (1.4.9)$$

(ii) If $((2N+1)h = 2\pi$, where N is a non-negative integer, then

$$f_h(b,x) = \sum_{k=-N}^{N} f((k-b)h) \, D_o(N, x-(k-b)h). \quad (1.4.10)$$

Proof of Part (i): Since $f(x + 2\pi) = f(x)$ for all $x \in \mathbb{R}$, for b an arbitrary real number, we have

$$f_h(b,x) = \sum_{k \in \mathbb{Z}} f((k-b)h) \, \frac{\sin\left\{\frac{\pi}{h}(x-(k-b)h)\right\}}{\frac{\pi}{h}(x-(k-b)h)}$$

$$= \sum_{s \in \mathbb{Z}} \sum_{k=-N+1}^{N} f(kh - bh + 2sNh) \, \frac{\sin\left\{\frac{\pi}{h}(x+bh-kh-2sNh)\right\}}{\frac{\pi}{h}(x+bh-kh-2sNh)}$$

$$= \sum_{k=-N+1}^{N} f((k-b)h) \, D_e(N, x-(k-b)h),$$

$$(1.4.11)$$

where, since $Nh = \pi$,

$$D_e(N, x - (k - b) h)$$

$$= \frac{h}{\pi} \sin\left\{\frac{\pi}{h}(x - (k - b) h)\right\} \sum_{s=-\infty}^{\infty} \frac{1}{x(k - b) h - 2s \pi}. \tag{1.4.12}$$

Hence, using (1.4.1), and the first of (1.4.6) we get (1.4.9).

Proof of Part (ii): The proof of this case is similar to the proof of Part (i), and is omitted. ∎

1.4.2   EXPLICIT SPECIAL CASES ON $[0, 2\pi]$

The most interesting cases of Theorem 1.4.2 are for $b = 0$ and for $b = 1/2$. Theorem 1.4.2 reduces to several different explicit formulas on $[-\pi, \pi]$ depending on whether $b = 0$ or $b = 1/2$, and depending on whether $2Nh = 2\pi$, or when $(2N+1)h = 2\pi$. Important special cases of these also occur depending on whether $f_h(b, x)$ is an even or an odd function on $\mathbb{R}$. Enumeration of these special cases requires two corollaries for the case of $Nh = \pi$: the first for interpolation at the points $\{k h\}_{k=0}^{2N}$; and the second for interpolation at the points $\{k h - h/2\}_{k=1}^{2N}$. Following this, we state two corollaries, this time for the case when $(2N + 1)h = 2\pi$: one for interpolation at the points $\{k h - h/2\}_{k=1}^{2N+1}$; and the second for interpolation at the points $\{k h\}_{k=0}^{2N+1}$. For sake of broader applicability, these theorems are stated to allow different values of f at $\pm\pi$. We shall return to f being both analytic on $\mathbb{R}$ as well as periodic in subsections of this section.

We omit the proofs of the four corollaries which follow, leaving them to the problems at the end of this section. The proofs follow readily using the results $f(x + 2\pi) = f(x) = (f(x) + f(x + 2\pi))/2$, and

$$\frac{1}{\tan(a + b)} + \frac{1}{\tan(a - b)} = \frac{2 \sin(2a)}{\cos(2b) - \cos(2a)},$$

$$\frac{1}{\sin(a + b)} + \frac{1}{\sin(a - b)} = \frac{4 \sin(b) \cos(a)}{\cos(2a) - \cos(2b)}.$$

Corollary 1.4.3 *If $Nh = \pi$, and $b = 0$ in (1.4.9):*

(i) Interpolating on $[-\pi, \pi]$:

$$f_h(0, x) = \sum_{k \in \mathbb{Z}} f(k\,h)\, S(k\,h)(x)$$

$$= \frac{\sin(N\,x)}{2\,N} \sum_{k=-N}^{N}{}' \frac{(-1)^k f\left(\frac{k\pi}{N}\right)}{\tan\left(\frac{1}{2}\left(x - \frac{k\pi}{N}\right)\right)}.$$

(1.4.13)

(ii) If $f(-x) = f(x)$ for all $x \in \mathbb{R}$, then (1.4.13) reduces to the expression

$$f_h(0, x) = \frac{\cos((N-1)\,x) - \cos((N+1)\,x)}{2\,N} \sum_{k=0}^{N}{}' \frac{(-1)^k f\left(\frac{k\pi}{N}\right)}{\cos(x) - \cos\left(\frac{k\pi}{N}\right)}.$$

(1.4.14)

for interpolation of f on $[0, \pi]$.

(iii) If $f(-x) = -f(x)$ for all $x \in \mathbb{R}$, then (1.4.13) reduces to the expression

$$f_h(0, x) = \frac{\sin(N\,x)}{N} \sum_{k=1}^{N-1} \frac{(-1)^k \sin\left(\frac{k\pi}{N}\right) f\left(\frac{k\pi}{N}\right)}{\cos(x) - \cos\left(\frac{k\pi}{N}\right)}.$$

(1.4.15)

for interpolation on $[0, \pi]$.

Corollary 1.4.4 *If $Nh = \pi$, and $b = 1/2$ in (1.4.9), then:*

(i) Interpolating on $[-\pi, \pi]$,

$$f_h\left(\frac{1}{2}, x\right) = \frac{\cos(N\,x)}{2\,N} \sum_{k=-N+1}^{N} \frac{(-1)^k f\left(\left(k - \frac{1}{2}\right)\frac{\pi}{N}\right)}{\tan\left(\frac{1}{2}\left(x - \left(k - \frac{1}{2}\right)\frac{\pi}{N}\right)\right)}.$$

(1.4.16)

(ii) If in addition we have $f(-x) = f(x)$ for $x \in \mathbb{R}$, then (1.4.16) reduces to the expression

$$f_h\left(\tfrac{1}{2}, x\right) = \frac{\cos(N x)}{N} \sum_{k=1}^{N} \frac{(-1)^k \sin\left(\left(k - \tfrac{1}{2}\right)\tfrac{\pi}{N}\right) f\left(\left(k - \tfrac{1}{2}\right)\tfrac{\pi}{N}\right)}{\cos(x) - \cos\left(\left(k - \tfrac{1}{2}\right)\tfrac{\pi}{N}\right)}.$$

$$(1.4.17)$$

for interpolation of f on $[0, \pi]$.

(iii) If $f(-x) = -f(x)$ for $x \in \mathbb{R}$, then (1.4.16) reduces to the expression

$$f_h\left(\tfrac{1}{2}, x\right) =$$

$$\frac{\sin((N+1)x) + \sin((N-1)x)}{2N} \sum_{k=1}^{N} \frac{(-1)^{k-1} f\left(\left(k - \tfrac{1}{2}\right)\tfrac{\pi}{N}\right)}{\cos(x) - \cos\left(\left(k - \tfrac{1}{2}\right)\tfrac{\pi}{N}\right)}.$$

$$(1.4.18)$$

for interpolation of f on $[0, \pi]$.

The formula (1.4.17) was obtained previously by Gabdulhaev (see [S1], Theorem 2.2.6).

Corollary 1.4.5 *If $(2N+1)h = 2\pi$, and $b = 0$ in (1.4.10),*

$$f_h(0, x) = \frac{\sin\left(\left(N + \tfrac{1}{2}\right)x\right)}{2N+1} \sum_{k=-N}^{N} \frac{(-1)^k f\left(\tfrac{2k\pi}{2N+1}\right)}{\sin\left(\tfrac{1}{2}\left(x - \tfrac{2k\pi}{2N+1}\right)\right)}. \qquad (1.4.19)$$

(ii) If in addition we have $f(-x) = f(x)$ for $x \in \mathbb{R}$, then (1.4.19) reduces to the expression

$$f_h(0, x) = \frac{\cos((N+1)x) - \cos(Nx)}{2N+1} \cdot$$

$$\cdot \left(\frac{f(0)}{\cos(x) - 1} + 2 \sum_{k=1}^{N} \frac{(-1)^k \cos\left(\tfrac{k\pi}{2N+1}\right) f\left(\tfrac{2k\pi}{2N+1}\right)}{\cos(x) - \cos\left(\tfrac{2k\pi}{2N+1}\right)} \right),$$

$$(1.4.20)$$

for interpolation of f on $[0, \pi]$.

(iii) If $f(-x) = -f(x)$ for $x \in \mathbb{R}$, then (1.4.19) reduces to the expression

$$f_h\left(0\,,x\right) \;=\; 2\,\frac{\sin(N\,x)+\sin((N+1)\,x)}{2\,N+1}\cdot$$

$$\cdot\sum_{k=1}^{N}\frac{(-1)^k\,\sin\left(\frac{k\,\pi}{2\,N+1}\right)\,f\left(\frac{2\,k\,\pi}{2\,N+1}\right)}{\cos(x)-\cos\left(\frac{2\,k\,\pi}{2\,N+1}\right)}\cdot$$

(1.4.21)

for interpolation of f on $[0,\pi]$.

Corollary 1.4.6 *If $(2\,N+1)\,h = 2\,\pi$, and $b = 1/2$ in (1.4.10), then:*

(i) Interpolating on $[-\pi\,,\pi]$,

$$f_h\left(\tfrac{1}{2},x\right) \;=\; \sum_{k\in\mathbb{Z}}f(k\,h-h/2)\,\mathrm{sinc}\left(\frac{x}{h}-k+1/2\right)$$

$$=\;\frac{\cos\left(\left(N+\tfrac{1}{2}\right)x\right)}{2\,N+1}\sum_{k=-N-1}^{N}{}'\,\frac{(-1)^{k-1}\,f\left(\frac{(2\,k+1)\,\pi}{2\,N+1}\right)}{\sin\left(\frac{1}{2}\left(x-\frac{(2\,k+1)\,\pi}{2\,N+1}\right)\right)}\cdot$$

(1.4.22)

(ii) If $f(-x) = f(x)$ for all $x \in \mathbb{R}$, then (1.4.22) reduces to the expression

$$f_h\left(\tfrac{1}{2},x\right) = \frac{\cos(N\,x)+\cos((N+1)\,x)}{2\,N+1}\cdot$$

$$\cdot\left(\frac{(-1)^N\,f(\pi)}{\cos(x)+1}+2\sum_{k=0}^{N-1}\frac{(-1)^k\,\sin\left(\frac{\left(k+\frac{1}{2}\right)\pi}{2\,N+1}\right)\,f\left(\frac{(2\,k+1)\,\pi}{2\,N+1}\right)}{\cos(x)-\cos\left(\frac{(2\,k+1)\,\pi}{2\,N+1}\right)}\right)\cdot$$

(1.4.23)

for interpolation of f on $[0,\pi]$.

(iii) If $f(-x) = -f(x)$ for all $x \in \mathbb{R}$, then (1.4.22) reduces to the expression

$$f_h\left(\tfrac{1}{2}, x\right) = 2\,\frac{\sin(N x) + \sin((N+1)\,x)}{2\,N + 1} \cdot$$

$$\cdot \sum_{k=0}^{N-1} \frac{(-1)^k \, \cos\left(\frac{(k+1/2)\,\pi}{2\,N+1}\right) f\left(\frac{(2\,k+1)\,\pi}{2\,N+1}\right)}{\cos(x) - \cos\left(\frac{(2\,k+1)\,\pi}{2\,N+1}\right)} . \tag{1.4.24}$$

for interpolation of f on $[0, \pi]$.

1.4.3 WAVELETS AND TRIGONOMETRIC POLYNOMIALS

The two Dirichlet kernels already introduced in (1.4.3) and (1.4.4) above are used regularly to prove the convergence of Fourier series approximation. We can also express these kernels for approximation on $(-\pi, \pi)$ in their complex variable forms:

$$D_e(N, x) = \frac{1}{2\,N} \sideset{}{'}\sum_{k=-N}^{N} e^{i\,k\,x} , \qquad D_o(N, x) = \frac{1}{2\,N+1} \sum_{k=-N}^{N} e^{i\,k\,x} . \tag{1.4.25}$$

Given f defined on $[-\pi, \pi]$, recall that its Fourier series is given by

$$F(x) = \sum_{k \in \mathbb{Z}} c_k \, e^{i\,k\,x} , \tag{1.4.26}$$

with

$$c_k = \frac{1}{2\,\pi} \int_{-\pi}^{\pi} F(x)\, e^{-i\,k\,x} \, dx . \tag{1.4.27}$$

It follows that

$$\frac{1}{2\,\pi} \int_{-\pi}^{\pi} F(x')\, D_e(N, x - x')\, dx'$$

$$= \frac{1}{2} c_{-N}\, e^{-i\,N\,x} + \sum_{k=-N+1}^{N-1} c_k\, e^{i\,k\,x} + \frac{1}{2} c_N\, e^{i\,N\,x} . \tag{1.4.28}$$

and also,

$$\frac{1}{2\pi} \int_{-\pi}^{\pi} F(x') D_o(N, x - x') \, dx' = \sum_{k=-N}^{N} c_k \, e^{i k x}, \qquad (1.4.29)$$

The formulas (1.4.28) and (1.4.29) as integrals of Dirichlet kernels are said to be *continuous wavelet transforms*.

The usual $(m + 1)$–point *Trapezoidal* integration rule $T_m(h, f)$ for approximate integration of f over $[-\pi, \pi]$ is, with $m \, h = 2 \pi$,

$$T_m(h, f) = h \left\{ \frac{1}{2} f(-\pi) + \sum_{k=1}^{m-1} f(-\pi + k \, h) + \frac{1}{2} f(\pi) \right\}. \qquad (1.4.30)$$

Likewise, the usual m–point *Midordinate* rule, is

$$M_m(h, f) = h \sum_{k=1}^{m} f(-\pi + (k - 1/2) \, h), \quad h = 2 \pi / m. \qquad (1.4.31)$$

It is easily seen that application of the Trapezoidal or Midordinate rules to the formulas ((1.4.28) and (1.4.29) yields the exact values of the integrals. Indeed:

– Application of the Trapezoidal rule to (1.4.28) yields the formula (1.4.13);

– Application of the Midordinate rule to (1.4.28) yields the formula (1.4.16);

– Application of the Midordinate rule to (1.4.29) yields the formula (1.4.19);

– Application of the Trapezoidal rule to (1.4.29) yields the formula (1.4.22); and

The formulas (1.4.14), (1.4.20), (1.4.16) and (1.4.22) are *discrete wavelet transforms*.

Remark. The approximations of the Corollaries 1.4.3 – 1.4.6 are trigonometric polynomials which interpolate f: The identities for

$f_h(0, \cdot)$ interpolate f at the points $\{k\,h\}$ for $k \in \mathbb{Z}$, while the identities for $f_h(1/2, \cdot)$ interpolate f at the points $\{k\,h + h/2\}$, $k \in \mathbb{Z}$. If f is even on $\mathbb{R}$, then we get interpolation of f by cosines, while if f is odd, we get interpolations of f by sines. We can thus state the following self–evident corollary.

Corollary 1.4.7 *There exist constants a_k^J and b_k^J, $J = 1,\,2,\,3$ and 4, such that:*

(i) If $N\,h = \pi$, then

$$f_h(0, x) = \sum_{k=0}^{N-1} a_k^1 \cos\left(\frac{2\pi k}{P} x\right) + \sum_{k=1}^{N} b_k^1 \sin\left(\frac{2\pi k}{P} x\right), \quad (1.4.32)$$

and

$$f_h\left(\frac{1}{2} x\right) = \sum_{k=0}^{N-1} a_k^2 \cos\left(\frac{2\pi k}{P} x\right) + \sum_{k=1}^{N} b_k^2 \sin\left(\frac{2\pi k}{P} x\right), \quad (1.4.33)$$

(ii) If $(2\,N + 1)h = 2\,\pi$, then

$$f_h(0, x) = \sum_{k=0}^{N} a_k^3 \cos\left(\frac{2\pi k}{P} x\right) + \sum_{k=1}^{N} b_k^3 \sin\left(\frac{2\pi k}{P} x\right), \quad (1.4.34)$$

and

$$f_h\left(\frac{1}{2} x\right) = \sum_{k=0}^{N} a_k^4 \cos\left(\frac{2\pi k}{P} x\right) + \sum_{k=1}^{N} b_k^4 \sin\left(\frac{2\pi k}{P} x\right), \quad (1.4.35)$$

Similar expressions for the cases of f even and odd, are also possible.

1.4.4 ERROR OF APPROXIMATION

In Sections 1.2 and 1.3 above we studied properties of the Cardinal function, which is an entire function of order 1 and type π/h and its uses for approximating other functions f. In particular, an infinite term Cardinal functions were used to approximate functions f for which $|\tilde{f}(y)| \le C \exp(-d|y|)$, $y \in \mathbb{R}$, where $\tilde{f}$ denotes the Fourier transform of f, and where C and d are positive numbers, and a finite term Cardinal function was used if f also satisfies the relation $|f(x)| \le C' \exp(-\alpha |x|)$, for $x \in \mathbb{R}$, where C' and α are positive constants (see Theorems 1.3.2 and 1.3.5 of [S1]).

In the present section we shall study relationships between the the above Cardinal series f_h, trigonometric polynomials and wavelets, and algebraic polynomials, as well as errors of approximation. To this end, we shall assume that f_h is not only defined on $\mathbb{R}$, but is analytic in the region $\mathcal{D}_d$ defined below, where d is a positive number. Let $R = e^d$, and let us define the following domains in the complex plane:

$$\mathcal{D}_d = \{z \in \mathbb{C} : |\Im z| < d\},$$

$$\mathcal{D}_{\pi,d} = \{z \in \mathbb{C} : |\Re z| \le \pi, \ |\Im z| < d\},$$

$$\mathcal{A}_R = \{\zeta \in \mathbb{C} : 1/R < |\zeta| < R\}$$

$$\mathcal{E}_R = \left\{w \in \mathbb{C} : \left(\frac{2\Re(w)}{R+1/R}\right)^2 + \left(\frac{2\Im(w)}{R-1/R}\right)^2 < 1\right\}.$$
$$(1.4.36)$$

We denote by $\mathbf{H}(\mathcal{B})$ the family of all functions that are analytic in a region $\mathcal{B}$ of the complex plane $\mathbb{C}$. We remark here, that $f \in \mathbf{H}(\mathcal{D}_d)$ us equivalent to the above assumption, that $|\tilde{f}(y)| \le C \exp(-d |y|)$ for $y \in \mathbb{R}$. Let $\mathbf{H}_\pi(\mathcal{D}_d)$ denote the subset of all periodic functions f in $\mathbf{H}(\mathcal{D}_d)$ such that for all $(x,y) \in \mathbb{R}^2$ and $x + iy \in \mathcal{D}_d$, we have $f(x + 2\pi + iy) = f(x + iy)$.

Two transformations will be useful for our purposes:

$$w = w(z) = e^{iz} \quad \Longleftrightarrow \quad z = -i \log(w), \quad \text{and}$$

$$\zeta = \zeta(w) = \frac{1}{2}\left(w + \frac{1}{w}\right) \quad \Longleftrightarrow \quad w = \zeta \pm i \sqrt{1 - \zeta^2}.$$
$$(1.4.37)$$

Given any $f \in \mathbf{H}_\pi(\mathcal{D}_d)$, the transformation $z = z(w)$ replaces $f(z)$ with $f(-i \log(w)) = F(w)$, with $F \in \mathbf{H}(\mathcal{A}_R)$, and with $R = e^d$. Similarly, given $F \in \mathbf{H}(\mathcal{A}_R)$, we can set $w = \exp(iz)$ to transform

F defined on $\mathcal{A}_R$ into f defined on $\mathcal{D}_d$, and where $f(x + 2\pi + iy) = f(x + iy)$ for all $z = x + iy \in \mathcal{D}_d$.

Note also, that the transformation $z = (1/i)\log(w)$ is a one–to–one transformation of $\mathcal{A}_R$ onto $\mathcal{D}_{\pi,d}$.

Every $f \in \mathbf{H}(\mathcal{A}_R)$ has a *Laurent expansion* of the form

$$F(w) = \sum_{k \in \mathbb{Z}} c_k\, w^k, \quad , w \in \mathcal{A}_R, \tag{1.4.38}$$

where the c_k can be obtained, e.g., using the formula

$$c_k = \frac{1}{2\pi} \int_{-\pi}^{\pi} F\left(e^{i\theta}\right) e^{-in\theta}\, d\theta, \quad n = 0, \pm 1, \pm 2, \dots. \tag{1.4.39}$$

Denote by $\mathbf{H}_e(\mathcal{A}_R)$ the set of all $F \in \mathbf{H}(\mathcal{A}_R)$ such $c_{-n} = c_n$, or equivalently, such that $F(1/w) = F(w)$ in (1.4.28).

Let us next invoke the mapping $\zeta = \zeta(w)$ of (1.4.37) above. This mapping satisfies the equation $\zeta(1/w) = \zeta(w)$, and although it is not a conformal map from $\mathcal{A}_R$ onto $\mathcal{E}_R$ (see Equation (1.4.36) above), it provides a one to one correspondence of functions $F \in \mathbf{H}_e(\mathcal{A}_R)$ with the functions $G \in \mathbf{H}(\mathcal{E}_R)$. Indeed, we have

$$\frac{1}{2}(w^n + w^{-n}) = T_n(\zeta) \tag{1.4.40}$$

with $T_n(\zeta)$ the Chebyshev polynomial. The polynomials T_n may also be defined either by the the equations $\xi = \cos(\theta)$ and $T_n(\xi) = \cos(n\,\theta)$, or recursively by means of the equations

$$\begin{cases} T_0(\zeta) = 1, \quad T_1(\zeta) = \zeta \\[2mm] T_{n+1}(\zeta) = 2\zeta\, T_n(\zeta) - T_{n-1}(\zeta), \quad n = 1, 2, \cdots. \end{cases} \tag{1.4.41}$$

Hence, it follows that if $F \in \mathbf{H}_e(\mathcal{A}_R)$, then the corresponding function G defined as above has the series representation

$$G(\zeta) = c_0 + 2 \sum_{n=1}^{\infty} c_n\, T_n(\zeta), \quad \zeta \in \mathcal{E}_R. \tag{1.4.42}$$

We shall use the following $\mathbf{H}^1$ norms in what follows:

$$\|f\|_{\mathcal{D}_d}^1 = \max_{m=0,1} \int_0^{2\pi} \left| f(x + (-1)^m \, i \, d^-) \, dx \right|$$

$$\|F\|_{\mathcal{A}_R}^1 = \max_{m=\pm 1} \int_{|w|=(R^-)^m} |F(w) \, dw| \qquad (1.4.43)$$

$$\|G\|_{\mathcal{E}_R}^1 = \lim_{\rho \to R^-} \int_{\zeta \in \partial \mathcal{E}_\rho} |G(\zeta) \, d\zeta|.$$

Of course these three norms have the same values, under the assumption that the three functions f, F, and G are related as above.

We next use complex variables to prove that under conditions of periodicity and analyticity of f, the wavelet polynomials of Theorem 1.4.2 and the Trapezoidal and Midordinate formulas converge exponentially, which is much better as compared with the $\mathcal{O}(h^2)$ convergence usually associated with these methods of approximation, in the absence of periodicity and analyticity.

Let d denote a positive constant, let f be periodic with period 2π on $\mathbb{R}$, and let f be analytic in the infinite strip $\mathcal{D}_d$ defined in (1.4.36). Let $0 < \varepsilon < 1$, and let us define the rectangle $\mathcal{D}_d^\varepsilon$ by

$$\mathcal{D}_d^\varepsilon = \{ z \in \mathbb{C} : |\Re z| < 1/\varepsilon, \ |\Im z| < (1 - \varepsilon)d \}. \qquad (1.4.44)$$

Under the assumption that f is also periodic with period 2π on $\mathbb{R}$, it follows that $f(x \pm i y)$, $|y| \leq d$, is as a function of x, also periodic with period 2π on $\mathbb{R}$. Let us assume furthermore that the integral of $|f|$ taken along the vertical sides of the boundary $\partial \mathcal{D}_d^\varepsilon$ of rectangle $\mathcal{D}_d^\varepsilon$ remains bounded as $\varepsilon \to 0$, and such that the norm $\|f\|_{\mathcal{D}_d}^1$ is finite. Let $x \in [-\pi, \pi]$, $h = \pi/N$, and consider the integral

$$E_\varepsilon(f) \equiv \frac{\sin\left\{ \frac{\pi}{h}(x + b\,h) \right\}}{2\pi i} \int_{\partial \mathcal{D}_d^\varepsilon} \frac{f(t)}{(t - x) \sin\left\{ \frac{\pi}{h}(t + b\,h) \right\}} \, dt, \quad (1.4.45)$$

see Equation (3.1.4) of [S1].

We then obtain the following results by proceeding as in the constructive proofs of Theorem 3.1.2 of [S1] and Theorem 1.4.2 above.

Theorem 1.4.8 *Let f be periodic on $\mathbb{R}$ with period 2π and analytic in $\mathcal{D}_d$, let $b \in \mathbb{R}$, and let $\|f\|_{\mathcal{D}_d}^1 < \infty$.*

(i) If $N h = \pi$, then for all $x \in \mathbb{R}$,

$$f(x) - f_h(b, x) = \frac{\sin(N(x + \pi b))}{2\pi i}.$$

$$\cdot \int_{-\pi}^{\pi} \left\{ \frac{f(t - i d)}{\tan\left(\frac{1}{2}(t - x - id)\right) \sin(N(t + \pi b - id))} \right.$$

$$\left. - \frac{f(t + i d)}{\tan\left(\frac{1}{2}(t - x + id)\right) \sin(N(t + \pi b + id))} \right\} dt,$$

$$(1.4.46)$$

and

$$|f(x) - f_h(b, x)| \le \frac{\|f\|_{\mathcal{D}_d}^1}{2\pi \, \tanh\left(\frac{d}{2}\right) \sinh(N d)}. \qquad (1.4.47)$$

(ii) If $(2N + 1) h = 2\pi$, then for all $x \in \mathbb{R}$,

$$f(x) - f_h(b, x) = \frac{\sin\left(\left(N + \frac{1}{2}\right)(x + \pi b)\right)}{2\pi i}.$$

$$\cdot \int_{-\pi}^{\pi} \left\{ \frac{f(t - i d)}{\sin\left(\frac{1}{2}(t - x - id)\right) \sin\left(\left(N + \frac{1}{2}\right)(t + \pi b - id)\right)} \right.$$

$$\left. - \frac{f(t + i d)}{\sin\left(\frac{1}{2}(t - x + id)\right) \sin\left(\left(N + \frac{1}{2}\right)(t + bh + id)\right)} \right\} dt,$$

$$(1.4.48)$$

and

$$|f(x) - f_h(b, x)| \le \frac{\|f\|_{\mathcal{D}_d}^1}{2\pi \sinh\left(\frac{d}{2}\right) \sinh\left(\left(N + \frac{1}{2}\right) d\right)}. \qquad (1.4.49)$$

Proof. Obtaining the above norm bounds from the integral (1.4.45) is straight forward. We use the inequalities $|\sin(\pi/h(t \pm i\,d)|) \geq \sinh(\pi\,d/h)$ (see [S1], Eq. (3.1.8)) and also, $|\tan((1/2)(t - x \pm i\,d))| \geq \tanh(d/2)$. ∎

It is interesting that we are able to get a single expression for the errors of each combination of quadratures, trapezoidal and Midordinate, for b either 0 or $1/2$, and with h being defined by either $h = \pi/N$ or by $h = 2\pi/(2N + 1)$. We prove only one of the formulas, since the proofs in all cases are similar.

We now use the notations of the Trapezoidal and Midordinate rules introduced in (1.4.30) and (1.4.31) above.

Theorem 1.4.9 *Let f satisfy the conditions of Theorem 1.4.8. If either $b = 0$ or $b = 1/2$, if $f_h(b,x)$ is defined as in (1.4.1), and if either $h = \pi/N$ or $h = 2\pi/(2N+1)$, then*

$$\int_{-\pi}^{\pi} (f(x) - f_h(b,x))\,dx = \frac{i}{\exp\left(\frac{\pi d}{h}\right)} \int_{-\pi}^{\pi} G(t)\,dt, \qquad (1.4.50)$$

with

$$G(t)$$
$$= \frac{f(t - id)\,\exp\left(\frac{-\pi i}{h}(t + bh)\right)}{\sin\left(\frac{\pi}{h}(t + bh - id)\right)} - \frac{f(t + id)\,\exp\left(\frac{\pi i}{h}(t + bh)\right)}{\sin\left(\frac{\pi}{h}(t + bh + id)\right)}.$$
$$(1.4.51)$$

Moreover,

$$\left| \int_{-\pi}^{\pi} \{f(x) - f_h(b,x)\}\,dx \right| \leq \frac{\|f\|_{\mathcal{D}_d}^1}{\exp\left(\frac{\pi d}{h}\right)\sinh\left(\frac{\pi d}{h}\right)}. \qquad (1.4.52)$$

where $\|f\|_{\mathcal{D}_d}^1$ is defined as in (1.4.43).

Proof. We sketch here the derivation of the error of quadrature using the formula (1.4.46) with $h = \pi/N$, omitting a similar proof for the derivation of the same result using the formula (1.4.48).

Considered as a function of x, the integrand (1.4.46) is periodic, with period 2π. Integrating both sides of (1.4.46), we consider first the evaluation of the integral

$$I_1 \equiv \frac{1}{2\pi i} \int_{-\pi}^{\pi} \frac{\sin\left\{\frac{\pi}{h}(x + bh)\right\}}{\tan\left\{\frac{\pi}{P}(x - t + id)\right\}} \, dx. \tag{1.4.53}$$

Note that, since both x and t are real, and since the integrand in (1.4.53) is periodic, of period 2π with respect to x, replacing x by $x + t$ does not change the value of this integral. Note also, that the value of the integral with respect to x of the second term in (1.4.46) can be readily deduced from the value of the integral of the first term, by just replacing id by $-id$.

We can deduce the expression

$$\frac{1}{\tan(\theta/2 + id/2)} = \frac{\sin(\theta) - i\,\sinh(d)}{\cosh(d) - \cos(\theta)}. \tag{1.4.54}$$

Moreover, it follows, using the identity

$$\frac{1}{\cosh(d) - \cos(\theta)} = \sum_{k=-\infty}^{\infty} \frac{e^{-ik\theta}}{R^{|k|}}, \tag{1.4.55}$$

in which $R = \exp(d)$, that for any integer ℓ,

$$\frac{1}{2\pi} \int_0^{2\pi} \frac{e^{i\ell\theta}}{\cosh(d) - \cos(\theta)} \, d\theta = \frac{1}{R^{|\ell|}}. \tag{1.4.56}$$

The result (1.4.52) can now be readily obtained using this identity. ∎

1.4.5   ALGEBRAIC INTERPOLATION AND QUADRATURE

We derive here the polynomial forms of the expressions (ii) and (iii) of Corollaries 1.4.3–1.4.6. To this end, we shall use the notation $R = \exp(d)$. Under the transformations (1.4.37), a function $f = f(x)$ which is periodic with period 2π, which is an even function on $\mathbb{R}$, and which is analytic in the region $\mathcal{D}_d$ of (1.4.36), then becomes a function $F = F(\zeta)$ that is analytic in the ellipse $\mathcal{E}_R$ defined in (1.4.36). Such a function F can be readily approximated by algebraic polynomials in ξ on $[-1, 1]$.

Similarly, if $f(x)$ is an odd function that is periodic with period 2π on $\mathbb{R}$ and analytic in $\mathcal{D}_d$, we can set $g(x) = f(x)\sin(x)$, and under the above stated transformations, $g(x)$ becomes a function $G(\zeta)$ which is again analytic in the ellipse $\mathcal{E}_R$. In this case, we shall consider the approximation of $\sqrt{1 - \xi^2}\, F(\xi)$ on the interval $[-1, 1]$. Finally, we shall assume that f is real–valued on $\mathbb{R}$, so that the corresponding functions F and G are real-valued on the interval $[-1, 1]$. In this case, the two $(\pm)$ integral terms of (1.4.46) have the same value.

Although the polynomials T_m have already been defined in (1.4.41) above, we now reintroduce them here, along with the Chebyshev polynomials of the second kind, U_m as well as their zeros, t_k and τ_k, such that $T_m(t_k) = 0$ and $U_m(\tau_k) = 0$:

$$
\begin{aligned}
&T_0(\xi) = 1; \quad T_1(\xi) = \xi = \cos(\theta), \\
&T_m(\xi) = \cos(m\,\theta), \quad t_k = \cos\left(\frac{k - 1/2}{m}\,\pi\right) \\
&U_m(\xi) = \frac{\sin\{(m+1)\,\theta\}}{\sin(\theta)}, \quad \tau_k = \cos\left(\frac{k}{m+1}\,\pi\right).
\end{aligned}
\tag{1.4.57}
$$

Algebraic Interpolations. We now examine each of the interpolation procedures of the (ii) and (iii)–parts of Corollaries 1.4.3–1.4.6, for algebraic polynomial interpolation. To this end, we shall use the notations

$$
\begin{aligned}
w_k &= \cos(k\,\pi/N), \\
x_k &= \cos((2k - 1)\,\pi/(2\,N)) \\
y_k &= \cos((2k\,\pi/(2\,N + 1)) \\
z_k &= \cos((2k - 1)\,\pi/(2\,N + 1)),
\end{aligned}
$$

throughout this section.

We omit the proofs of the following theorems, since they involve simple trigonometric identities, the most complicated of these being

$$
\begin{aligned}
\frac{1}{\tan\left(\frac{u-v}{2}\right)} &= -\frac{\sin(u) + \sin(v)}{\cos(u) - \cos(v)} \\[2ex]
\frac{1}{\sin\left(\frac{u-v}{2}\right)} &= -\frac{2\sin\left(\frac{u+v}{2}\right)}{\cos(u) - \cos(v)}.
\end{aligned}
\tag{1.4.58}
$$

These identities enable us to obtain the necessary monomial terms $\xi - w_k$ and $\xi - \tau_k$ in the denominators of the interpolation polynomials.

Theorem 1.4.10 *Let f be periodic on $\mathbb{R}$, with period 2π. Then:*

(a) If f is an even function defined on $\mathbb{R}$, the function $f_h(0,x)$ of (1.4.14) can also be represented in the form of a polynomial of degree N, i.e.,

$$f_h(0,x) = \sum_{k=0}^{N} \frac{P_{N+1}(\xi)}{(\xi - w_k)\, P'_{N+1}(w_k)}\, F(w_k), \qquad (1.4.59)$$

with $w_k = \cos(k\,\pi/N)$, and with $P_{N+1}(\xi) = T_{N-1}(\xi) - T_{N+1}(\xi)$.

(b) If f is an odd function defined on $\mathbb{R}$, the function $f_h(0,x)$ of (1.4.15) can also be represented on $[-1,1]$ in the form

$$f_h(0,x) = \sum_{k=1}^{N-1} \frac{Q(\xi)}{(\xi - w_k)\, Q'(\tau_k)}\, F(w_k), \qquad (1.4.60)$$

with $w_k = \cos(k\,\pi/N)$, and with $Q(\xi) = \sqrt{1 - \xi^2}\, U_{N-1}(\xi)$.

Theorem 1.4.11 *Let f be periodic, with period 2π on $\mathbb{R}$. Then:*

(a) If f is an even function defined on $\mathbb{R}$, the function $f_h(1/2,x)$ of (1.4.17) can also be represented in the form of a polynomial of degree $N - 1$, i.e.,

$$f_h\left(\frac{1}{2},x\right) = \sum_{k=1}^{N} \frac{T_N(\xi)}{(\xi - x_k)\, T'_N(x_k)}\, F(x_k), \qquad (1.4.61)$$

with $x_k = \cos((2\,k - 1)\,\pi/(2\,N))$.

(b) If f is an odd function defined on $\mathbb{R}$, the function $f_h\left(\frac{1}{2},x\right)$ of (1.4.18) can also be represented on $[-1,1]$ in the form

$$f_h\left(\frac{1}{2},x\right) = \sum_{k=1}^{N} \frac{Q(\xi)}{(\xi - x_k)\, Q'(x_k)}\, F(x_k), \qquad (1.4.62)$$

with $x_k = \cos((2\,k-1)\,\pi/(2\,N))$, and with $Q(\xi) = \sqrt{1 - \xi^2}\, (U_{N-1}(\xi) + U_{N+1}(\xi))$.

Theorem 1.4.12 *Let f be periodic, with period 2π. Then:*

(a) If f is an even function defined on $\mathbb{R}$, the function $f_h(1/2, \xi)$ of (1.4.20) is also represented in the form of a polynomial of degree N, i.e.,

$$f_h\left(\tfrac{1}{2}, x\right) = P_{N+1}(\xi) \cdot$$

$$\cdot \left(\frac{F(1)}{2\,(\xi - 1)\,P'_{N+1}(1)} + \sum_{k=1}^{N} \frac{F(y_k)}{(\xi - y_k)\,P'_{N+1}(y_k)}\right),$$

$$(1.4.63)$$

with $y_k = \cos((2\,k\,\pi/(2\,N+1)))$, and with $P_{N+1}(\xi) = T_{N+1}(\xi) - T_N(\xi)$.

(b) If f is an odd function defined on $\mathbb{R}$, the function $f_h(1/2, x)$ of (1.4.21) is also represented on $[-1, 1]$ in the form

$$f_h\left(\frac{1}{2}, x\right) = \sum_{k=1}^{N} \frac{Q(\xi)}{(\xi - y_k)\,Q'(y_k)}\,F(y_k), \qquad (1.4.64)$$

with $y_k = \cos(2\,k\,\pi/(2\,N+1))$, and with

$$Q(\xi) = \sqrt{1 - \xi^2}\,\left(U_N(\xi) + U_{N-1}(\xi)\right).$$

Theorem 1.4.13 *Let f be periodic, with period $2\pi = (2\,N+1)\,h$. Then:*

(a) If f is an even function defined on $\mathbb{R}$, the function $f_h(1/2, x)$ of (1.4.23) is also represented in the form of a polynomial of degree N, i.e.,

$$f_h\left(\tfrac{1}{2}, x\right) = P_{N+1}(\xi) \cdot$$

$$\cdot \left(\frac{F(-1)}{(\cos(x) + 1)\,P'(-1)} + \sum_{k=0}^{N-1} \frac{F(z_k)}{(\xi - z_k)\,P'_{N+1}(z_k)}\right),$$

$$(1.4.65)$$

with $z_k = \cos((2\,k+1)\,\pi/(2\,N+1))$, and with $P_{N+1}(\xi) = T_N(\xi) + T_{N+1}(\xi)$.

(b) If f is an odd function defined on $\mathbb{R}$, the function $f_h(1/2, x)$ of (1.4.24) is also represented on $[-1, 1]$ in the form

$$f_h\left(\frac{1}{2}, x\right) = \sum_{k=1}^{N} \frac{Q(\xi)}{(\xi - z_k) Q'(z_k)} F(z_k), \qquad (1.4.66)$$

with $z_k = \cos((2k+1)\pi/(2N+1))$, and with

$$Q(\xi) = \sqrt{1 - \xi^2} \, (U_{N+1}(\xi) + U_N(\xi)).$$

Algebraic Interpolation Errors.

Using the transformations (1.4.37), we replace x in Theorem 1.4.8 by ζ, taking $\xi = \Re\zeta$, and recalling that $R = e^d$, , $\xi = \Re\zeta$, we get the following result:

Theorem 1.4.14 *(a) Let $\Phi(\xi)$ denote either one of the polynomials on the right hand side of (1.4.59) or (1.4.61). Then we have, for $-1 \leq \xi \leq 1]$,*

$$|F(\xi) - \Phi(\xi)| \leq \frac{R^{1/2} + R^{-1/2}}{\pi \, (R^{1/2} - R^{-1/2}) \, (R^N - R^{-N})} \, \|F\|_{\mathcal{E}_R}^1. \quad (1.4.67)$$

(b) Let $\Psi(\xi)$ denote either of the polynomials on the right hand sides of (1.4.63) or (1.4.65). Then we have, for $-1 \leq \xi \leq 1$,

$$|F(\xi) - \Psi(\xi)| \leq \frac{1}{\pi \, (R^N - R^{-N})} \, \|F\|_{\mathcal{E}_R}^1. \quad (1.4.68)$$

If f is an odd function of period 2π defined on $\mathbb{R}$, which is analytic in the region $\mathcal{D}_d$ of (1.4.36), then g defined by $g(x) = f(x)\sin(x)$ is an even function of period 2π on $\mathbb{R}$ which is analytic in the region $\mathcal{D}_d$ of (1.4.36). By using the transformations (1.4.37) we transform the functions $f(x)$ and $g(x)$ into functions $F(\zeta)$ and $G(\zeta)$ defined on the ellipse $\mathcal{E}_R$, on which G is analytic and normed as in (1.4.43). We arrive at the following theorem.

Theorem 1.4.15 *Set $G(\xi) = F(\xi)\sqrt{1 - \xi^2}$, and let G be analytic in $\mathcal{E}_R$.*

(a) Let Φ denote the term in the sum on the right hand side of either (1.4.60) or (1.4.62). Then $|F(\xi) - \Phi(\xi)|$ is bounded for all $\xi \in [-1, 1]$ by the right hand side of (1.4.67).

(b) Let Ψ denote the term in the sum on the right hand side of either (1.4.64) or (1.4.66). Then $|F(\xi) - \Psi(\xi)|$ is bounded for all $\xi \in [-1, 1]$ by the right hand side of (1.4.68).

Connections with Quadratures on $[-1, 1]$. The above interpolation schemes enable a variety of quadratures. All of these formulas, including, e.g., the well known *Gauss–Chebyshev* and the *Clenshaw–Curtis* procedures can be readily obtained from the above interpolation results.

Let us state three theorems, the proofs of which are straight–forward consequences of the results already obtained above. As stated, there appears to be little difference between the three theorems; the difference occurs when we assume that not only the function f, but also the function g in Theorems 1.4.18 and 1.4.19 is analytic on the region $\mathcal{D}_d$ defined in (1.4.45). Analogously, we also have that $F(\xi) = \sqrt{1 - \xi^2}\, G(\xi)$, where $G \in \mathbf{H}^1(\mathcal{E}_R)$.

It is convenient to use the following notations for the variety of formulas that are possible:

1. *Gauss–Chebyshev and Related Quadratures.* The case when $f \in \mathbf{H}^1(\mathcal{D}_d)$, and when f is an even function of period 2π on $\mathbb{R}$; equivalently, let F be defined on $\mathcal{E}_R$ with $F = f$ based on the transformations (1.4.37), so that $F \in \mathbf{H}^1(\mathcal{E}_R)$, where $F(\zeta) = \mathcal{F}\left(e^{i\theta}\right) = f(\theta)$. In this case, we have

$$I(f) = \int_0^\pi f(\theta)\, d\theta \quad \Longleftrightarrow \quad \mathcal{I}(F) = \int_{-1}^1 \frac{F(\xi)}{\sqrt{1 - \xi^2}}\, d\xi. \quad (1.4.69)$$

For this case, we have four quadrature formula approximations corresponding to the explicit special cases of Corollary 1.4.3 (ii), with $h = \pi/N$, Corollary 1.4.4 (ii), with $h = 2\pi/(2N+1)$, Corollary 1.4.5 (ii), with $h = \pi/N$ and Corollary 1.4.6 (ii), with $h = 2\pi/(2N+1)$:

$$Q_h^I(f) = h\left\{\frac{1}{2}f(0) + \frac{1}{2}f(\pi) + \sum_{k=1}^{N-1} f(k\,h)\right\},$$
$$h = \pi/N\,,$$

$$Q_h^{II}(f) = h\sum_{k=1}^{N} f\left(\left(k - \frac{1}{2}\right)h\right),\quad h = \pi/N\,,$$

$$Q_h^{III}(f) = h\left\{\frac{1}{2}f(0) + \sum_{k=1}^{N} f(k\,h)\right\},$$
$$h = \pi/(N + 1/2).$$

$$Q_h^{IV}(f) = h\left\{\frac{1}{2}f(\pi) + \sum_{k=1}^{N} f\left(\left(k - \frac{1}{2}\right)h\right)\right\},$$
$$h = \pi/(N + 1/2)\,,$$

(1.4.70)

These quadratures become the following, under the transformations (1.4.37):

$$\mathcal{Q}_N^I(F) = h\left\{\frac{1}{2}F(-1) + \frac{1}{2}F(1) + \sum_{k=1}^{N-1} F(w_k)\right\},$$
$$h = \pi/N\,,$$

$$\mathcal{Q}_N^{II}(F) = h\sum_{k=1}^{N} F(x_k),\quad h = \pi/N\,,$$

$$\mathcal{Q}_N^{III}(F) = h\left\{\frac{1}{2}F(1) + \sum_{k=1}^{N} F(y_k)\right\},$$
$$h = \pi/(N + 1/2)$$

$$\mathcal{Q}_N^{IV}(F) = h\left\{\frac{1}{2}F(-1) + \sum_{k=1}^{N} F(z_k)\right\},$$
$$h = \pi/(N + 1/2).$$

(1.4.71)

We can then state the following

Theorem 1.4.16 *Let* $f \in \mathbf{H}(\mathcal{D}_d)$, *and let* f *be an even function of period* 2π *defined on* $\mathbb{R}$. *Then*

$$I(f) = \int_0^\pi f(x)\,dx$$

$$
\begin{aligned}
&= Q_N^I(f) + E_h^I(f) \\
&= Q_N^{II}(f) + E_h^{II}(f) \\
&= Q_N^{III}(f) + E_h^{III}(f) \\
&= Q_N^{VI}(f) + E_h^{IV}(f),
\end{aligned}
\tag{1.4.72}
$$

with each of the errors $E_h^J(f)$ *bounded as in Theorem 1.4.9.*

Equivalently, let $F \in \mathbf{H}^1(\mathcal{E}_R)$. *Then*

$$\mathcal{I}(F) = \int_{-1}^1 \frac{F(\xi)}{\sqrt{1 - \xi^2}}\,d\xi$$

$$
\begin{aligned}
&= \mathcal{Q}_N^I(F) + \varepsilon_N^I(F) \\
&= \mathcal{Q}_N^{II}(F) + \varepsilon_N^{II}(F) \\
&= \mathcal{Q}_N^{III}(F) + \varepsilon_N^{III}(F) \\
&= \mathcal{Q}_N^{IV}(F) + \varepsilon_N^{IV}(F),
\end{aligned}
\tag{1.4.73}
$$

where, ε_N^I *and* ε_N^{II} *are bounded by*

$$\frac{2\,\|F\|_{\mathcal{E}_R}^1}{R^N\left(R^N - R^{-N}\right)} \tag{1.4.74}$$

whereas, ε_N^{III} *and* ε_N^{IV} *are bounded by*

$$\frac{2\,\|F\|_{\mathcal{E}_R}^1}{R^{N+1/2}\left(R^{N+1/2} - R^{-N-1/2}\right)}, \tag{1.4.75}$$

and where $\|F\|_{\mathcal{E}_R}^1$ *is defined as in (1.4.43).*

2. *Clenshaw–Curtis Type Formulas.* These are also readily obtainable, via use of the above approach. Consider, for example, the following theorem, in which we again assume that f is a periodic function of period 2π defined on $\mathbb{R}$, that $f \in \mathbf{H}^1(\mathcal{D}_d)$,

and furthermore, that $f(x) = g(x) \sin(x)$. Equivalently, we assume that $F \in \mathbf{H}(\mathcal{E}_R)$, with $F(\xi) = G(\xi)\sqrt{1 - \xi^2}$. Note, it is obvious at this point, that the functions g and G need not be bounded at the end-points of integration, whereas the values $f(0)$, $f(\pi)$, and $F(\pm 1)$ must, in fact, exist and be finite. If also $g \in \mathbf{H}(\mathcal{D}_d)$ then the terms $f(0)$, $f(\pi)$, and $F(\pm 1)$ must all vanish, as was tacitly assumed by Clenshaw and Curtis [ClCu] and Imhof [Im].

Hence, paraphrasing the steps that we took preceding Theorem 1.4.16 above, we get the following results, for which the points w_k, x_k, y_k, and z_k are the same as above:

$$Q_h^I(g) = h\left\{\frac{1}{2}f(0) + \frac{1}{2}f(\pi) + \sum_{k=1}^{N-1}\sin(k\,h)\,g(k\,h)\right\},$$
$$h = \pi/N,$$
$$Q_h^{II}(g) = h\sum_{k=1}^{N}\sin\left(\left(k - \frac{1}{2}\right)h\right)g\left(\left(k - \frac{1}{2}\right)h\right),$$
$$h = \pi/N,$$
$$Q_h^{III}(g) = h\left\{\frac{1}{2}f(0) + \sum_{k=1}^{N}\sin(k\,h)\,g(k\,h)\right\},$$
$$h = \pi/(N + 1/2),$$
$$Q_h^{IV}(g) = \begin{aligned}&\frac{h}{2}f(\pi)\\ &+h\sum_{k=1}^{N}\sin\left(\left(k - \frac{1}{2}\right)h\right)g\left(\left(k - \frac{1}{2}\right)h\right),\end{aligned}$$
$$h = \pi/(N + 1/2).$$

$$(1.4.76)$$

Analogously, the quadratures over $(-1, 1)$ take the form

$$Q_N^I(G) = h \left\{ \frac{1}{2} F(-1) + \frac{1}{2} F(1) + \sum_{k=1}^{N-1} \sin(kh) \, G(w_k) \right\},$$
$$h = \pi/N \,,$$

$$Q_N^{II}(G) = h \sum_{k=1}^{N} \sin\left(\left(k - \frac{1}{2} \right) h \right) G(x_k), \quad h = \pi/N \,,$$

$$Q_N^{III}(G) = h \left\{ \frac{1}{2} F(1) + \sum_{k=1}^{N} \sin(k \, h) \, G(y_k) \right\},$$
$$h = \pi/(N + 1/2) \,,$$

$$Q_N^{IV}(G) = h \left\{ \frac{1}{2} F(-1) + \sum_{k=1}^{N} \sin\left(\left(k - \frac{1}{2} \right) h \right) G(z_k) \right\},$$
$$h = \pi/(N + 1/2).$$

$$(1.4.77)$$

We can then state the following

Theorem 1.4.17 *Let* $f(x) = \sin(x) \, g(x)$, *let* $f \in \mathbf{H}(\mathcal{D}_d)$, *and let* f *be an even function of period* 2π *defined on* $\mathbb{R}$. *Then*

$$\int_0^\pi \sin(x) \, g(x) \, dx$$

$$\begin{aligned}
&= Q_N^I(g) + E_h^I(f) \\
&= Q_N^{II}(g) + E_h^{II}(f) \\
&= Q_N^{III}(g) + E_h^{III}(f) \\
&= Q_N^{IV}(g) + E_h^{IV}(f)
\end{aligned}$$

$$(1.4.78)$$

with each of the errors $E_h^J(f)$ *bounded as in Theorem 1.4.9.*

Equivalently, let $F(\xi) = \sqrt{1 - \xi^2} \, G(\xi)$, *and let* $F \in \mathbf{H}^1(\mathcal{E}_R)$. *Then*

$$\int_{-1}^1 G(\xi) \, d\xi$$

$$\begin{aligned}
&= Q_N^I(G) + \varepsilon_N^I(F) \\
&= Q_N^{II}(G) + \varepsilon_N^{II}(F) \\
&= Q_N^{III}(G) + \varepsilon_N^{III}(F) \\
&= Q_N^{IV}(G) + \varepsilon_N^{IV}(F),
\end{aligned}$$

$$(1.4.79)$$

where, ε_N^I *and* ε_N^{II} *are bounded by*

$$\frac{2\,\|F\|_{\mathcal{E}_R}^1}{R^N\,(R^N - R^{-N})} \tag{1.4.80}$$

whereas, ε_N^{III} *and* ε_N^{IV} *are bounded by*

$$\frac{2\,\|F\|_{\mathcal{E}_R}^1}{R^{N+1/2}\,(R^{N+1/2} - R^{-N-1/2})}\,, \tag{1.4.81}$$

and where $\|F\|_{\mathcal{E}_R}^1$ *is defined as in (1.4.79).*

3. *Other Formulas.* Similar results could easily be deduced for approximation of integrals of the form

$$\int_{-1}^{1} G(x)\,\sqrt{1 - x^2}\,dx \tag{1.4.82}$$

under the assumption that $F \in \mathbf{H}(\mathcal{E}_R)$, with $F(x) = (1 - x^2)\,G(x)$. Equivalently, we assume that $f(x) = \sin^2(x)\,g(x)$ is an even function of period $2\,\pi$ defined on $\mathbb{R}$, that belongs to $\mathbf{H}(\mathcal{D}_d)$. It is then easily seen, that the $\sin(\cdot)$ terms of the previous theorem are then just replaced by $\sin^2$ terms.

Hence, paraphrasing the above *Clenshaw-Curtis*–type presentation, we get the following results, for which the points w_k, x_k, y_k, and z_k are the same as in (1.4.71):

$$Q_h^I(g) = h \left\{ \frac{1}{2} f(0) + \frac{1}{2} f(\pi) + \sum_{k=1}^{N-1} \sin^2(k\,h)\, g(k\,h) \right\},$$
$$h = \pi/N,$$
$$Q_h^{II}(g) = h \sum_{k=1}^{N} \sin^2\left(\left(k - \frac{1}{2}\right) h\right) g\left(\left(k - \frac{1}{2}\right) h\right),$$
$$h = \pi/N,$$
$$Q_h^{III}(g) = h \left\{ \frac{1}{2} f(0) + \sum_{k=1}^{N} \sin^2(k\,h)\, g(k\,h) \right\},$$
$$h = \pi/(N + 1/2),$$
$$Q_h^{IV}(g) =$$
$$h \left\{ \frac{1}{2} f(\pi) + \sum_{k=1}^{N} \sin^2\left(\left(k - \frac{1}{2}\right) h\right) g\left(\left(k - \frac{1}{2}\right) h\right) \right\},$$
$$h = \pi/(N + 1/2).$$
$$(1.4.83)$$

Analogously, the quadratures over $(-1, 1)$ take the form

$$\mathcal{Q}_N^I(G) = h \left\{ \frac{1}{2} F(-1) + \frac{1}{2} F(1) + \sum_{k=1}^{N-1} \sin^2(kh) G(w_k) \right\},$$
$$h = \pi/N,$$
$$\mathcal{Q}_N^{II}(G) = h \sum_{k=1}^{N} \sin^2\left(\left(k - \frac{1}{2}\right) h\right) G(y_k), \quad h = \pi/N,$$
$$\mathcal{Q}_N^{III}(G) = h \left\{ \frac{1}{2} F(1) + \sum_{k=1}^{N} \sin^2(k\,h) G(z_k) \right\},$$
$$h = \pi/(N + 1/2)$$
$$\mathcal{Q}_N^{IV}(G) = h \left\{ \frac{1}{2} F(-1) + \sum_{k=1}^{N} \sin^2\left(\left(k - \frac{1}{2}\right) h\right) G(x_k) \right\},$$
$$h = \pi/(N + 1/2).$$
$$(1.4.84)$$

We can then state the following

Theorem 1.4.18 *Let* $f(x) = \sin^2(x)\, g(x)$, *let* $f \in \mathbf{H}(\mathcal{D}_d)$, *and let* f *be an even function of period* 2π *defined on* $\mathbb{R}$. *Then*

$$\int_0^\pi \sin^2(x)\, g(x)\, dx$$

$$
\begin{aligned}
&= Q_N^I(g) + E_h^I(f) \\
&= Q_N^{II}(g) + E_h^{II}(f) \\
&= Q_N^{III}(g) + E_h^{III}(f) \\
&= Q_N^{IV}(g) + E_h^{IV}(f)
\end{aligned}
\tag{1.4.85}
$$

with each of the errors $E_h^J(f)$ *bounded as in Theorem 1.4.9.*

Equivalently, let $F(\xi) = (1 - \xi^2)\, G(\xi)$, *and let* $F \in \mathbf{H}(\mathcal{E}_R)$. *Then*

$$\int_{-1}^1 G(\xi)\, d\xi$$

$$
\begin{aligned}
&= \mathcal{Q}_N^I(G) + \varepsilon_N^I(F) \\
&= \mathcal{Q}_N^{II}(G) + \varepsilon_N^{II}(F) \\
&= \mathcal{Q}_N^{III}(G) + \varepsilon_N^{III}(F) \\
&= \mathcal{Q}_N^{IV}(G) + \varepsilon_N^{IV}(F),
\end{aligned}
\tag{1.4.86}
$$

where, ε_N^I *and* ε_N^{III} *are bounded by*

$$\frac{2\, \|F\|_{\mathcal{E}_R}^1}{R^N\, (R^N - R^{-N})} \tag{1.4.87}$$

whereas, ε_N^{II} *and* ε_N^{IV} *are bounded by*

$$\frac{2\, \|F\|_{\mathcal{E}_R}^1}{R^{N+1/2}\, (R^{N+1/2} - R^{-N-1/2})}, \tag{1.4.88}$$

and where $\|F\|$ *is defined as in (1.4.79).*

1.4.6   Wavelet Differentiation

We shall discuss here, the approximation of derivatives via use of the derivatives of wavelet approximations. To this end, we shall use the notations

$$D_e(N,P,x)$$

$$= \frac{1}{2N}\left\{1 + \cos(2N\pi x/P) + 2\sum_{k=1}^{N-1}\cos(2k\pi x/P)\right\}$$

$$= \frac{\sin\{\frac{2N\pi x}{P}\}}{2N\tan\{\frac{\pi x}{P}\}},$$

$$D_o(N,P,x) = \frac{1}{2N+1}\left\{1 + 2\sum_{k=1}^{N}\cos(2k\pi x/P)\right\}$$

$$= \frac{\sin\{\frac{(N+1/2)\pi x}{P}\}}{(2N+1)\sin\{\frac{\pi x}{P}\}}.$$

$$(1.4.89)$$

which can be used for approximation on any interval $[a, a + P]$.

Differentiating the terms D_e and D_o with respect to x denoting these by $D'_{e/o}(M,x)$, we get

$$f_h(x) = \sum_k f(x_k)\, D_{e/o}(M,P,x-x_k),$$

$$(1.4.90)$$

$$f'_h(x) = \sum_k f(x_k)\, D'_{e/o}(M,P,x-x_k),$$

with $D_{e/o}$ as in $(1.4.30)$ and $(1.4.31)$, i.e.,

$$D'_e(N,P,x)$$
$$= -\frac{\pi}{P}\sin\left(\frac{2N\pi x}{P}\right) - \frac{2\pi}{NP}\sum_{k=1}^{N}k\sin\left(\frac{2k\pi x}{P}\right),$$

$$(1.4.91)$$

$$D'_o(N,P,x) = -\frac{4\pi}{(2N+1)P}\sum_{k=1}^{N}k\sin\left(\frac{2k\pi x}{P}\right),$$

and where, in $(1.4.91)$,

$$\sum_k = \sum_{k=0}^{2N}, \quad x_k = a + \frac{(k-1)P}{2N}, \qquad \text{for} \quad D_e$$

$$\sum_k = \sum_{k=1}^{2N+1}, \quad x_k = a + \frac{(k-1/2)P}{2N+1}, \quad \text{for} \quad D_o.$$

(1.4.92)

These formulas are not accurate in neighborhoods of end–points if the function f is not periodic of period P. And they are of course not accurate in neighborhoods of points where f is not smooth.

1.4.7   Wavelet Indefinite Integration

In this case, we integrate the first formula of (1.4.89) and then insert the indefinite integral notations

$$E_e(N,P,x) = \int^x D_e(N,P,t)\,dt$$

$$= \frac{1}{2N}\left\{ x + \frac{P}{2N\pi} \sin\left(\frac{2N\pi x}{P}\right) \right.$$

$$\left. + \frac{P}{\pi}\sum_{k=1}^{N-1}\frac{1}{k}\sin\left(\frac{2k\pi x}{P}\right) \right\}$$

$$E_o(N,P,x) = \int^x D_o(N,P,t)\,dt$$

$$= \frac{1}{(2N+1)}\left\{ x + \frac{P}{\pi}\sum_{k=1}^{N}\frac{1}{k}\sin\left(\frac{2k\pi x}{P}\right) \right\}.$$

(1.4.93)

This yields

$$\int_a^x f_h(t)\,dt$$
$$= \sum_k \alpha_k \int_a^x D_{e/o}(M,P,t-x_k)\,f(x_k)\,dt$$

(1.4.94)

$$= \sum_k \alpha_k \left\{ E_{e/o}(M,P,x-x_k) - E_{e/o}(M,P,a-x_k) \right\}.$$

where the α_k are either 0 or $1/2$, and with the x_k suitably defined as in (1.4.92) above, depending which of the interpolation schemes of (1.4.90) is used.

Several integration matrices are therefore possible, corresponding to Corollaries 1.4.4–1.4.7. Suppose for example, we have the representation (1.4.14) for f_h, over the interval $[a, b]$, so that $P = b - a$, $h = P/(2N)$ and $x_k = a + (k-1)h$, for $k = 1, \ldots, 2N+1$. We can then form an indefinite integration matrix $A = [a_{j,k}]$ of order $2N+1$ with

$$
\begin{aligned}
a_{j,k} &= \int_a^{x_j} D_e(N, P, x - x_k)\, dx \\[2mm]
&= E_e(N, P, x_j - x_k) - E_e(N, P, a - x_k), \\
&\quad j, k = 1, \ldots, 2N+1.
\end{aligned}
$$
(1.4.95)

If we then form the vectors

$$
\begin{aligned}
\mathbf{f} &= (f(x_1), \ldots, f(x_{2N+1}))^T \\
\mathbf{F} &= (F_1, \ldots, F_{2N+1})^T = A\mathbf{f},
\end{aligned}
$$
(1.4.96)

then, under the assumption that f is smooth and of period P on $\mathbb{R}$, we get the approximation

$$
\int_a^x f(t)\, dt \approx \frac{1}{2} F_1 D_e(N, P, x - a) + \frac{1}{2} F_{2N} D_e(N, P, x - b)
$$

$$
+ \sum_{k=1}^{2N-1} F_k D_e(N, P, x - x_k).
$$
(1.4.97)

1.4.8   HILBERT TRANSFORMS

Let us present here some Hilbert transform results for periodic functions. The following result, in which "P.V." denotes the principal value of an integral, is readily deduced starting with the identity (1.3.11) and proceeding as in the proof of Theorem 1.4.2 (see Problem 1.4.3).

Theorem 1.4.19 *Let f be defined, and uniformly bounded on the real line $\mathbb{R}$, let f be periodic, with period P on $\mathbb{R}$, and for arbitrary $b \in \mathbb{R}$, define $f_h(b,x)$ by*

$$f_h(b,x) = \sum_{k \in \mathbb{Z}} f(k\,h - b\,h) \frac{\sin\left\{\frac{\pi}{h}(x + b\,h - k\,h)\right\}}{\frac{\pi}{h}(x + b\,h - k\,h)}. \qquad (1.4.98)$$

(i) If h is defined by $h = P/(2\,N)$, where N is a positive integer, then

$$\lim_{X \to \infty} \frac{P.V.}{\pi\,i} \int_{-X}^{X} \frac{f_h(b,t)}{t - x}\, dt$$

$$= \sum_{k=0}^{2\,N-1} f(k\,h - b\,h) \frac{\cos\left\{\frac{\pi}{h}(x + b\,h - k\,h)\right\} - 1}{2\,N \tan\left\{\frac{\pi}{P}(x + b\,h - k\,h)\right\}}. \qquad (1.4.99)$$

(ii) If h is defined by $h = P/(2\,N+1)$, where N is a non-negative integer, then

$$\lim_{X \to \infty} \frac{P.V.}{\pi\,i} \int_{-X}^{X} \frac{f_h(b,t)}{t - x}\, dt$$

$$= \sum_{k=0}^{2\,N} f(k\,h - b\,h) \frac{\cos\left\{\frac{\pi}{h}(x + b\,h - k\,h)\right\} - 1}{(2\,N+1) \sin\left\{\frac{\pi}{P}(x + b\,h - k\,h)\right\}}. \qquad (1.4.100)$$

We can deduce a similar result for the case when the Hilbert transform is taken over just one period. To this end, let us assume a complex plane setting, with U the unit disc, and with the Hilbert transform $\mathcal{S}\,f$ of a function f taken over the boundary, ∂U and evaluated at $z \in \partial U$. We then get

$$\begin{aligned}
(\mathcal{S}\,f)(z) &= \frac{P.V.}{\pi\,i} \int_{\partial U} \frac{f(\zeta)}{\zeta - z}\, d\zeta - c_0, \\
&= \frac{P.V.}{2\,\pi\,i} \int_{-\pi}^{\pi} \frac{f\left(e^{i\,\theta'}\right)}{\tan\left(\frac{\theta' - \theta}{2}\right)}\, d\theta'
\end{aligned} \qquad (1.4.101)$$

where c_0 is defined as in (1.4.42). If we take $f(\zeta) = \zeta^n$, with n an integer, we get the well known result ([S1, Problem 1.3.5])

$$(\mathcal{S}\,\zeta^n)(z) = z^n \, \mathrm{sgn}(n).$$
(1.4.102)

Theorem 1.4.20 *Let D_e and D_o be defined as in (1.4.3)–(1.4.4). Then,*

$$\frac{P.V.}{2\pi i} \int_{-\pi}^{\pi} \frac{D_e(N,t)}{\tan\left(\frac{t-x}{2}\right)}\, dt \;=\; -\frac{\sin^2\left(\frac{Nx}{2}\right)}{N\,\tan\left(\frac{x}{2}\right)},$$

$$\frac{P.V.}{2\pi i} \int_{-\pi}^{\pi} \frac{D_o(N,t)}{\tan\left(\frac{t-x}{2}\right)}\, dt \;=\; -\frac{\sin\left(\frac{Nx}{2}\right)\sin\left(\frac{(N+1)x}{2}\right)}{(N+1/2)\,\sin\left(\frac{x}{2}\right)}.$$
(1.4.103)

The Hilbert transforms of the functions f_h in Theorem 1.4.2 can thus be explicitly expressed, based on these formulas.

1.4.9   DISCRETE FOURIER TRANSFORM

Let us here derive DFT approximations based on two one dimensional methods, Sinc series and wavelet series. We start with $2N+1$ function values $f_k = f(k\,h_x)$, of a function f, for $k = -N,\dots,\ N$ and $h_x = X/(2N)$ given on the interval $[-X/2, X/2]$. These values may be used to approximate f on $\mathbb{R}$ with $v(x)$, where v denotes the sum in (1.3.28), but with the $k = \pm N$ terms halved. The Fourier transform $\hat{v}$ of v, a Fourier polynomial on the interval $(-\pi/h_x, \pi/h_x)$ and zero on the outside of this interval can then be used to approximate the Fourier transform $\hat{f}$ of f on $\mathbb{R}$ – see (1.3.30). Here the Fourier and inverse transforms are defined by the integrals

$$\hat{f}(y) \;=\; \int_{\mathbb{R}} e^{i\,x\,y}\, f(x)\, dx,$$

$$f(x) \;=\; \int_{\mathbb{R}} e^{-i\,x\,t}\, \hat{f}(y)\, dy.$$
(1.4.104)

That is, the values $\hat{f}_\ell = \hat{f}(\ell\,h_x)$, with $h_y = \pi/(N\,h_x) = \pi/X$, and with $\ell = -N,\ \dots,\ N$ define $\hat{v}$. Similarly, the given the values $\hat{v}_\ell = \hat{v}(l\,h_y)$, $\ell = -N,\dots N$ define a function $\hat{v}$ that is a Fourier polynomials on $[-N\,h_y, N\,h_y] = [-\pi/h_x, \pi/h_x]$ and zero on the remainder of the real line, and by taking the inverse Fourier transform $\hat{v}$ of this function we get a finite Sinc series v that approximates

f at the Sinc points $\{k\,h_x\}_{k=-N}^{N}$. We thus derive here two explicit matrices A and B which relate the two vectors

$$\mathbf{f} \approx \mathbf{v} = (f_{-N}, \ldots, f_N)^T$$

$$\hat{\mathbf{f}} \approx \hat{\mathbf{v}} = (\hat{v}_{-N}, \ldots, \hat{v}_N)$$

(1.4.105)

by the formulas

$$\hat{\mathbf{v}} = A\,\mathbf{v} \quad \text{and} \quad \mathbf{v} = B\,\hat{\mathbf{v}}.$$

(1.4.106)

Given a set of numbers c_{-N}, c_{-N+1}, $\ldots$, c_N, we shall use the notation $\sum_{k=-N}^{N}{}' c_k = \frac{1}{2}\,c_{-N} + c_{-N+1} + \ldots + c_{N-1} + \frac{1}{2}\,c_N$, i.e., the prime on the sum indicates that the first and last terms are to be halved.

(i) *Fourier Transform of a Finite Sinc Sum.*

Given f defined at the Sinc points $\{k\,h_x\}_{k=-N}^{N}$, with $h_x > 0$, define v by the finite Sinc sum,

$$v(x) = \sum_{k=-N}^{N}{}' f(k\,h_x)\,S(k,h_x)(x).$$

(1.4.107)

Notice that $v(k\,h_x) = f(k\,h_x)$. The Fourier transform of v taken over $\mathbb{R}$ is just (see Problem 1.2.6)

$$
\hat{v}(y) = \int_{\mathbb{R}} e^{i\,x\,y}\,v(x)\,dx
$$

$$
= \begin{cases} h_x \displaystyle\sum_{k=-N}^{N}{}' v(k\,h_x)\,e^{i\,k\,h_x\,y} & \text{if} \quad |y| \le \frac{\pi}{h_x} \\[2mm] 0 & \text{if} \quad |y| > \frac{\pi}{h_x}. \end{cases}
$$

(1.4.108)

As discussed in the above subsections, this Fourier polynomial in y is determined by its $m = 2\,N+1$ values $v(\ell\,h_y)$ on $-\pi/h_x \le y \le \pi/h_x$, with $h_y = (\pi/h_x)/N$. If we now evaluate $\hat{v}(y)$ at $y = j\,h_y$, we get

$$\hat{v}(j\,h_y) = h_x \sum_{\ell=-N}^{N}{}' e^{i\,j\,k\,h_x\,h_y}\,v(\ell\,h_x).$$

(1.4.109)

The values $\hat{v}(k\,h_y)$ approximate the values $\hat{f}(k\,h_y)$ (see [S1, §3.3.3]). The equation (1.4.109) thus determines the matrix $\mathbf{A}$, i.e., noting that $h_x\,h_y = \pi/N$, we get

$$A = [a_{jk}] \quad \text{with} \quad a_{jk} = h_x\,\{\exp(i\,j\,k\,\pi/N)\}', \qquad (1.4.110)$$

where the prime indicates that the $ell = \pm N$ terms are to be halved.

(ii) *Inverse Fourier Transform of Trigonometric Polynomial.*

Next let $v(x)$ and $\hat{v}(y)$ be related as in (1.4.104) and (1.4.108) above, and let us assume that we are given the data $\{\hat{v}_\ell\} = \{\hat{v}(\ell\,h_y)\}_{\ell=-N}^{N}$, and with $h_y = \pi/(N\,h_x)$. We combine this data with the results of Corollary 1.4.3 (i), to form the function $\hat{v}$ that is a Fourier polynomial on $[-\pi/h_x,\pi/h_x]$ and zero on the remainder of $\mathbb{R}$, i.e.,

$$\hat{v}(y) = \begin{cases} \displaystyle\sum_{\ell=-N}^{N}{}' D_e(N,2\,\pi/h_x,y-\ell\,h_y)\,\hat{v}_\ell & \text{if} \quad -\pi/h_x \le y \le \pi/h_x \\[2mm] 0 & \text{if} \quad y \in \mathbb{R}, |y| > \pi/h_x. \end{cases}$$

$$(1.4.111)$$

with

$$D_e(N,2\,\pi/h_x,y) = \frac{1}{2\,N}\sum_{k=-N}^{N}{}' e^{i\,k\,h_x\,y}. \qquad (1.4.112)$$

The inverse transform of the function $D_e(N,2\,\pi/h_x,\cdot)$ taken over $(-\pi/h_x,\pi/h_x)$ is

$$\frac{1}{2\,\pi}\int_{-\pi/h_x}^{\pi/h_x} D_e(N,2\,\pi/h_x,y-\ell\,h_y)\,e^{-i\,x\,y}\,dy$$

$$(1.4.113)$$

$$= \frac{1}{2\,N\,h_x}\sum_{k=-N}^{N}{}' S(k,h_x)(x)\,e^{-i\,k\,\ell\,h_x\,h_y},$$

so that

$$v(x) = \frac{1}{2\pi} \int_{-\pi/h_x}^{\pi/h_x} e^{-ixy}\, \hat{v}(y)\, dy$$

$$= \frac{1}{2N\,h_x} \sum_{k=-N}^{N}{}' S(k,h_x)(x) \sum_{\ell=-N}^{N}{}' e^{-ik\ell h_x h_y}\, \hat{v}(\ell\, h_y).$$

$$(1.4.114)$$

Taking $x = k\,h_x$ and $h_x\,h_y = \pi/N$ in this equation, we get

$$v(k\,h) = \frac{1}{2N\,h_x} \sum_{\ell=-N}^{N}{}' e^{-ik\ell\pi/N}\, \hat{v}(\ell\, h_y). \qquad (1.4.115)$$

This equation determines the matrix B:

$$B = [a_{jk}] \quad \text{with} \quad b_{jk} = 1/(2N\,h_x)\{\exp(-ijk\pi/N)\}', \quad (1.4.116)$$

where the prime indicates that the $k = \pm N$ terms are to be halved.

In effect, we used the Part (i) of Corollary 1.4.3 to obtain our approximations above. Similar (but not identically the same) matrix relations obtain using the Parts (i) of Corollaries 1.4.4, 1.4.5 and 1.4.6.

PROBLEMS FOR SECTION 1.4

1.4.1 Let f be analytic in the region $\mathcal{D}_d$ defined in (1.4.43), and let f be periodic on $\mathbb{R}$, with period P.

(a) Prove that if $h > 0$, then the expression

$$\lim_{c\to\infty} \frac{1}{h} \int_{-c}^{c} \text{sinc}\left(\frac{x-t}{h}\right) f(t)\, dt$$

exists, and defines a function $F_h(x)$ for all $x \in \mathbb{R}$.

(b) Prove that if $h = P/(2N)$, then the function F_h defined in the (a)–Part of this problem can also be expressed via the continuous wavelet formula

$$F_h(x) = \frac{1}{P} \int_{-P/2}^{P/2} \frac{\sin\left(\frac{\pi}{h}(x-t)\right)}{\tan\left(\frac{\pi}{P}(x-t)\right)} f(t)\, dt.$$

(c) Prove that if $h = P/(2N+1)$, then the function F_h defined in the (a)–Part of this problem can also be expressed via the continuous wavelet formula

$$F_h(x) = \frac{1}{P} \int_{-P/2}^{P/2} \frac{\sin\left(\frac{\pi}{h}(x-t)\right)}{\sin\left(\frac{\pi}{P}(x-t)\right)} f(t)\, dt.$$

1.4.2 Let f satisfy the conditions stated in Problem 1.4.1.

(a) Apply the Trapezoidal rule over $\mathbb{R}$,

$$\int_{\mathbb{R}} g(t)\, dt \approx T_h(g) \equiv h \sum_{k \in \mathbb{Z}} g(k\,h)$$

to the integral expression for F_h do deduce that

$$F_h(x) \approx f_h(x) \equiv \sum_{k \in \mathbb{Z}} f(k\,h)\, \mathrm{sinc}\left(\frac{x - k\,h}{h}\right)$$

(b) Apply the Trapezoidal and Midordinate rules (1.4.37) and (1.4.39) to the integral expression in the (b)–Part of Problem 1.4.1, to get (1.4.14) and (1.4.18).

(c) Apply the Trapezoidal and Midordinate rules (1.4.37) and (1.4.39) to the (c)–Part of Problem 1.4.1, to get (1.4.22) and (1.4.26).

1.4.3 Give a detailed proof of Corollary 1.4.3.

1.4.4 Give a detailed proof of Corollary 1.4.4.

1.4.5 Give a detailed proof of Corollary 1.4.5.

1.4.6 Give a detailed proof of Corollary 1.4.6.

1.4.7 Give a detailed proof of Theorem 1.4.20.

1.5   Sinc Methods on Γ

In this section we shall consider using Sinc methods to approximate functions as well as calculus operations of functions over finite, semi–infinite, and infinite intervals, indeed, more generally, over planar arcs, Γ. A particular class of *one–to–one* transformations φ of Γ to $\mathbb{R}$ are ideal for this purpose. These transform the class of functions $\mathbf{L}_{\alpha,d}(\mathrm{id})$ and $\mathbf{M}_{\alpha,d}(\mathrm{id})$ that we have introduced in the previous section to convenient classes over Γ, and thus they yield similar accurate approximations over Γ that were obtained for $\mathbb{R}$.

For practical usefulness, it is preferable to select functions φ which can be explicitly expressed, and for which it is also possible to explicitly express the inverse functions, φ^{-1}.

1.5.1   SINC APPROXIMATION ON A FINITE INTERVAL

Consider first the case of a finite interval, (a, b). The mapping

$$w = \varphi(z) = \log\left(\frac{z-a}{b-z}\right), \quad z \in (a,b)$$

$$\Longleftrightarrow z = \varphi^{-1}(w) = \frac{a + b\,e^w}{1 + e^w}, \quad w \in \mathbb{R}$$

(1.5.1)

provides a one–to–one transformation of the interval (a,b) onto $\mathbb{R}$. The Sinc points are defined for $h > 0$ and $k \in \mathbb{Z} = 0, \pm 1, \pm 2, \cdots$, by $z_k = \varphi^{-1}(kh) = (a + b\,e^{kh})/(1 + e^{kh})$.

Example 1.5.1 Let $(a,b) = (0,1)$; we wish to approximate the function $F(x) = x^{1/3}(1-x)^{1/2}$. In this case, $\varphi(x) = \log(x/(1-x))$, and $\varphi^{-1}(w) = e^w/(1 + e^w)$. Thus

$$f(w) = F\left(\varphi^{-1}(w)\right) = \frac{e^{w/3}}{(1 + e^w)^{5/6}}$$

We would like to have this function belong to the class $\mathbf{L}_{\alpha,d}(\mathrm{id})$, for some positive numbers α and d. We may note first, that $f(w) = \mathcal{O}\left(e^{-|w|/3}\right)$, as $w \to -\infty$, while $f(w) = \mathcal{O}\left(e^{-|w|/2}\right)$, as $w \to \infty$, and we may therefore take $\alpha = 1/3$.

Next, the Fourier transform $\tilde{f}$ of f is given by (see Problem 1.5.1)

$$\tilde{f}(y) = \int_{\mathbb{R}} f(w) e^{i y w} \, dw = \frac{\Gamma(1/3 + i\,y)\,\Gamma(1/2 + i\,y)}{\Gamma(5/6)},$$

and using *Stirling's formula*, i.e.,

$$\Gamma(x+1) = (2\pi x)^{1/2}(x/e)^x \left[1 + \mathcal{O}(1/x)\right], \quad x \to \infty, \tag{1.5.2}$$

we find that

$$\tilde{f}(y) = \mathcal{O}\left(\exp(-\pi|y|)\,|y|^{-1/6}\right), \quad y \to \pm\infty,$$

so that $d = \pi$. Hence $f \in \mathbf{L}_{1/3,\pi}(\mathrm{id})$.

Upon selecting a positive integer N, defining h via (1.3.27), and selecting the Sinc points as above, we find, using (1.3.28), that

$$\sup_{x \in (0,1)} \left| x^{1/3} \left(1 - x\right)^{1/2} - \sum_{k=-N}^{N} z_k^{1/3} \left(1 - z_k\right)^{1/2} S(k,h) \circ \varphi(x) \right|$$

$$= \mathcal{O}\left(N^{1/2} e^{-\pi\,(N/3)^{1/2}}\right). \tag{1.5.3}$$

This improves considerably the $\mathcal{O}\left(N^{-1/3}\right)$ error given in standard approximation theory texts, of best approximation of $x^{1/3}(1 - x)^{1/2}$ on $(0,1)$ via polynomials of degree $2N + 1$ in x.

Let us observe that the derivatives of the function $f(x) = x^{1/3}\,(1 - x)^{1/2}$ are unbounded at the end–points of the interval $(0,1)$, i.e., f has singularities at these points. This type of singularity is a frequently occurring one when one solves partial differential equations. In the case when the nature of the singularity is known, then suitable boundary element bases can be selected to produce an approximation that converges rapidly. On the other hand, in the solution of three dimensional partial differential equation problems it is often difficult to deduce *a priori* the nature of the singularity, and in such cases methods based on polynomial or finite element approximations converge very slowly. However, we see from Corollary 1.3.7 that Sinc methods still converge at an exponential rate, even in such cases.

We now introduce two important spaces of functions, $\mathbf{L}_{\alpha,d}(\varphi)$ and $\mathbf{M}_{\alpha,d}(\varphi)$ associated with *Sinc* approximation on the finite interval (a,b). These spaces are related to the spaces $\mathbf{L}_{\alpha,d}(\text{id})$ and $\mathbf{M}_{\alpha,d}(\text{id})$ that were introduced in the previous section. For the case of $\mathbf{L}_{\alpha,d}(\varphi)$, we assume that α, and d are arbitrary fixed positive numbers. The space $\mathbf{L}_{\alpha,d}(\varphi)$ consists of the family of all functions f such that,

$$f(z) = \begin{cases} \mathcal{O}(|z-a|^\alpha), & \text{as } z \to a, \\ \mathcal{O}(|z-b|^\alpha), & \text{as } z \to b, \end{cases} \qquad (1.5.4)$$

and such that the Fourier transform $\tilde{F} = \{f \circ \varphi^{-1}\}^\sim$ satisfies the relation

$$\hat{F}(y) = \mathcal{O}\left(\exp(-d|y|)\right) \quad y \to \pm\infty. \qquad (1.5.5)$$

Thus the function F from Example 1.5.1 belongs to $\mathbf{L}_{1/3,\pi}(\varphi)$.

In order to define the second space, $\mathbf{M}_{\alpha,d}(\varphi)$, it is convenient to assume that α, and d are restricted such that $0 < \alpha \le 1$, and $0 < d < \pi$). Then, $\mathbf{M}_{\alpha,d}(\varphi)$ denotes the family of all functions g that are continuous and uniformly bounded on Γ, such that $g - L\,g \in \mathbf{L}_{\alpha,d}(\varphi)$, where

$$L\,g(z) = \frac{(b-z)\,g(a) + (z-a)\,g(b)}{b-a}. \qquad (1.5.6)$$

The class $\mathbf{M}_{\alpha,d}(\varphi)$ includes functions which are bounded on $[a,b]$ but which need not vanish at the end–points a or b of (a,b). For example, $\mathbf{M}_{\alpha,d}(\varphi)$ includes all those functions g with non-smooth ($\mathbf{Lip}_\alpha$ i.e., singular) behavior at the end–points of (a,b). Such "$\mathbf{Lip}_\alpha$" functions frequently arise in the solution of partial differential equations. Let us state here, that *a function f is said to be of class $\mathbf{Lip}_\alpha$ on a closed interval $[a,b]$ if there exists a constant C such that $|f(x) - f(y)| \le C\,|x-y|^\alpha$ for all points x and y on $[a,b]$.*

1.5.2   SINC SPACES FOR INTERVALS AND ARCS

Let us at once introduce some basic definitions, for approximation on an arc Γ, including finite, semi-infinite and infinite intervals, as well as genuine non–interval arcs, such as an arc of a circle.

Definition 1.5.2 *Let φ denote a smooth one–to–one transformation of an arc Γ, with end–points a and b onto $\mathbb{R}$, such that $\varphi(a) = -\infty$, and $\varphi(b) = \infty$. Let $\psi = \varphi^{-1}$ denote the inverse map, so that*

$$\Gamma = \{z \in \mathbb{C} : z = \psi(u), u \in \mathbb{R}\}. \tag{1.5.7}$$

Given φ, ψ and a positive number h, define the Sinc points z_k by

$$z_k = z_k(h) = \psi(kh), \ \ k = 0, \pm 1, \pm 2, \cdots, \tag{1.5.8}$$

and a function ρ, by

$$\rho(z) = e^{\varphi(z)}. \tag{1.5.9}$$

Observe that $\rho(z)$ increases from 0 to ∞ as z traverses Γ from a to b.

Corresponding to positive numbers α and d, let $\mathbf{L}_{\alpha,d}(\varphi)$ denote the family of all functions F defined on Γ for which

$$F(z) = \begin{cases} \mathcal{O}\left(\rho(z)^\alpha\right) & \text{as } z \to a, \\ \\ \mathcal{O}\left(\rho(z)^{-\alpha}\right) & \text{as } z \to b. \end{cases} \tag{1.5.10}$$

and such that the Fourier transform $\{F \circ \varphi^{-1}\}^\sim$ satisfies the relation

$$\left\{F \circ \varphi^{-1}\right\}^\sim (\zeta) = \mathcal{O}\left(e^{-d|\zeta|}\right) \tag{1.5.11}$$

for all $\zeta \in \mathbb{R}$.

Another important family of functions is $\mathbf{M}_{\alpha,d}(\varphi)$, with $0 < \alpha \leq 1$, and $0 < d < \pi$. It consists of all those functions F defined on Γ such that

$$G = F - LF \in \mathbf{L}_{\alpha,d}(\varphi), \tag{1.5.12}$$

and where LF is defined by

$$LF(x) = \frac{F(a) + \rho(x) F(b)}{1 + \rho(x)}. \tag{1.5.13}$$

The Fourier exponential decay condition (1.5.11) is equivalent to the more easily checked analyticity condition that F is analytic on the region

$$\mathcal{E}_d = \{z \in \mathbb{C} : |\arg(\varphi(z))| < d\}.$$

Remark. It may be shown that $\mathbf{M}_{\alpha,d}(\varphi) \subset \mathbf{Lip}_\alpha(\overline{\Gamma})$ for arcs Γ of finite length. We also remark, with L defined as in (1.5.13), the term $L F$ satisfies the relations

$$\begin{aligned}
&L F \in \mathbf{M}_{1,d}(\varphi) \\
&L F(z) - L F(a) = \mathcal{O}(\rho(z)), \quad z \to a \\
&L F(z) - (L F)(b) = \mathcal{O}(1/\rho(z)), \quad z \to b,
\end{aligned} \tag{1.5.14}$$

and this is why we require $0 < \alpha \le 1$ in the definition of the class $\mathbf{M}_{\alpha,d}(\varphi)$. That we cannot have $d \ge \pi$ was already shown in Example 1.5.1. This result will again be demonstrated in the proof of Theorem 1.5.3 below.

The classes $\mathbf{L}_{\alpha,d}(\varphi)$ and $\mathbf{M}_{\alpha,d}(\varphi)$ are both convenient for and well suited to Sinc methods, because they are easy to recognize, because they house the solution to most differential and integral equation problems which arise from applications, and because they guarantee the rapid convergence of Sinc approximation described in the theorems which follow. Justification for this assertion can be based on theoretical considerations. The purpose of this handbook is to provide justification of this assertion based on examples born from the theory presented in Chapters 1 and 2.

Remark. Instead of the spaces $\mathbf{L}_{\alpha,d}(\varphi)$ and $\mathbf{M}_{\alpha,d}(\varphi)$ we could define more general ones, $\mathbf{L}_{\alpha,\beta,d}(\varphi)$ and $\mathbf{M}_{\alpha,\beta,d}(\varphi)$, by replacement of (1.5.10) above with

$$F(z) = \begin{cases} \mathcal{O}\left(\rho(z)^\alpha\right) & \text{as } z \to a, \\[2mm] \mathcal{O}\left(\rho(z)^{-\beta}\right) & \text{as } z \to b. \end{cases}$$

In this case, we would be able to replace the sum $\sum_{k=-N}^{N}$ in (1.3.28) above, as we did in (1.5.27) below with a possibly more efficient form, $\sum_{k=-M}^{N}$. In fact, all of the programs of this handbook are written for this more general case. We have simply chosen $M = N$ in our theoretical presentations in order to simplify the proofs.

Note also, that if α, α', β, β', d, and d' are suitable positive numbers selected as above, and if

$$\alpha \le \alpha', \quad \beta \le \beta' \text{ and } d \le d',$$

then $\mathbf{M}_{\alpha',\beta',d'} \subseteq \mathbf{M}_{\alpha,\beta,d}$.

In the next theorem we state and prove some properties of the spaces $\mathbf{L}_{\alpha,d}(\varphi)$ and $\mathbf{M}_{\alpha,d}(\varphi)$ which are consequences of the above definitions. The statements of this theorem mainly involve operations of calculus. However we also wish to include Cauchy and Hilbert transforms, and to this end, we now define these operations. With apologies to the reader, we deviate here from our commitment to not use complex variables.

Let Γ denote an oriented analytic arc in the complex plane, with end–points a and b, let $x \in \Gamma$, and let $z \in \mathbb{C}$, $z \notin \Gamma$. The Hilbert and Cauchy transforms of a function F taken over Γ are defined by the equations

$$(\mathcal{S} F)(x) = \frac{\text{P.V.}}{\pi i} \int_\Gamma \frac{F(t)}{t-x} \, dt, \quad x \in \Gamma$$

$$(\mathcal{C} F)(z) = \frac{1}{2\pi i} \int_\Gamma \frac{F(t)}{t-z} \, dt, \quad z \notin \Gamma. \tag{1.5.15}$$

Remark. The following results are useful for determining the nature of singularities following operations of calculus, in solutions of differential equations, etc.

Theorem 1.5.3 *Let* $\alpha \in (0,1]$, $d \in (0,\pi)$, *and take* $d' \in (0,d)$.

i. *If* $F \in \mathbf{M}_{\alpha,d}(\varphi)$, *then* $F'/\varphi' \in \mathbf{L}_{\alpha,d'}(\varphi)$;

ii. *If* $F \in \mathbf{M}_{\alpha,d}(\varphi)$, *then* $F^{(n)}/(\varphi')^n \in \mathbf{L}_{\alpha,d'}(\varphi)$, $n = 1, 2, 3, \cdots$;

iii. *If* $F'/\varphi' \in \mathbf{L}_{\alpha,d}(\varphi)$, *then* $F \in \mathbf{M}_{\alpha,d}(\varphi)$;

iv. *If* $F \in \mathbf{L}_{\alpha,d}(\varphi)$, *then* $\int_\Gamma |\varphi'(x) F(x) \, dx| < \infty$; *and*

v. *If* $F \in \mathbf{L}_{\alpha,d}(\varphi)$, *then both* $\mathcal{S} F$ *and* $\mathcal{C} F$ *belong to* $\mathbf{M}_{\alpha,d'}(\varphi)$.

Proof of Part i: Set $G = F - LF$, with LF defined as in (1.5.12) above. Then

$$(1/\varphi')\,(L\,F)' = \frac{(F(b) - F(a))\,\rho}{(1 + \rho)^2}\,,$$

which obviously satisfies the asymptotic relations in (1.5.14). Next, the identities $\rho \circ \psi(w) = e^w$, and

$$\left\{\frac{F'}{\varphi'} \circ \varphi^{-1}\right\}^{\sim}(w) = \int_{\mathbb{R}} \frac{e^w}{(1 + e^w)^2}\,e^{ixw}\,dx = \frac{\pi x}{\sinh(\pi x)}\,,\qquad (1.5.16)$$

together with Theorem 1.5.3 with $0 < d \leq \pi$, and Problem 1.3.3 show that $(1/\varphi')\,(L\,F)' \in \mathbf{L}_{\alpha,d}(\varphi) \subseteq \mathbf{L}_{\alpha,d'}(\varphi)$.

Define $H = \{(-1)/(2\,\pi)\,(G \circ \varphi^{-1})\}^{\sim}(-t)$, and since $G \in \mathbf{L}_{\alpha,d}(\varphi)$ we have $|H(t)| \leq C'\exp(-d|t|)$, $t \in \mathbb{R}$.

Hence, with $x = \varphi^{-1}(w)$,

$$\frac{G'(x)}{\varphi'(x)} = G' \circ \psi(w)\,\psi'(w)$$

$$= \int_{\mathbb{R}} e^{iwt}\,H(t)\,(it)\,dt.$$

But since $|H(t)| \leq C'\exp(-d|t|)$, $t \in \mathbb{R}$, it follows that $|(it)H(t)| \leq C'\exp(-(d-\varepsilon)|t|) \leq C'\exp(-d'\,|t|)$ for arbitrary positive $\varepsilon < d - d'$. We have shown that G'/φ' and $(LF)'/\varphi'$ are in $\mathbf{L}_{\alpha,d'}(\varphi)$. Hence their sum, $F'/\varphi' \in \mathbf{L}_{\alpha,d'}(\varphi)$.

This completes the proof of Part i.

Proof of Part ii. This is similar to the proof of Part i., and we omit it.

Proof of Part iii. By Problem 1.5.3 (c) it suffices to consider the case of $\varphi(x) = x$. Now

$$|F(\infty)| = \left|F(0) + \int_0^\infty F'(t)\,dt\right| < \infty\,,$$

Therefore $F(\infty)$ (and similarly, $F(-\infty)$) exists. Let $G = F - LF$, i.e., for $t \in \mathbb{R}$,

$$G(t) = (F - LF)(t) = F(t) - \frac{F(-\infty) + e^t\,F(\infty)}{1 + e^t}\,.$$

so that $G(\pm\infty) = 0$, and we obtain

$$G(t) = -\int_t^\infty G'(\tau)\,d\tau = \int_{-\infty}^t G'(\tau)\,d\tau. \qquad (1.5.17)$$

Now

$$G'(t) = F'(t) - \frac{(F(\infty) - F(-\infty))\,e^t}{(1+e^t)^2},$$

which clearly belongs to $\mathbf{L}_{\alpha,d'}(\mathrm{id})$.

From the first equality in (1.5.17) we obtain a constant C such that $G(t) \le C\,e^{-\alpha t}$ as $t \to \infty$. That is $G(t) = \mathcal{O}(e^{-\alpha t})$ as $t \to \infty$, and likewise $G(t) = \mathcal{O}\left(e^{-\alpha|t|}\right)$ as $t \to -\infty$. Hence the Fourier transform $\tilde{G}$ exists,

$$G(t) = \frac{1}{2\pi}\int_{\mathbb{R}} \tilde{G}(x)\,e^{-i\,x\,t}\,dx.$$

This equation yields

$$G'(t) = \frac{1}{2\pi}\int_{\mathbb{R}}(ix)\tilde{G}(x)\,e^{-i\,x\,t}\,dx,$$

and by our assumption that $F' \in \mathbf{L}_{1,d}(id)$, it follows that there exists a constant C_1, such that $|x\,\tilde{g}(x)| \le C_1\exp(-d\,|x|)$, and also, that $|\tilde{G}(x)| < \infty$, $x \in \mathbb{R}$. However, taken together, these two inequalities imply that there exists a constant C_2, such that $|\tilde{G}(x)| \le C_2\exp(-d\,|x|)$, $x \in \mathbb{R}$.

Proof of Part iv. By our assumption on F, there exists a constant C such that

$$|F(x)| \le C\frac{|\rho(x)|^\alpha}{(1+|\rho(x)|)^{2\alpha}}. \qquad (1.5.18)$$

Hence,

$$\int_\Gamma |F(x)\,\varphi'(x)\,dx| \le \int_{\mathbb{R}} C\frac{e^{\alpha x}}{(1+e^x)^{2\alpha}}\,dx = C\frac{(\Gamma(\alpha))^2}{\Gamma(2\,\alpha)}. \qquad (1.5.19)$$

where $\Gamma(\alpha) = \int_0^\infty t^{\alpha-1}\,e^{-t}\,dt$ denotes Euler's Gamma function. This last inequality shows that the left hand side of (1.5.19) is bounded.

Proof of Part v. It is shown in [Gk], that if C is a closed contour, if $G \in \mathbf{Lip}_\alpha(C)$, and if the Hilbert transform of G is taken over C, then this Hilbert transform also belongs to $\mathbf{Lip}_\alpha(C)$. Hence, if Γ' is another arc from b to a, which does not intersect Γ, we set $C = \Gamma \cup \Gamma'$, and we extend the definition of F to C by setting $G = F$ on Γ, while $G = 0$ on Γ'. Then G and also its Hilbert transform taken over C belong to $\mathbf{Lip}_\alpha(C)$. In view of the relations

$$\frac{P.V.}{\pi i} \int_C \frac{G(t)}{t - x} \, dt = \frac{P.V.}{\pi i} \int_\Gamma \frac{F(t)}{t - x} \, dt = (\mathcal{S}F)(x) \, ,$$

it thus follows that $\mathcal{S} F \in \mathbf{Lip}_\alpha(C)$.

Let $z = z^L$ and $z = z^R$ denote points respectively to the left and right of Γ, as one traverses Γ from a to b, and set

$$
\begin{aligned}
F^L(x) &= \lim_{z^L \to x} (\mathcal{C} F)(z^L) \\
F^R(x) &= \lim_{z^R \to x} (\mathcal{C} F)(z^R)
\end{aligned}
\tag{1.5.20}
$$

where the limits are taken non-tangentially. The *Plemelj* formulas [Gk] state that

$$
\begin{aligned}
F^L(x) &= \frac{1}{2}\left(f(x) + \mathcal{S}(x)\right), \\
F^R(x) &= \frac{1}{2}\left(f(x) - \mathcal{S}(x)\right).
\end{aligned}
\tag{1.5.21}
$$

Next, for any fixed $\tau \in (-d, d)$, let $\Gamma_\tau = \{z \in \mathbb{C} : \Im\varphi(z) = \tau\}$. By Cauchy's theorem, $F^L(z^L)$ is independent of τ for any z^L on the left–hand side of Γ_τ. It thus follows from the definition (1.5.15) of $\mathcal{C} F$ that F^L is analytic and bounded in the region $\mathcal{D}_{d'} = \{z \in \mathbb{C} : |\Im\varphi(z)| < d'\}$, for any $d' \in (0, d)$. Moreover, $F \in \mathbf{Lip}_\alpha(\Gamma_\tau)$, from which it follows that $F \in \mathbf{Lip}_\alpha(\mathcal{D}_{d'})$. Similarly, F^R is also analytic and bounded and of class $\mathbf{Lip}_\alpha$ in $\mathcal{D}_{d'}$. By assumption F is also analytic, bounded, and of class $\mathbf{Lip}_\alpha$ in $\mathcal{D}_{d'}$. By inspection of the above Plemelj formulas (1.5.21), it follows that each of the functions, F, F^L, F^R, and $\mathcal{S}F$ are analytic, bounded, and of class $\mathbf{Lip}_\alpha$ in $\mathcal{D}_{d'}$, i.e., that these functions belong to $\mathbf{M}_{\alpha,d'}(\varphi)$. ∎

1.5.3   IMPORTANT EXPLICIT TRANSFORMATIONS

Let us next describe some transformations for intervals and arcs frequently encountered in applications. These transformations cover all of the different types of singular behavior arising in applications.

Example 1.5.4   Transformation #1: *The Identity Map over* $\mathbb{R} = (-\infty, \infty)$. If $\Gamma = \mathbb{R}$, and we wish to approximate functions with exponential decay and whose Fourier transforms have exponential decay, take $\varphi(z) = z$. With L as above, and $g \in \mathbf{M}_{\alpha,d}(\mathrm{id})$, we have $L\, g(z) = [g(-\infty) + e^z\, g(\infty)]/(1 + e^z)$, and $f = g - Lg \in \mathbf{L}_{\alpha,d}(\varphi)$, the set of all functions f such that for some positive constants C and α, the inequality $|f(z)| \leq C\, \exp(-\alpha|z|)$ holds for all $z \in \mathbb{R}$, and such that the Fourier transform, $\tilde{f}$, satisfies the inequality $|\tilde{f}(w)| \leq C'\, e^{-d\,|w|}$ for all $w \in \mathbb{R}$. The Sinc points z_j are defined by $z_j = j\, h$, and $1/\varphi'(z_j) = 1$.

Example 1.5.5 Transformation #2: *Finite Interval* (a, b). For the cased of $\Gamma = (a, b)$, we have $\varphi(z) = \log((z - a)/(b - z))$, $\psi(w) = \varphi^{-1}(w) = (a + b\, e^w)/(1 + e^w)$, $L\, g(z) = ((b - z)\, g(a) + (z - a)\, g(b))/(b - a)$. $\mathbf{L}_{\alpha,d}(\varphi)$ is the class of all functions f for which we have $|f(z)| < c|z - a|^\alpha |b - z|^\alpha$ on Γ, and such that for some positive constants C and d, the Fourier transform $\tilde{f}$ of $f \circ \psi$ satisfies the inequality $|\tilde{f}(x)| \leq C\, e^{-d\,|x|}$ for all $x \in \mathbb{R}$. The Sinc points z_j are $z_j = \psi(jh) = (a + b\, e^{j\,h})/(1 + e^{j\,h})$, and $1/\varphi'(z_j) = (b - a)\, e^{j\,h}/(1 + e^{j\,h})^2$. For most examples of this handbook, we shall simply take $a = 0$ and $b = 1$, i.e., for the case of $\Gamma = (0, 1)$.

Example 1.5.6   Transformation #3: *Approximating Functions over* $(0, \infty)$ *having Algebraic Behavior for Large and Small Values of the Argument.* If $\Gamma = (0, \infty)$, take $\varphi(z) = \log(z)$, so that $\psi(w) = \varphi^{-1}(w) = e^w$, and $L\, g(z) = [g(0) + z\, g(\infty)]/(1 + z)$. $\mathbf{L}_{\alpha,d}(\varphi)$ is the family of all functions f such that for some positive constants C and α we have $f(z) = \mathcal{O}(|z|^\alpha)$ as $z \to 0$, $f(z) = \mathcal{O}(|z|^{-\alpha})$ as $z \to \infty$, and such that for some positive constant d, the Fourier transform $\tilde{f}$ of $f \circ \psi$ satisfies the inequality $|\tilde{f}(x)| \leq C\, e^{-d\,|x|}$ for all $x \in \mathbb{R}$. The Sinc points z_j are defined by $z_j = \psi(j\, h) = e^{j\,h}$, and $1/\varphi'(z_j) = e^{j\,h}$.

Example 1.5.7   Transformation #4: *Approximating Functions over* $(0, \infty)$ *having Algebraic Behavior for Small and Exponential Behavior for Large Values of the Argument.* If $\Gamma = (0, \infty)$, let us take

$\varphi(z) = \log(\sinh(z))$, so that $\psi(w) = \varphi^{-1}(w) = \log\left(e^w + \sqrt{1 + e^{2w}}\right)$, while L is given by $L\,g(z) = [g(0) + \sinh(z)\,g(\infty)]/(1 + \sinh(z))$. $\mathbf{L}_{\alpha,d}(\varphi)$ is the class of all functions f such that for some positive constants C and α we have $f(z) = \mathcal{O}(|z|^\alpha)$ as $z \to 0$, $f(z) = \mathcal{O}(\exp(-\alpha z))$ as $z \to \infty$, and such that for some positive constant d, the Fourier transform $\tilde{f}$ of $f \circ \psi$ satisfies the inequality $|\tilde{f}(x)| \le C\,e^{-d|x|}$ for all $x \in \mathbb{R}$. The Sinc points z_j are defined by $z_j = \psi(h\,h) = \log[e^{j\,h} + (1 + e^{2j\,h})^{1/2}]$, and $1/\varphi'(z_j) = (1 + e^{-2j\,h})^{-1/2}$.

Example 1.5.8   Transformation #5: *Approximating Functions over* $\mathbb{R} = (-\infty, \infty)$ *having Algebraic Decay at* $\pm\infty$. If $\Gamma = \mathbb{R}$, we take $\varphi(z) = \log[z + (1 + z^2)^{1/2}]$, so that $\psi(w) = \varphi^{-1}(w) = \sinh(w)$ and $L\,g(z) = [g(-\infty) + (z + (1 + z^2)^{1/2})\,g(\infty)]/[1 + z + (1 + z^2)^{1/2}]$. Then $\mathbf{L}_{\alpha,d}(\varphi)$ is the class of all functions f such that $f(z) = \mathcal{O}(|z|^{-\alpha})$ as $z \to \pm\infty$ on $\mathbb{R}$, and such that for some positive constant d, the Fourier transform $\tilde{f}$ of $f \circ \psi$ satisfies the inequality $|\tilde{f}(x)| \le C\,e^{-d|x|}$ for all $x \in \mathbb{R}$. The Sinc points z_j are defined by $z_j = \sinh(j\,h)$, and $1/\varphi'(z_j) = \cosh(j\,h)$.

Example 1.5.9   Transformation #6: *Approximating Functions over* $\mathbb{R} = (-\infty, \infty)$, *having Algebraic Behavior at* $-\infty$, *Exponential Behavior at* ∞. If $\Gamma = \mathbb{R}$, we let $\varphi(z) = \log\{\sinh[z + (1 + z^2)^{1/2}]\}$, so that, with $t(w) = \log\left(e^w + \sqrt{1 + e^{2w}}\right)$, we have $\psi(w) = \varphi^{-1}(w) = (1/2)\,(t(w) - 1/t(w))$. Then $L\,g(z) = [g(-\infty) + \sinh(z + (1 + z^2)^{1/2})\,g(\infty)]/[1 + \sinh(z + (1 + z^2)^{1/2})]$, and the class $\mathbf{L}_{\alpha,d}(\varphi)$ is the class of all functions f such that $f(z) = \mathcal{O}(|z|^{-\alpha})$, $z \to -\infty$, such that $f(z) = \mathcal{O}(\exp(-\alpha z))$, $z \to \infty$, and such that for some positive constant d, the Fourier transform $\tilde{f}$ of $f \circ \psi$ satisfies the inequality $|\tilde{f}(x)| \le C\,e^{-d|x|}$ for all $x \in \mathbb{R}$. The Sinc points z_j are defined by $z_j = (1/2)[t_j - 1/t_j]$, where $t_j = \log[e^{j\,h} + (1 + e^{2j\,h})^{1/2}]$, and $1/\varphi'(z_j) = (1/2)(1 + 1/t_j^2)(1 + e^{-2j\,h})^{-1/2}$.

Actually, the transformation #5 & #6 are special cases of *double exponential transformations* which was discovered by Takahasi and Mori [Mo1, Mo2] and extensively studied by Sugihara [Su1, Su2, Su3]; it may happen, after making a transformation from Γ to $\mathbb{R}$, the Fourier transform of the resulting function decays faster than at the rate $\exp(-d|x|)$ as $x \to \pm\infty$ on $\mathbb{R}$, in which case, we may be able to make an additional transformation from $\mathbb{R}$ to $\mathbb{R}$, and thus achieve even faster convergence. For example, suppose that, α and β are positive constants, and that we wish to approximate the integral

$$I = \int_{-1}^{1} (1+x)^{\alpha-1} (1-x)^{\beta-1}\, dx.$$

By setting $x = \varphi^{-1}(w)$, where φ is defined as in Example 1.5.5 above, with $a = -1$ and $b = 1$, we get the integral

$$I = 2^{\alpha+\beta-2} \int_{\mathbb{R}} \frac{e^{\alpha w}}{(1+e^w)^\alpha\,(1+e^w)^\beta}\, dw.$$

Now, it may be shown, by proceeding as in Example 1.5.1 above, that the Fourier transform $\hat{f}$ of the integrand f in this last integral satisfies $\hat{f}(x) = \mathcal{O}(\exp(-(\pi-\varepsilon)|x|))$ as $x \to \pm\infty$ on $\mathbb{R}$, where $\varepsilon > 0$ but is otherwise arbitrary. Hence we can make a second *exponential transformation*, $w = \sinh(v)$ (see Example 1.5.8) in the integral above, to get

$$I = 2^{\alpha+\beta-2} \int_{\mathbb{R}} \frac{\exp(\alpha\,\sinh(v))\,\cosh(v)}{(1+\exp(\sinh(v)))^\alpha\,(1+\exp(\sinh(v)))^\beta}\, dv\,,$$

At this point it may be shown that the Fourier transform of the integrand is $\mathcal{O}(\exp(-(\pi/2-\varepsilon)|x|))$ as $x \to \pm\infty$ on $\mathbb{R}$. However, the new integrand decays much more rapidly than exponentially, and thus requires relatively few points for accurate evaluation.

Example 1.5.10   Transformation #7: *Approximation on a Circular Arc.* If Γ is the arc $\{z \in \mathbb{C} : z = e^{i\theta}, u < \theta < v\}$, with $0 < v - u < 2\pi$, take $\varphi(z) = i(v-u)/2 + \log[(z - e^{iu})/(e^{iv} - z)]$, so that $\psi(w) = \varphi^{-1}(w) = [e^{w+iv} + e^{i(u+v)/2}]/[e^w + e^{i(v-u)/2}]$, and $L\,g(z) = [(e^{iv} - z)f(e^{iu}) + (z - e^{iu})f(e^{iv})]/(e^{iv} - e^{iu})$. The space $\mathbf{L}_{\alpha,d}(\varphi)$ is the class of all functions f such that for some positive constants c and α, we have, $|f(z)| < c|z - e^{iu}|^\alpha|e^{iv} - z|^\alpha$, and such that for some positive constant d, the Fourier transform $\tilde{f}$ of $f \circ \psi$ satisfies the inequality $|\tilde{f}(x)| \leq C\,e^{-d|x|}$ for all $x \in \mathbb{R}$. The Sinc points z_j are $z_j = [e^{jh+iv} + e^{i(u+v)/2}]/[e^{jh} + e^{i(v-u)/2}]$, and $1/\phi'(z_j) = 2\,i\,e^{jh+iv}\,\sin((v-u)/2)/[e^{jh} + e^{i(v-u)/2}]^2$.

Example 1.5.11   Transformation #8: *Approximation on more General Arcs* Γ. We define an arc Γ of finite length in the (x,y)–plane by a pair of coordinates

$$\Gamma = \{(x(t), y(t)) : 0 < t < 1\}, \qquad (1.5.22)$$

and with the added requirement that:

(i) $(\dot{x}(t))^2 + (\dot{y}(t))^2 > 0$ for all $t \in (0,1)$, and

(ii) Setting $t = e^w/(1 + e^w)$, the functions X and Y defined by

$$X(w) = x\left(\frac{e^w}{1 + e^w}\right), \quad Y(w) = y\left(\frac{e^w}{1 + e^w}\right)$$

should both belong to the space $\mathbf{M}_{\alpha,d}(\mathrm{id})$.

We are able to deal with problems over such arcs Γ via Sinc methods applied to the interval $(0,1)$. For example, if ds denotes an element of arc length, we can transform an integral over Γ to an integral over $(0,1)$ as follows:

$$\int_{\Gamma} f(x, y)\, ds = \int_0^1 f(x(t), y(t)) \sqrt{(\dot{x}(t))^2 + (\dot{y}(t))^2}\, dt.$$

The Sinc points of such an arc Γ are the points $\{(X(kh), Y(kh))\}$, for $k \in \mathbb{Z}$.

On the other hand, if the arc has infinite length, and one finite end–point, it may be better to define such an arc as a map from the arc to $(0, \infty)$, and in this case, we could use one of the $\Gamma = (0, \infty)$ maps to define the Sinc points. Finally, if the arc starts at a point at infinity, e.g., at the point $\infty\, e^{i a_1}$ and ends up at another point at infinity, e.g., $\infty\, e^{i a_2}$, then it may be more convenient to define such an arc as a map from $\mathbb{R}$ to the arc, and depending on the problem to be solved[1]. We may then be able to use one of the $\Gamma = \mathbb{R}$ maps to define the Sinc points.

1.5.4   INTERPOLATION ON Γ

Our aim here is the approximation of functions belonging to the space $\mathbf{M}_{\alpha,d}(\varphi)$ defined in Definition 1.5.2 above. In the first sub-section we derive a basis for interpolation at the Sinc points on the

---

[1]We could, e.g., select one of the above three $\mathbb{R} \to \mathbb{R}$ Sinc maps of Examples 1.5.7 to 1.5.9, whichever yields the most suitable space for solution of the problem via Sinc methods.

interior of an interval. While this basis generates each of the methods of approximation in the other subsections, the basis is seldom used in computation, except for evaluation of the final computed answer at points other than Sinc points. We shall see that the elimination of interpolation at end–points of Γ is a definite advantage, an obvious one being that it enables approximation of functions which are either undefined, or unbounded at end–points of Γ. If for example, we require a boundary value of a computed approximation, then the computed value at the nearest Sinc point is sufficiently accurate for practical purposes.

Let us first establish a notation which we shall use throughout the remainder of this handbook. We now use notations for approximation using bases $\{\omega_j\}_{-M}^N$ rather than for $\{\omega_j\}_{-N}^N\}$, since this is the manner in which they are used in the programs of this handbook. We also assume in force here, the definitions of $\mathbf{L}_{\alpha,\beta,d}(\varphi)$ and $\mathbf{M}_{\alpha,\beta,d}(\varphi)$, and such that either $M = [\beta/\alpha\, N]$, or $N = [\alpha/\beta\, M]$, where $[\cdot]$ denotes the greatest integer function.

Definition 1.5.12   *For given a positive integers M and N, let D and V denote linear operators acting on functions u defined on Γ given by*

$$\begin{aligned} D\,u &= \operatorname{diag}[u(z_{-M}), \ldots, u(z_N)] \\ V\,u &= (u(z_{-M}), \ldots, u(z_N))^T \end{aligned} \qquad (1.5.23)$$

where $z_j = \varphi^{-1}(j\,h)$ denote the Sinc points. Set

$$\begin{aligned} h &= \left(\frac{\pi d}{\beta N}\right)^{1/2}, \\ \gamma_j &= S(j, h) \circ \varphi, \quad j = -M, \ldots, N, \\ \omega_j &= \gamma_j, \quad j = -M+1, \ldots, N-1, \\ \omega_{-M} &= \frac{1}{1+\rho} - \sum_{j=-M+1}^{N} \frac{1}{1+e^{j\,h}}\gamma_j, \\ \omega_N &= \frac{\rho}{1+\rho} - \sum_{j=-M}^{N-1} \frac{e^{j\,h}}{1+e^{j\,h}}\gamma_j, \\ \varepsilon_N &= N^{1/2}e^{-(\pi d\beta N)^{1/2}}. \end{aligned} \qquad (1.5.24)$$

The ω_j are the basis functions, and we may define the row vector of basis functions

$$\mathbf{w} = (\omega_{-M}, \dots, \omega_N). \tag{1.5.25}$$

Finally, let $\| \cdot \|$ denote the uniform norm on Γ, i.e.,

$$\|f\| = \sup_{x \in \Gamma} |f(x)|. \tag{1.5.26}$$

This norm will be used through the remainder of this handbook.

For given f defined in Γ, we can now form the *Sinc* approximation,

$$f(x) \approx \sum_{k=-M}^{N} f(z_k)\, \omega_k(x), \tag{1.5.27}$$

or in terms of the notation defined above,

$$f \approx \mathbf{w}Vf.$$

Let us now discuss several features of the important approximation (1.5.27) – features that are exhibited in the proof of the next theorem.

1. Given $G \in \mathbf{L}_{\alpha,\beta,d}(\varphi)$, then $G \circ \varphi^{-1} \in \mathbf{L}_{\alpha,\beta,d}(\mathrm{id})$, and this function can then be approximated on $\mathbb{R}$ as described in Theorem 1.3.5, i.e.,

$$\left\| G - \sum_{k=-M}^{N} G(z_k)\, S(k,h) \right\| = \mathcal{O}\left(\varepsilon_N\right), \quad N > 1. \tag{1.5.28}$$

2. We may note that the basis function $\omega_{-M}(x)$ (resp., $\omega_N(x)$) approaches the value 1 as x approaches the end–point a (resp., b) of Γ, and it approaches the value 0 as x approaches the end–point b (resp., a) of Γ. Whereas uniform approximation on $\mathbb{R}$ via the basis $\{S(k,h)\}$ is necessarily restricted to functions that vanish at the end–points of Γ, we will now be able to approximate functions which remain bounded, but need not vanish at the end–points of Γ.

3. Let us note that the basis function $\omega_{-M}(x)$ (resp., $\omega_N(x)$) is close to 1 at the Sinc point $x = z_{-M}$ (resp., $x = z_N$), and yet vanishes at all of the other Sinc points z_j, $j = -M+1, \dots N$

(resp., $j = -M, \ldots, N - 1$). Also, for $j \neq -M$ or for $j \neq N$, $\omega_j(z_k) = 1$ for $k = j$, and 0 for $k \neq j$. The bases $\{\omega_j\}_{-M}^{N}$ is therefore *discrete orthogonal*, at least for purposes of application. Simple multiplicative factors would have enabled us to have $\omega_{-M}(z_{-M}) = \omega_N(z_N) = 1$, but such a factors would introduce an added complication in the definition of these functions, complications that are not worth bothering with, since when approximating functions in $\mathbf{M}_{\alpha,\beta,d}(\varphi)$, the resulting difference is negligible.

4. If $f \in \mathbf{M}_{\alpha,\beta,d}(\varphi)$, the differences $f(a) - f(z_{-M})$ and $f(b) - f(z_N)$ are of the order of $e^{-\beta N h}$. These terms are negligible compared with the error of approximation, enabling us to make the substitutions $f(z_{-M}) = f(a)$, and $f(z_N) = f(b)$ (see Equations (1.5.10)–(1.5.13) and our remark following (1.5.14)). This feature is a very convenient; it eliminates having to evaluate functions whose boundary values are defined only in terms of limits, as is often the case in applications. In addition the computer evaluation of the Sinc function $S(j, h) \circ \varphi(x)$ poses problems at the end–points a and b of Γ, since φ is not finite at these points.

We state and prove the following theorem, the proof of which touches on all of the above points. We remind the reader that either $M = [\beta/\alpha\, N]$, or $N = [\alpha/\beta\, M]$.

Theorem 1.5.13 *If $F \in \mathbf{M}_{\alpha,\beta,d}(\varphi)$, then, as $N \to \infty$,*

$$\|F - \mathbf{w}\, V\, F\| = \mathcal{O}(\varepsilon_N). \qquad (1.5.29)$$

Proof. Recall that $F - LF \in \mathbf{L}_{\alpha,\beta,d}(\varphi)$. By proceeding as in the proof of (1.3.28), it follows, in view of our definition of h in (1.5.24) that

$$F = LF + \sum_{j=-M}^{N} (F - LF)(z_j)\, \gamma_j + E,$$

$$\qquad (1.5.30)$$

$$E = \mathcal{O}(\varepsilon), \quad N \to \infty,$$

where here and for the remainder of this proof we shall write ε for ε_N, δ for $\varepsilon_N/\sqrt{N}$, and F_k for $F(z_k)$. Moreover, by assumption, both α and β are on the interval $(0,1]$, so $\delta_L = e^{-\alpha M h} \geq e^{-Mh}$ and similarly, $\delta_R = e^{-\beta N h} \geq e^{-Nh}$, and we therefore have

$$e^{-Mh} = \rho(z_{-M}) = \mathcal{O}(1/\rho(z_N)) = \mathcal{O}(\delta), \quad M, N > 1, \quad (1.5.31)$$

with δ either δ_L or δ_R.

Also note by (1.5.14), that since $F \in \mathbf{M}_{\alpha,\beta,d}(\varphi)$, we have

$$\begin{aligned} F(a) - F_{-M} &= \mathcal{O}(\delta_L) \\ F(b) - F_N &= \mathcal{O}(\delta_R) \end{aligned} \qquad (1.5.32)$$

Furthermore, since $x \in \Gamma$, the results of Problem 1.2.11 yield the inequalities

$$|S(j,h) \circ \varphi(x)| \leq 1, \quad x \in \mathbb{R},$$
$$\sum_{j=-M}^{N} |S(j,h) \circ \varphi(x)| = \mathcal{O}(\log(N)) \quad M, N > 1. \qquad (1.5.33)$$

Expanding the approximation gotten from (1.5.30), and recalling the way ω_j was defined in Definition 1.5.12, we have

$$F = LF + \sum_{j=-M}^{N} (F_j - LF(z_j))\, \gamma_j + E \qquad (1.5.34)$$

$$= \sum_{j=-M}^{N} F_j\, \omega_j \qquad (1.5.35)$$

$$+ F_{-M} \left\{ \frac{\rho(z_{-M})}{1 + \rho(z_{-M})} \right\} \gamma_{-M} \qquad (1.5.36)$$

$$+ F_N \left\{ \frac{1}{1 + \rho(z_N)} \right\} \gamma_N \qquad (1.5.37)$$

$$+ (F_{-M} - F(a)) \sum_{j=-M}^{N} \frac{\gamma_j}{1 + \rho(z_j)} \qquad (1.5.38)$$

$$+ (F_N - F(b)) \sum_{j=-M}^{N} \frac{\gamma_j\, \rho(z_j)}{1 + \rho(z_j)} \qquad (1.5.39)$$

$$+ E. \qquad (1.5.40)$$

Let us now consider in turn the error terms (1.5.36)–(1.5.40).

Eq. (1.5.36) : Since F is bounded on Γ, since $|\gamma_{-M}| \leq 1$, and since $\rho(z_{-M}) = \mathcal{O}(\delta)$, the term in (1.5.36) is of the order of δ.

Eq. (1.5.37) : This term is also of the order of δ, since F is bounded on Γ, since $1/(1 + \rho(z_N)) = \mathcal{O}(\delta)$, and since $|\gamma_N| \leq 1$.

Eqs. (1.5.38) and (1.5.39) : Both of these terms are $\mathcal{O}(\delta \log(N)) = \mathcal{O}(\varepsilon)$ by (1.5.32) and (1.5.33).

Since the term E in (1.5.40) has already been shown to be $\mathcal{O}(\varepsilon)$ in (1.5.30), the statement of Theorem 1.5.13 now follows. ∎

Remark. We remark, that if $G \in \mathbf{L}_{\alpha,\beta,d}(\varphi)$, then it is convenient to take $w_j = \mathrm{sinc}\{[\varphi - jh]/h\}$, $j = -M, \dots, N$, instead of as defined in Definition 1.5.12, since the corresponding approximation of G as defined in (1.5.12)–(1.5.13) then also vanishes at the end points of Γ, just as F then vanishes at the end points of Γ.

1.5.5   SINC APPROXIMATION OF DERIVATIVES

The formulas given here yield approximations of derivatives which may be used to obtain approximate solutions of ordinary and partial differential equations. Notice, the results of the theorem which follows are uniformly accurate, even when derivatives of f are unbounded at end–points of Γ.

Theorem 1.5.14 *Let* $f \in \mathbf{M}_{\alpha,\beta,d}(\varphi)$, *and let* k *be any non–negative integer. Then*

$$\left\| \left(\frac{h}{\varphi'}\right)^k \left[f^{(k)} - \mathbf{w}^{(k)} V f\right] \right\| = \mathcal{O}(\varepsilon_N), \quad N > 1. \qquad (1.5.41)$$

Proof. The proof follows easily from Theorem 1.5.3, and is omitted. ∎

We remark that the row vector of functions $(h/\varphi')^k \, \mathbf{w}^{(k)}$ generates an $m \times m$ matrix when evaluated at the Sinc points z_i, i.e.,

$$\left[\left(\frac{h}{\varphi'(z_i)}\right)^k w_j^{(k)}(z_i) \right]_{i,j=-M,\dots,N}.$$

For purposes of applying Sinc theory to solving boundary value problems it is convenient to use the notation $I^{(k)}$ for the $m \times m$ matrix with $(i,j)^{th}$ entry $\delta_{i-j}^{(k)}$, where $\delta_\ell^{(k)}$ is defined as in (1.3.15).

In this case, if $p(x)$ with $p \in \mathbf{L}_{\alpha,d}(\varphi)$ is approximated on Γ by $P(h,x)$, with

$$P(h,x) = \sum_{j=-M}^{N} p(z_j)\, S(j,h) \circ \varphi(x)\,,$$

then by differentiating both sides of the approximation $p \approx P(h,\cdot)$, we get the approximation $p'(z_i) \approx P'(h,z_i)$, or, by means of the above defined operator V and diagonal matrix $D(u)$,

$$V\, P' \approx D(\varphi'/h)\, I^{(1)}\, V\, p.$$

We advise the user to use caution when applying this formula to the approximation of a derivative, since it is inaccurate in the neighborhood of a finite end–point of Γ due to the fact that φ' becomes unbounded in the neighborhood of such points. Derivatives of functions in $\mathbf{M}_{\alpha,d}(\varphi)$ may also be unbounded at such points.

There are many instances of when a function f has several bounded higher order derivatives on a closed and bounded interval Γ, finite or semi–infinite, when we wish to approximate derivatives of f knowing only function values at Sinc points. We shall return to this problem again, in §1.7 below, where we derive some novel formulas based on polynomial methods that can yield more accurate approximations on the closed arc Γ.

As shown above the matrices $D\left((\varphi'/h)^k\right) I^{(k)}$ arise when we form the kth derivative in a Sinc approximation of f. The matrices $I^{(k)}$ have eigenvalues bounded by π^k (see [S1, §7.2]). They are symmetric for k even, and skew symmetric for k odd. They are all well conditioned for even k, and this is also the case for k odd, provided that the size m of $I^{(k)}$ is even. In such cases the eigenvalues of smallest magnitude of all of these matrices of size m is approximately equal to $(\pi/N)^k)$ ([S1, Lemma 7.2.4]). If k is odd, and if $I^{(k)}$ is of odd order, then $I^{(k)}$ has a zero eigenvalue. These features show that Sinc methods are robust and easily adapted to solution of ordinary differential equations via collocation (i.e., solution obtained through function evaluation at Sinc points – see the next subsection). Hence, although the above approximation of $V P'$ is inaccurate in the neighborhoods of finite end–points of Γ, the use of the inverse of the matrix

$$A_k = D\left(\left(\frac{\varphi'}{h}\right)^k\right) I^{(k)}$$

in a differential equation is nevertheless satisfactory (see §7.2 of [S1]) if $I^{(k)}$ is non-singular. That is, whereas the use of $A_k V f$ is not an accurate approximation, each of the approximations $A_k^{-1} A_j V f$ is accurate whenever $0 \le j \le k$ (and where A_0 here denotes the unit matrix), and when applied to functions f with $f/\varphi^{k-j} \in \mathbf{L}_{\alpha,d}(\varphi)$ (see Theorem 1.5.3).

Several methods of approximation of derivatives are possible via use of Sinc bases. One method that we have not mentioned in this text, is the use of integration by parts (see §7.2 of [S1]) based on Theorem 1.5.14 above. Other methods, for approximating the first derivative involve the use of the inverse of an indefinite integration matrix. Sinc-Pack has programs for all of these, including programs based on the formulas of §1.7 below. Illustrations are provided in Chapter 3.

1.5.6 SINC COLLOCATION

We know from Theorem 1.5.13, that if $\|f - \mathbf{w} V f\| = \mathcal{O}(\varepsilon_N)$, or equivalently, for some constant C, $\|f - \mathbf{w} V f\| \le C \varepsilon_N$, then Theorem 1.5.15 follows. This guaranteeing an accurate final approximation of f on Γ provided that we know a good approximation to f at the Sinc points. We leave its proof which follows directly via the triangle inequality and Problem 1.2.11 to the reader. It tells us that accurate approximations of f at Sinc points imply accurate approximation of f on all of Γ.

Theorem 1.5.15 If $f \in \mathbf{M}_{\alpha,\beta,d}(\varphi)$, and $\mathbf{c} = (c_{-M}, \ldots, c_N)^T$ is a complex vector of order m, such that for some $\delta > 0$,

$$\left(\sum_{j=-M}^{N} |f(z_j) - c_j|^2\right)^{1/2} < \delta, \qquad (1.5.42)$$

then

$$\|f - \mathbf{w}\,\mathbf{c}\| < C\varepsilon_N + \delta. \qquad (1.5.43)$$

Similar results of the type (1.5.43) hold if the least squares error of
(1.5.42) is replaced by a ℓ^p error, for $1 \le p \le \infty$.

1.5.7   Sinc Quadrature

We record next the standard Sinc quadrature formula. It belongs to
the Sinc family of powerful tools for solving differential and integral
equations.

Theorem 1.5.16 *If $f/\varphi' \in \mathbf{L}_{\alpha,\beta,d}(\varphi)$, then*

$$\left| \int_a^b f(x)\,dx - h\,\{V(1/\varphi')\}^T(Vf) \right| = \mathcal{O}\left(\varepsilon_N\right)\,, \quad M,\ N > 1. \quad (1.5.44)$$

Remarks.
1. Note that (1.5.44) follows at once from (1.5.29) upon setting $x = \psi(t)$ in the integral in (1.5.44) and then applying the Trapezoidal rule to the resulting integral (see Equations (1.3.8), (1.3.9)).

2. An excellent example is to consider the approximation of the integral $\int_0^1 f(x)\,dx$, in which we take $f(x) = x^{-2/3}(1-x)^{-1/\sqrt{\pi}}$. Notice that f is infinite at both ends of the interval of integration. In this case, we take $\varphi(x) = \log(x/(1-x))$, so that $1/\varphi'(x) = x(1-x)$, giving $f(x)/\varphi'(x) = x^{1/3}(1-x)^{1-1/\sqrt{\pi}}$. Hence. $f/\varphi' \in \mathbf{L}_{\alpha,\beta,d}(\varphi)$, with $\alpha = 1/3$, with $\beta = 1 - 1/\sqrt{\pi}$,, and with $d = \pi$.

This illustrates that the condition $f/\varphi' \in \mathbf{L}_{\alpha,\beta,d}(\varphi)$, rather than being difficult to apply, allows instead for the efficient numerical approximation of integrals that may be unbounded at one or both end-points of a finite interval. If the condition is satisfied, it guarantees that we can accurately approximate an integral over a finite interval, over a semi-infinite interval, over $\mathbb{R}$, or even over an arc, i.e., over any of the regions of the transformations of §1.5.3.

Example 1.5.17 *A Quadrature Algorithm.* The following is an algorithm based on the transformation of Example 1.5.6 for computing $\int_0^\infty f(x)\,dx$ to within any error ε (with error as small as 10^{-16} in double precision). Rapid convergence assumes that f satisfies the conditions of Theorem 1.5.16. While more efficient versions of this algorithm are known [ShSiS], what we have here, is short and easy to program, and moreover, it contains the essential ideas of how to achieve arbitrary accuracy in approximating an integral using

(1.5.44). Following this algorithm, we give explicit transformations that allow the use of this algorithm for each of the transformations described in Examples 1.5.4 to 1.5.11.

Algorithm 1.5.18 *Algorithm for Approximating $\int_0^\infty f(x)\,dx$ to within an Error of ε.*

User writes $f(x)$ routine
input ε
$\varepsilon_1 \leftarrow \sqrt{\varepsilon}$
$h \leftarrow 2$
$e \leftarrow \exp(h)$
$q \leftarrow e$
$T \leftarrow f(1)$
1 $p \leftarrow 1/q$
$r \leftarrow p * f(p) + q * f(q)$
$T \leftarrow T + r$
$r_1 \leftarrow |r|$
if $r_1 < \varepsilon$ go to 2
$q \leftarrow e * q$
go to 1
2 $T \leftarrow h * T$
6 $M \leftarrow 0$
$e \leftarrow \sqrt{e}$
$q \leftarrow e$
go to 3
7 output ε, h, T

3 $p \leftarrow 1/q$
$r \leftarrow p * f(p) + q * f(q)$
$M \leftarrow M + r$
$r_1 \leftarrow |r|$
if $r_1 < \varepsilon$ go to 4
$q \leftarrow e * q$
go to 3
4 $M \leftarrow h * M$
$T_1 \leftarrow (T + M)/2$
$\delta \leftarrow |T - M|$
if $\delta < \varepsilon_1$ go to 5
$T \leftarrow T_1$
$e \leftarrow \sqrt{e}$
$h \leftarrow h/2$
go to 6
5 $T \leftarrow T_1$
go to 7

Let us now illustrate a sequence of transformations $\varphi : E \to \mathbb{R}$, as well as functions $q : E \to (0, \infty)$ in integrals e.g., of the form

$$I = \int_E f(y)\,dy, \tag{1.5.45}$$

The transformation $y = q(x)$ then transforms I into

$$J = \int_0^\infty g(x)\,dx, \tag{1.5.46}$$

where

$$g(x) = f(q(x))\,q'(x), \tag{1.5.47}$$

TABLE 1.1. Intervals and transformations for Algorithm 1.5.18

E	φ	q
$(0, \infty)$	$\log(y)$	x
$(0, 1)$	$\log\left(\dfrac{y}{1-y}\right)$	$\dfrac{x}{1+x}$
$(0, \infty)$	$\log(\sinh(y))$	$\log(x + \sqrt{1+x^2})$
$(-\infty, \infty)$	y	$\log(x)$
$(-\infty, \infty)$	$\log\left\{y + \sqrt{1+y^2}\right\}$	$(1/2)(x - 1/x)$
$(-\infty, \infty)$	$\log\left\{\sinh\left[y + \sqrt{1+y^2}\right]\right\}$	$(1/2)(\zeta - 1/\zeta),$ $\zeta = \log\left(x + \sqrt{1+x^2}\right)$

TABLE 1.2. Continuation of Table 1.1

E	q	q'
$(0, \infty)$	x	1
$(0, 1)$	$\dfrac{x}{1+x}$	$(1+x)^{-2}$
$(0, \infty)$	$\log(x + \sqrt{1+x^2})$	$1/\sqrt{1+x^2}$
$(-\infty, \infty)$	$\log(x)$	$1/x$
$(-\infty, \infty)$	$(1/2)(x - 1/x)$	$(1/2)(1 + 1/x^2)$
$(-\infty, \infty)$	$(1/2)(\zeta - 1/\zeta)$ $\zeta = \log\left(x + \sqrt{1+x^2}\right)$	$(1 + 1/\zeta^2)/\sqrt{1+x^2}$

and where $g/\varphi_1'(x) = \log(x)$. We can now apply the Algorithm 1.5.18 to approximate the integral I to arbitrary accuracy. The specific functional forms are given in Table 1.1.

1.5.8   SINC INDEFINITE INTEGRATION

Let us next describe Sinc indefinite integration over an interval or an arc. A detailed derivation and proof of Sinc indefinite integration is given in [S1, §3.6 and §4.5].

At the outset, we specify numbers σ_k and e_k, by

$$\sigma_k = \int_0^k \text{sinc}(x)\, dx, \quad k \in \mathbb{Z},$$

$$(1.5.48)$$

$$e_k = 1/2 + \sigma_k,$$

and we then define an $m \times m$ (*Toeplitz*) matrix $I^{(-1)}$ by $I^{(-1)} = [e_{i-j}]$, with e_{i-j} denoting the $(i,j)^{th}$ element of $I^{(-1)}$.

Additional definitions are the operators $\mathcal{J}^+$, $\mathcal{J}^-$, $\mathcal{J}_m^+$ and $\mathcal{J}_m^-$, and $m \times m$ matrices A^+ and A^- :

$$(\mathcal{J}^+ f)(x) = \int_a^x f(t)\, dt,$$

$$(\mathcal{J}^- f)(x) = \int_x^b f(t)\, dt,$$

$$(\mathcal{J}_m^+ f)(x) = \mathbf{w}(x)\, A^+ V f, \quad A^+ = h\, I^{(-1)} D(1/\varphi'),$$

$$(\mathcal{J}_m^- f)(x) = \mathbf{w}(x)\, A^- V f, \quad A^- = h\, (I^{(-1)})^T D(1/\varphi'),$$

$$(1.5.49)$$

where $D(\cdot)$ and $V(\cdot)$ are defined as in (1.5.23), and where the dependence of $\mathcal{J}_m^\pm$ on m is through the dependence of $A^\pm$, V and $\mathbf{w}$ on m.

The following theorem then enables us to collocate (linear or non-linear, non–stiff or stiff) initial value problems over an interval or an arc.

Theorem 1.5.19 *If $f/\varphi' \in \mathbf{L}_{\alpha,\beta,d}(\varphi)$, then, for all $N > 1$,*

$$\|\mathcal{J}^+ f - \mathcal{J}_m^+ f\| = \mathcal{O}(\varepsilon_N),$$

$$\|\mathcal{J}^- f - \mathcal{J}_m^- f\| = \mathcal{O}(\varepsilon_N).$$

$$(1.5.50)$$

1.5.9   SINC INDEFINITE CONVOLUTION

We consider here the approximation of the two indefinite convolution integrals on a finite or possibly infinite interval $(a, b) \subseteq \mathbb{R}$:

$$p(x) = \int_a^x f(x - t)\, g(t)\, dt,$$

$$q(x) = \int_x^b f(t - x)\, g(t)\, dt.$$

$$(1.5.51)$$

The approximation of these integrals is a very important part of this Sinc–Pack handbook. As we illustrate in the later chapters, our methods of approximating these integrals enable new formulas for approximating Laplace transform inversions (§1.5.10), Hilbert and Cauchy transforms (§1.5.12), for approximating solutions to ordinary differential equation problems (§3.9), for approximating solutions to Wiener–Hopf integral equations (§1.5.15), for evaluating more general convolution–type integrals of the form

$$r(x) = \int_a^x k(x, x - t, t) \, dt, \qquad (1.5.52)$$

(in §1.5.11), and for approximating Abel–type integrals (§3.6). In addition, our methods of solving integrals of the form (1.5.51) are far simpler to implement than those in current use to solve control theory problems. Our methods of approximating the integrals p and

q are original and powerful methods for solving partial differential equations, which frequently converge faster by orders of magnitude than conventional in use for solving such equations.

We shall use the notation of Definition 1.5.12, with the exception of modifying the basis to the following:

$$
\begin{aligned}
\gamma_j(x) &= S(j, h) \circ \phi(x), \quad j = -M, \cdots, N \\
\omega_j(x) &= \gamma_j(x), \quad j = -M + 1, \cdots, N - 1 \\
\omega_{-M} &= \left(1 + e^{-Mh}\right) \left[\frac{1}{1 + \rho} - \sum_{j=-(M-1)}^{N} \frac{\gamma_j}{1 + e^{jh}} \right] \\
\omega_N &= \left(1 + e^{-Nh}\right) \left[\frac{\rho}{1 + \rho} - \sum_{j=-M}^{N-1} \frac{e^{jh}\gamma_j}{1 + e^{jh}} \right].
\end{aligned}
\qquad (1.5.53)
$$

This basis is the same as that in (1.5.24), except for ω_{-M} and ω_N, which have been modified slightly for purposes of this proof, to ensure that these functions attain the exact value of 1 at the Sinc points where they interpolate, i.e., $\omega_{-M}(z_{-M}) = \omega_N(z_N) = 1$. We shall comment further on this modification in the proof of Theorem 1.5.20 below. In addition we shall use the definitions and notations of the previous section (§1.5.8) for the indefinite integral operators $\mathcal{J}^{\pm}$, and for their Sinc approximations, $\mathcal{J}_m^{\pm}$, with the exception, that these approximating operators are here defined via the basis $\mathbf{w} = (\omega_{-M}, \ldots, \omega_N)$, with the functions ω_j as in (1.5.53) rather than the ω_j in (1.5.24).

We shall also limit our proof to the approximation only of p, in that the procedure for approximating q is nearly identical. The interested reader should consult the references [S7] or [S1, §4.6] for original derivations of Sinc convolution.

We introduce here the equivalent of a Laplace transform of f. To this end, we shall assume that the usual Laplace transform $\hat{f}$ of f given by

$$\hat{f}(s) = \int_0^c e^{-st} f(t)\, dt\,, \qquad c \geq (b - a)\,,$$

exists for all $\Re s > 0$. We mention here that the Laplace transform of f is customarily defined over $(0, \infty)$. However, we have chosen a somewhat more general form $(0, c)$, in order to also be able to approximate integrals over finite intervals (a, b), with functions f that do not have a Laplace transform over $(0, \infty)$, e.g., for $f(x) = \exp(x^2)$. Furthermore, instead of using $\hat{f}$, it is more convenient for us to use F, defined by $F(s) = \hat{f}(1/s)$. Hence

$$F(s) = \int_0^c e^{-t/s} f(t)\, dt\,, \qquad c \geq (b - a)\,. \qquad (1.5.54)$$

We shall also assume for simplicity of presentation that the definition of F has been extended to the closed right half plane, and thus avoid writing limits in the inversion formulas. Since the transformation $s \to 1/s$ is a conformal map of the right half plane onto itself, the Bromwich inversion formula is transformed as follows:

$$f(t) \;=\; \frac{1}{2\pi i} \int_{-i\infty}^{i\infty} \hat{f}(s)\, e^{st}\, ds$$

$$\;=\; -\frac{1}{2\pi i} \int_{-i\infty}^{i\infty} \frac{F(s)\, e^{t/s}}{s^2}\, ds.$$

Let us now prove the following result, where as usual $\|\cdot\|$ denotes the sup norm on (a, b), as defined in (1.5.26), and in which the function $\varphi : (a, b) \to \mathbb{R}$ is as in Definition 1.5.2:

Theorem 1.5.20   (a) Given f and g in (1.5.51), let indefinite integrals $p(x)$ and $q(x)$ exist and are uniformly bounded on the finite or possibly infinite interval (a, b), and let F be defined as in (1.5.54). Then the following operator identities hold:

$$p = F\left(\mathcal{J}^+\right) g, \quad q = F\left(\mathcal{J}^-\right) g. \tag{1.5.55}$$

(b) *Suppose $g/\varphi' \in \mathbf{L}_{\alpha,\beta,d}(\varphi)$, and let M, N, h and ε_N be defined as in Definition 1.5.12. If for some positive c' independent of N we have $|F'(s)| \leq c'$ for all $\Re s \geq 0$, then there exists a constant C independent of N such that*

$$\|p - F\left(\mathcal{J}_m^+\right) g\| \leq C\,\varepsilon_N$$
$$\tag{1.5.56}$$
$$\|q - F\left(\mathcal{J}_m^-\right) g\| \leq C\,\varepsilon_N.$$

Remarks. (i) Suppose that $A^\pm$ (see (1.5.49)) can be diagonalized with $A^+ = X\,S\,X^{-1}$, with $A^- = Y\,S\,Y^{-1}$, and with S a diagonal matrix. Then, with $\mathcal{J}_m^\pm = \mathbf{w}(x)\,A^\pm V$, we have for $j = 0,\ 1,\ \ldots$,

$$\left((\mathcal{J}_m^+)^j\right)(x) \;=\; \mathbf{w}(x)\,(A^+)^j\,V \;=\; \mathbf{w}(x)\,X\,S^j\,X^{-1}\,V,$$

$$\left((\mathcal{J}_m^-)^j\right)(x) \;=\; \mathbf{w}(x)\,(A^-)^j\,V \;=\; \mathbf{w}(x)\,Y\,S^j\,Y^{-1}\,V,$$

and therefore, we have, with the basis (1.5.53), that

$$\left(F\left(\mathcal{J}_m^+\right) g\right)(x) \;=\; \mathbf{w}(x)\,X\,F(S)\,X^{-1}\,V\,g$$
$$\tag{1.5.57}$$
$$\left(F\left(\mathcal{J}_m^-\right) g\right)(x) \;=\; \mathbf{w}(x)\,Y\,F(S)\,Y^{-1}\,V\,g.$$

That is, whereas $F(\mathcal{J}^\pm)\,g$ is abstract symbolic and not readily applicable, we are able to evaluate the approximations, $F\left(\mathcal{J}_m^\pm\right) g$.

It follows, as a result of the definition of the basis given in (1.5.53), that whereas the formulas (1.5.56) are exact using the basis (1.5.57), in practice we would use the less cumbersome, basis (1.5.24), thus introducing only a negligible error, as discussed in the remarks preceding the statement of Theorem 1.5.13.

(ii) Consider for case of $0 < b - a < \infty$. Although the function F is different for every $c \geq (b-a)$, with c as in (1.5.54), the functions p and q in (1.5.55) and their approximations $F\left(\mathcal{J}_m^\pm\right) g$ are nevertheless independent of c. On the other hand, $F'(s)$ may not be bounded if $c = \infty$. For example, if $(a,b) = (0,1)$, and if $f(t) = t$, then

$$F(s) = \begin{cases} s^2 & \text{if } c = \infty \\ s^2\left(1 - e^{-1/s}\right) - s\,e^{-1/s} & \text{if } c = b - a. \end{cases}$$

In this case, $F'(s)$ is unbounded on the right half plane for the case of $c = \infty$, and bounded for the case of a finite c. Hence taking the smallest possible c may yield a better bound for the error of approximation in the (b)–Part of Theorem 1.5.20.

(iii) We have assumed in the theorem that $(\mathcal{J}^{\pm}g)(x)$, $p(x)$ and $q(x)$ exist and our uniformly bounded for all $x \in (a, b)$. This need not always be the case – see our worked example in §3.6.

   Proof of Part (a) of Theorem 1.5.20. Let us start with the well known result

$$((\mathcal{J}^{+})^n g)(x) = \int_a^x \frac{(x - t)^{n-1}}{(n-1)!}\, g(t)\, dt\,, \qquad n = 0,\, 1,\, \ldots,$$

and the easily verifiable result that with any u such that $u/\varphi' \in \mathbf{L}_{\alpha,\beta,d}(\varphi)$ we have

$$\Re\left(\mathcal{J}^{+}u\,,u\right) = \Re\left(\int_a^b (\mathcal{J}^{+}u)(x)\,\overline{u(x)}\,dx\right) = \frac{1}{2}\left|\int_a^b u(x)\,dx\right|^2 \geq 0.$$

Notice by Theorem 1.5.16, that the integral of u taken over (a, b) exists and is bounded. This "inner product" inequality shows that the real part of the spectrum of the operator $\mathcal{J}^{+}$ is non–negative. Since $\int_a^b |g(t)|\,dt$ is finite, under our assumption that $g/\varphi' \in \mathbf{L}_{\alpha,\beta,d}(\varphi)$ it follows that the operator $\mathcal{J}^{+}$ is then bounded by $b - a$. (Infinite intervals (a, b) are handled via a sequence of nested finite intervals approaching (a, b).)

The following "Dunford operator integral approach" (see [S1, pp. 225–227] yields

$$p(x) = -\frac{1}{2\pi i} \int_a^x \int_{-i\infty}^{i\infty} \int_a^x \exp((x-t)/s) \, F(s) \, ds \, g(t) dt$$

$$= -\frac{1}{2\pi i} \int_{-i\infty}^{i\infty} \sum_{n=0}^{\infty} \frac{(x-t)^n}{s^{n+2} \, n!} \, g(t) \, dt F(s) \, ds$$

$$= -\frac{1}{2\pi i} \int_{-i\infty}^{i\infty} \frac{\mathcal{J}^+}{s^2} \sum_{n=0}^{\infty} \frac{(\mathcal{J}^+)^n}{s^n} \, g \, F(s) \, ds$$

$$= -\frac{1}{2\pi i} \int_{-i\infty}^{i\infty} \left(\frac{1}{s} - \frac{1}{s - \mathcal{J}^+} \right) F(s) \, ds \, g$$

$$= (F(\mathcal{J}^+) - F(0^+)) \, g = F(\mathcal{J}^+) \, g \,,$$

where the last line obtains as a result of $F(0^+) = \hat{f}(\infty) = 0$. Notice also, that this operator valued contour integral can be evaluated by contour integration, via use of the contour $\mathcal{C} = \{z = x + i \, y : -R \le y \le R\} \cup \{z = R e^{i\theta} : -\pi/2 \le \theta \le \pi/2\}$. As R increases, this contour contains the spectrum of $\mathcal{J}$ and furthermore, the part of the contour integral along the circular part of this contour approaches zero as $R \to \infty$.

Proof of Part (b) of Theorem 1.5.20. Let us first show that the spectrum of $\mathcal{J}_m^{\pm}$ is also on the closed right half plane. Consider the case of $\mathcal{J}_m^+$. By definition (1.5.48) we have $I^{(-1)} = H + K$, where each of the m^2 entries of H is $1/2$, whereas K is a skew symmetric matrix. Hence, if $\mathbf{c} = (c_{-M}, \dots, c_N)^T$ is an arbitrary complex column vector, then

$$\mathbf{c}^* I^{(-1)} \mathbf{c} = \frac{1}{2} \left| \sum_{j=-M}^{N} c_j \right|^2 + \mathbf{c}^* K \mathbf{c}.$$

Since $\Re \mathbf{c}^* K \mathbf{c} = 0$, it follows from this identity that the the eigenvalues of $I^{(-1)}$ are all in the closed right half plane. Next, from (1.5.49) we have $A^+ = I^{(-1)} D$, where $D = D(h/\varphi') = \mathrm{diag}(d_{-M}, \dots, d_N)$ is a diagonal matrix with positive entries d_j. Since the eigenvalues of A^+ are the same as the eigenvalues of $D^{1/2} A^+ D^{-1/2} = D^{1/2} I^{(-1)} D^{1/2}$, it follows, with $\mathbf{c}$ as above, that

$$\mathbf{c}^* \, D^{1/2} \, I^{(-1)} \, D^{1/2} \, \mathbf{c} = \frac{1}{2} \left| \sum_{j=-M}^{N} c_j \, d_j \right|^2 + \mathbf{c}^* \, D^{1/2} \, K \, D^{1/2} \, \mathbf{c}.$$

(Notice, if we set $c_j = u(z_j)$ with u the function used in the above inner product examination $(\mathcal{J}^+ u, u)$, and with $z_j = \varphi^{-1}(j\,h)$, a Sinc–point, then the above sum $\sum_{j=-M}^{N} c_j \, d_j$ is just the Sinc quadrature approximation of $\int_a^b u(x)\,dx$ – see §1.5.7). This identity confirms again that the eigenvalues of A^+ are all on the closed right half plane. Indeed, it has been tested[2] by direct computation, that the eigenvalues of $I^{(-1)}$ are, in fact, on the open right half plane, for all orders m up to 1024. This means that the matrices $F(A^\pm)$ are well defined for all these values of m.

We then have

$$\|(F(\mathcal{J}^\pm) - F(\mathcal{J}_m^\pm))\,g\|$$

$$= \left\| \int_0^1 F'\left(t\,\mathcal{J}^\pm + (1-t)\,\mathcal{J}_m^\pm\right) dt \, (\mathcal{J}^\pm - \mathcal{J}_m^\pm)\,g \right\|$$

$$\leq \left\| \int_0^1 F'\left(t\,\mathcal{J}^\pm + (1-t)\,\mathcal{J}_m^\pm\right) dt \right\| \, \|(\mathcal{J}^\pm - \mathcal{J}_m^\pm)\,g\| \,.$$

Since the spectra of the operators $\mathcal{J}^\pm$ and $\mathcal{J}_m^\pm$ lie on the closed and convex right half plane, the same is true of their convex linear combination, $t\,\mathcal{J}^\pm + (1-t)\,\mathcal{J}_m^\pm$. The assertion of Theorem 1.5.20 (b) then follows by our assumption of the boundedness of F' on the right half plane, and by Theorem 1.5.19. ∎

Remark. We have tacitly assumed that A^+ and A^- can be diagonalized for every m. Although this has not been proved to date, it has so far been the case for all problems that we have attempted. Now the matrices Y which diagonalize $I^{(-1)}$ are well conditioned for $m = 1, 2, \ldots, 1024$. On the other hand if the condition number of a matrix X such that $A^+ = h I^{(-1)} D(1/\varphi') = X S X^{-1}$ is large,

---

[2] A still unsolved problem, is a proof or disproof, for which the author is offering \$300, that all eigenvalues of $I^{(-1)}$ lie on the open right half plane, for every positive integer m.

then a slight alteration of the parameter h (which does not appreciably change the accuracy of approximation) will usually lower this condition number.

1.5.10   LAPLACE TRANSFORM INVERSION

Let $x \in (0, a)$, with $a > 0$. Starting with the identity

$$f(x) - f(0) = \int_0^x f'(t)\, dt\,,$$

$$= \int_0^x f'(x - t)\, g(t)\, dt\,, \quad g(t) \equiv 1\,,$$

(1.5.58)

which is a convolution of the form (1.5.51), in which $g(t) = 1$. The (ordinary) Laplace transform equivalent of f is

$$\hat{f}(s) = \int_0^c \exp(-x\, s)\, f(x)\, dx\,, \quad c \geq a$$

$$= \{s\, \hat{f}(s)\}\{1/s\}\,,$$

(1.5.59)

or, replacing s by $1/s$ and $\hat{f}(1/s)$ by $\mathcal{F}(s)$, and then setting $\mathcal{G}(s) = \mathcal{F}(s)/s$, we get (see (1.5.49) for the definition of $\mathcal{J}^+$)

$$f = \mathcal{G}(\mathcal{J}^+)\, g$$

$$f(x) \approx \mathbf{w}(x)\, \mathcal{G}(A^+)\, \mathbf{1}\,,$$

(1.5.60)

where $\mathbf{1} = (1,\, 1,\, \ldots,\, 1)^T$.

It may be shown that this approximation is valid (in the sense of a relative error of the order of ε) even when $f/\varphi' \in \mathbf{L}_{\alpha,\beta,d}(\varphi)$.

Remark. Typically, inversion of the Laplace transform is numerically difficult. However, the approximation result (1.5.60) now makes it possible for engineers to solve problems in control theory that in the past have proven to be difficult, or intractable. Using Sinc, we see that the problem is reduced to finding the eigenvalues and eigenvectors of a single matrix, namely A^+.

1.5.11   MORE GENERAL $1 - d$ CONVOLUTIONS

We illustrate here the approximation of the following three one dimensional integrals, using the results of §1.5.9.

$$p(x) = \int_a^x k(x - t, t) \, dt, \quad x \in (a, b),$$

$$q(x) = \int_a^x k(x, x - t) \, dt, \quad x \in (a, b), \qquad (1.5.61)$$

$$r(x) = \int_a^x k(x, x - t, t) \, dt, \quad x \in (a, b).$$

The approximation of these integrals, particularly the approximation of $p(x)$ and $r(x)$, have important ramifications regarding the solution of PDE over curvilinear regions via Sinc convolution. We shall in fact use these results in §2.4.3 and in §2.4.4.

1. *The integral $p(x)$.* Let us assume here that the Laplace transform

$$K(s, t) = \int_0^c k(x, t) \, \exp(-x/s) \, dx, \quad c \geq (b - a) \qquad (1.5.62)$$

is known. If it is not known, then we shall have to approximate this integral using Sinc quadrature, for each Sinc point z_n on (a, b), and for each eigenvalue s_k of the indefinite integration matrix A^+ (see (1.5.49) and the last line of (1.5.64) below).

We assume that there exist $\mathbf{L}^1((a, b) \times (a, b))$ approximations to $k(x, t)$ as in the first line of (1.5.63) below, so that

$$k(x,t) = \lim_{\nu \to \infty} \sum_{j=1}^{\nu} a_j^{(\nu)}(x)\, b_j^{(\nu)}(t)$$

$$p_j^{(\nu)}(x) \equiv \int_a^x a_j^{(\nu)}(x-t)\, b_j^{(\nu)}(t)\, dt$$

$$\widehat{k_j^{(\nu)}}(s,t) \equiv \int_0^c \exp(-x/s)\, a_j^{(\nu)}(x)\, b_j^{(\nu)}(t)\, dx$$

$$K(s,t) = \lim_{\nu \to \infty} \sum_j \widehat{k_j^{(\nu)}}(s,t) \qquad (1.5.63)$$

$$p_j^{(\nu)}(x) = \int_{\frac{b}{\nu}}^x \widehat{k_j^{(\nu)}}(s,t)\, dt$$

$$p^{(\nu)}(x) = \sum_{j=1}^{\nu} p_j^{\nu}(x)$$

$$p(x) = \lim_{\nu \to \infty} p^{(\nu)}(x).$$

Then, by proceeding as in (1.5.55)–(1.5.57) above, we get, in the notation of (1.3.57), with $X = [x_{ij}]$ and $X^{-1} = [x^{ij}]$, and with $z_{-M}, \ldots, z_N$ denoting the Sinc points,

$$p_j^{(\nu)}(z_n) = \sum_{k=-M}^{N} x_{nk} \sum_{\ell=-M}^{N} x^{k\ell}\, \widehat{a_j^{(\nu)}}(s_k)\, b_j^{(\nu)}(z_\ell)$$

$$= \sum_{k=-M}^{N} x_{nk} \sum_{\ell=-M}^{N} x^{k\ell}\, \widehat{k_j^{(\nu)}}(s_k, z_\ell)$$

$$p^{(\nu)}(z_n) = \sum_{k=-M}^{N} x_{nk} \sum_{\ell=-M}^{N} x^{k\ell}\, \widehat{k^{(\nu)}}(s_k, z_\ell), \quad \text{and}$$

$$p(z_n) = \lim_{\nu \to \infty} p^{(\nu)}(z_n) = \sum_{k=-M}^{N} x_{nk} \sum_{\ell=-M}^{N} x^{k\ell} K(s_k,\, z_\ell).$$

$$(1.5.64)$$

It is convenient by way of evaluating this result to let $\mathbf{u}_k$ denote the kth row of $K = [K(k,\ell)] = [K(s_k, z_\ell)]$, $\mathbf{v}_k$ the kth row of $Xi = X^{-1}$, to then form a column vector $\mathbf{w}$ with k^{th} component $w_k = \mathbf{u}_k * (\mathbf{v}_k)^T$. We can then evaluate the column vector $\mathbf{p} = (p_{-M}, \ldots, p_n)^T$ via use of the expression

$$\mathbf{p} = X\,\mathbf{w}. \qquad (1.5.65)$$

This then yields the accurate approximation $p(z_k) = p_k$.

2. *The Integral $q(x)$.* This is similar to the case of $p(x)$ above. Instead of (1.3.58) and (1.3.59), we now write

$$K(x, s) = \int_0^c k(x, u) \exp(-u/s) \, du, \quad c \geq (b - a)$$

$$= \lim_{\nu \to \infty} \sum_{k=1}^{\nu} e_k^{(\nu)}(x) f_k^{(\nu)}(s),$$

$$(1.5.66)$$

and so, we get

$$q(x) = K(x, \mathcal{J}^+) \, 1$$

$$= \lim_{\nu \to \infty} \sum_{k=1}^{\nu} e_k^{(\nu)}(x) f_k^{(\nu)}(\mathcal{J}^+) \, 1$$

$$(1.5.67)$$

$$\approx \lim_{\nu \to \infty} \sum_{k=1}^{\nu} e_k^{(\nu)}(x) f_k^{(\nu)}(\mathcal{J}_m^+) \, 1$$

$$= \mathbf{w} \, X \lim_{\nu \to \infty} \sum_{k=1}^{\nu} e_k^{(\nu)}(x) f_k^{(\nu)}(S) \, X^{-1} \, \mathbf{1},$$

where $\mathbf{1}$ is a column vector of m ones. That is, setting

$$X = [x_{jk}], \qquad X^{-1} = [x^{jk}],$$

$$(1.5.68)$$

$$y^{\ell} = \sum_{n=1}^{m} x^{\ell n},$$

we get

$$q(z_j) \approx \sum_{\ell=1}^{m} x_{j,\ell} \, K(z_j, s_\ell) \, y^{\ell}. \qquad (1.5.69)$$

3. *The integral $r(x)$.* The treatment of this integral is also similar to that of $p(x)$ above. Instead of (1.5.62) and (1.5.63), we now write

$$K(x, s, t) = \int_0^c k(x, u, t) \exp(-u/s)\, du, \quad c \geq (b - a)$$

$$= \lim_{\nu \to \infty} \sum_{k=1}^{\nu} e_k^{(\nu)}(x)\, f_k^{(\nu)}(s, t).$$

$$(1.5.70)$$

Hence, by proceeding as for $p(x)$ above, we set

$$e^{(\nu)}(x)\, f^{(\nu)}(s, t) = \sum_{k=1}^{\nu} e_k^{(\nu)}(x)\, f_k^{(\nu)}(s, t),$$

$$(1.5.71)$$

$$\rho^{(\nu)}(z_n) = e^{(\nu)}(z_n) \sum_{k=-M}^{N} x_{nk} \sum_{\ell=-M}^{N} x^{k\ell} f^{(\nu)}(s_k, z_\ell).$$

to get

$$r(z_n) \approx \lim_{\nu \to \infty} \rho^{(\nu)}(z_n)$$

$$= \sum_{k=-M}^{N} x_{nk} \sum_{\ell=-M}^{N} x^{k\ell} K(z_n, s_k, z_\ell). \qquad (1.5.72)$$

We thus get the approximation $r = (r_{-M}, \ldots, r_N)$, with $r(z_n) \approx r_n$.

Remarks.

1. The reader may note that if in the above equations he/she makes the replacements (see §1.5.8)

$$\begin{array}{rcl}
\mathcal{J}^+ & \to & \mathcal{J}^- \\
\mathcal{J}_m^+ & \to & \mathcal{J}_m^- \\
A^+ & \to & A^-,
\end{array}$$

then the above theory applies verbatim to the integrals

$$p(x) = \int_x^b k(t-x,t)\,dt\,, \quad x \in (a,b),$$

$$q(x) = \int_x^b k(x,t-x)\,dt\,, \quad x \in (a,b), \qquad (1.5.73)$$

$$r(x) = \int_x^b k(x,t-x,t)\,dt\,, \quad x \in (a,b).$$

2. As already mentioned, the equations of this section have important applications for solving PDE, since they enable us to use *separation of variables* to solve partial differential equations over curvilinear regions. Of course the student of engineering is already familiar with this technique applied to cylindrical and spherical regions where the existence of specific orthogonal bases make such separation possible. Chapters 2 & 4 of this handbook contain illustrations of variables using Sinc convolution over significantly more general regions, namely, those for which the boundary can be expressed as a union of a finite number of explicit functional expressions.

1.5.12 Hilbert and Cauchy Transforms

We consider here some methods of approximating the *Cauchy* and *Hilbert* transforms. The Cauchy transform of a function g defined on an arc Γ, taken Γ and evaluated at $z \in \mathbb{C}$, with $z \notin \Gamma$, takes the form

$$(\mathcal{C}g)(z) = \frac{1}{2\pi i} \int_\Gamma \frac{g(t)}{t-z}\,dt. \qquad (1.5.74)$$

The Hilbert transform of a function g defined on Γ, taken over Γ, and evaluated at $x \in \Gamma$ is defined by

$$(\mathcal{S}g)(x) = \frac{P.V.}{\pi i} \int_\Gamma \frac{g(t)}{t-x}\,dt\,, \qquad (1.5.75)$$

where "P.V." denotes the principal value of the integral. Initially we state some approximations derived in [§5.2, S1] which have already been successfully used in [SSc, Bia3, S1]. See also Theorem 1.3.2, Equation (1.3.13) above. Finally, we describe a method for approximating these integrals via use of Sinc convolution.

1. *Use of Explicit Basis Functions.* Let $z = x + iy$, with x and y real, and let us define the following functional expressions, which are deduced via inspection of the results of Problems 1.2.4 and 1.2.5:

$$s_k(x) \;=\; \frac{h}{\pi} \, \frac{\sin\left\{\frac{\pi}{h}\left(\varphi(x) - kh\right)\right\}}{\varphi'(z_k)(x - z_k)} \tag{1.5.76}$$

$$t_k(x) \;=\; \frac{h}{\pi i} \, \frac{\cos\left\{\frac{\pi}{h}\left(\varphi(x) - kh\right)\right\} - 1}{\varphi'(z_k)(x - z_k)} \tag{1.5.77}$$

$$c_k(z) \;=\; \frac{h}{2\pi i} \, \frac{\exp\left\{\frac{\pi i}{h}\left(\varphi(z) - kh\right)\right\} - 1}{\varphi'(z_k)(z - z_k)}. \tag{1.5.78}$$

Note that for $z = x$, we have $c_k(x) = \frac{1}{2}\left(s_k(x) + t_k(x)\right)$. If k and ℓ are integers, then

$$s_k(z_\ell) \;=\; \begin{cases} 0 \text{ if } k \neq \ell \\ 1 \text{ if } k = \ell \end{cases} \tag{1.5.79}$$

$$t_k(z_\ell) \;=\; \begin{cases} \dfrac{h}{\pi i} \dfrac{1 - (-1)^{k-\ell}}{\varphi'(z_k)(z_k - z_\ell)} & \text{if } \;\; k \neq \ell \\[2mm] 0 & \text{if } \;\; k = \ell. \end{cases} \tag{1.5.80}$$

$$c_k(z_\ell) \;=\; \begin{cases} \dfrac{h}{2\pi i} \dfrac{1 - (-1)^{k-\ell}}{\varphi'(z_k)(z_k - z_\ell)} & \text{if } \;\; k \neq \ell \\[2mm] \frac{1}{2} & \text{if } \;\; k = \ell \end{cases} \tag{1.5.81}$$

We then have the following theorem, the proof of which can be found in [ES] and also in [S1, §5.2], and is omitted here. Part (c) of this theorem involves complex variables. In the theorem, the function φ is not only a one–to–one map of Γ to the real line $\mathbb{R}$, but is also a one–to–one map of a region $\mathcal{D}$ of the complex plane to the strip $\mathcal{D}_d = \{z \in \mathbb{C} : |\Im z| < d\}$. The domain $\mathcal{D}^+$ then denotes the part of $\mathcal{D}$ to the left of Γ as one traverses Γ from a to b. The number $N(g, \mathcal{D}, z)$ in the theorem is defined by

$$N(g, \mathcal{D}, z) = \int_{\partial \mathcal{D}} \left| \frac{F(t)}{t - z} dt \right|. \tag{1.5.82}$$

Theorem 1.5.21 (a) If $g \in \mathbf{M}_{\alpha,\beta,d}(\varphi)$, then, for all $x \in \Gamma$,

$$|\varepsilon_1(x)| \equiv \left| g(x) - \sum_{k=-\infty}^{\infty} g(z_k) s_k(x) \right|$$

(1.5.83)

$$\leq \frac{1}{2\pi \sinh(\pi d/h)} N(g, \mathcal{D}, x).$$

(b) If $g \in \mathbf{M}_{\alpha,\beta,d}(\varphi)$, then, for all $x \in \Gamma$,

$$\varepsilon_2(x) \equiv \left| \frac{P.V.}{\pi i} \int_{\Gamma} \frac{g(t)}{t-x} dt - \sum_{k=-\infty}^{\infty} g(z_k) t_k(x) \right|$$

(1.5.84)

$$\leq \frac{e^{-\pi d/h} + 1}{2\pi \sinh(\pi d/h)} N(g, \mathcal{D}, x).$$

(c) If $g \in \mathbf{M}_{\alpha,\beta,d}(\varphi)$, then for all $z \in \mathcal{D}^+$, we have

$$\varepsilon_3(z) \equiv \left| \frac{1}{2\pi i} \int_{\Gamma} \frac{g(\zeta)}{\zeta - z} d\zeta - \sum_{j=-\infty}^{\infty} g(z_j) c_j(z) \right|$$

(1.5.85)

$$\leq \frac{e^{-\pi d/h} + 1}{2\pi \sinh(\pi d/h)} N(g, \mathcal{D}, z).$$

For example, the part (b) of this theorem yields the approximation

$$(\mathcal{S}g)(z_\ell) \equiv \frac{P.V.}{\pi i} \int_{\Gamma} \frac{g(t)}{t - z_\ell} dt$$

(1.5.86)

$$\approx \frac{h}{\pi i} \sum_{k \neq \ell} \frac{g(z_k)}{\varphi'(z_k)} \frac{1 - (-1)^{k-\ell}}{z_k - z_\ell}.$$

The above bounds for $\varepsilon_1(x)$, $\varepsilon_2(x)$, and ε_3, and as well as the bounds in (1.5.88) (1.5.88) on the basis functions s_k, t_k and c_k defined in (1.5.76)–(1.5.78) were established in [ES] (see also [S1, §5.2]).

We may note that under our assumptions on g, the terms $N(g, \mathcal{D}, x)$ and $N(g, \mathcal{D}, z)$ may become unbounded as z approaches a point of

$\partial \mathcal{D}$. In such a case the above sums must be interpreted in the sense of a relative error. On the other hand, a uniform bound may be possible, if these equations hold for all $x \in \Gamma$, or for all $z \in \mathcal{D}'$ for the case of the part (c) of Theorem 1.5.22. For example, if we have $g \in \mathbf{L}_{\alpha,\beta,d}(\varphi)$, instead of $g \in \mathbf{M}_{\alpha,\beta,d}(\varphi)$, then the errors ε_j in Theorem 1.5.22 are uniformly bounded, and in particular,

$$N(g, \Gamma) \equiv \sup_{x \in \Gamma} N(g, \mathcal{D}, x) < \infty$$

$$N(g, \mathcal{D}) \equiv \sup_{z \in \mathcal{D}} N(g, \mathcal{D}, z) < \infty.$$

(1.5.87)

In order to truncate the sums, we also use the bounds

$$\sup_{x \in \Gamma} |s_k(x)| \le e^h$$

$$\sup_{x \in \Gamma} |t_k(x)| \le e^h$$

(1.5.88)

$$\sup_{z \in \mathcal{D}^+} |c_k(z)| \le \sqrt{2}\, e^h.$$

We assume in the following corollary, that M and N are related by $M = [\alpha/\beta\, N]$, where $[\cdot]$ denotes the greatest integer function.

Corollary 1.5.22 *If $g \in \mathbf{L}_{\alpha,\beta,d}(\varphi)$, then there exist positive constants C_i, $i = 1, 2, 3$, that are independent of N, such that*

$$\sup_{x \in \Gamma} \left| g(x) - \sum_{j=-M}^{N} g(z_j)\, s_j(x) \right| < C_1\, \varepsilon_N, \quad x \in \Gamma$$

$$\sup_{x \in \Gamma} \left| \frac{P.V.}{\pi i} \int_{\Gamma} \frac{g(t)}{t - x} - \sum_{j=-M}^{N} g(z_j)\, t_j(x) \right| < C_2\, \varepsilon_N, \quad x \in \Gamma$$

$$\sup_{z \in \mathcal{D}^+} \left| \frac{1}{2\pi i} \int_{\Gamma} \frac{g(t)}{t - z} - \sum_{j=-M}^{N} g(z_j)\, c_j(z) \right| < C_3\, \varepsilon_N, \quad z \in \mathcal{D}^+.$$

(1.5.89)

2. *Cauchy and Hilbert Transforms via Sinc Convolution.* In what
follows here, we derive our most preferable methods of approximating Cauchy and Hilbert transforms. Initially, we present the Cauchy
transform, which can be effectively approximated provided that we
can approximate the special function $E_1(z)$ for complex values of z.
We then use this function, together with Plemelj's formula [Gv], to
derive a method of approximating the Hilbert transform. This latter
formula does not require evaluation of the special function E_1. Initially, we derive the formulas for the interval $(a, b) \subseteq \mathbb{R}$. In 3. below,
we extend these formulas to the case of an analytic arc.

Let $(a, b) \subseteq \mathbb{R}$, and let $y > 0$. The Cauchy integral in the part (c)
of Theorem 1.5.22,

$$g(x + i\,y) = (\mathcal{C}\,f)(x + i\,y) = \frac{1}{2\,\pi\,i} \int_a^b \frac{f(t)}{t - x - i\,y}\,dt\,, \qquad (1.5.90)$$

is of course also a convolution. Next, illustrate an alternative way of
approximating Hilbert and Cauchy integrals using Sinc convolutions.

We first rewrite the integral expressing g in the form

$$(\mathcal{C}\,f)(x + i\,y) = -\frac{1}{2\,\pi\,i} \int_a^x \frac{f(t)}{x - t + i\,y}\,dt + \frac{1}{2\,\pi\,i} \int_x^b \frac{f(t)}{t - x - i\,y}\,dt.$$
$$(1.5.91)$$

We thus need two Laplace transforms,

$$F^+(s\,,i\,y) \;=\; \frac{-1}{2\,\pi\,i} \int_0^\infty \frac{e^{-t/s}}{t + i\,y}\,dt\,,$$

$$(1.5.92)$$

$$F^-(s\,,i\,y) \;=\; \frac{1}{2\,\pi\,i} \int_0^\infty \frac{e^{-t/s}}{t - i\,y}\,dt.$$

Inspecting these two integrals we see immediately that if y and s are
real, then

$$F^-(s\,,i\,y) = \overline{F^+(s\,,i\,y)}. \qquad (1.5.93)$$

Setting $t + i\,y = \tau$ in the above integral expression for $F^+(s\,,i\,y)$,
we get

$$F^+(s,i\,y) \;=\; -\frac{e^{i\,y/s}}{2\,\pi\,i} \int_{iy/s}^{\infty} \frac{e^{-\tau}}{\tau}\, d\tau = -\frac{e^{i\,y/s}}{2\,\pi\,i}\, E_1(i\,y/s)\,,$$

$$(1.5.94)$$

$$F^-(s,i\,y) \;=\; \frac{e^{-i\,y/s}}{2\,\pi\,i}\, E_1(-i\,y/s)\,,$$

where E_1 denotes the well known special function expression (see [Na, Eq. 5.1.5])

$$E_1(z) = \int_z^{\infty} \frac{e^{-t}}{t}\, dt\,, \quad |\arg(z)| < \pi. \qquad (1.5.95)$$

Theorem 1.5.23 *If $\mathcal{J}^\pm$, and $A^\pm$ are defined corresponding to an interval $(a,b) \subseteq \mathbb{R}$ as in §1.5.8, the operator V and basis function vector $\mathbf{w}$ as in Definition 1.5.12, and if $F^\pm$ are defined as in (1.5.94) above, then the Cauchy transform of f defined on (a,b) is approximated as follows:*

$$(\mathcal{C}\,f)(x+i\,y) \;=\; \frac{1}{2\pi i} \int_a^b \frac{f(t)}{t-x-i}\, dt$$

$$= \;(F^+\,(\mathcal{J}^+,i\,y) + F^-(\mathcal{J}^-,i\,y))\, f)\,(x)\,,$$

$$\approx \; \mathbf{w}(x)\,(F^+(A^+,i\,y) + F^-(A^-,i\,y))\,V f\,,$$

$$(1.5.96)$$

If $f \in \mathbf{L}_{\alpha,\beta,d}(\varphi)$, then the error of this approximation is $\mathcal{O}(\varepsilon_N)$, whereas if $f \in \mathbf{M}_{\alpha,\beta,d}(\varphi)$, then $g(x,y)$ may become unbounded as $x+i\,y \to a$ or as $x+i\,y \to b$, and in this case the approximation (1.5.96) is accurate to within a relative error of the order of ε_N.

We must of course exercise care in using this approximation when y is small. To this end, the representation of E_1 given below, in terms of the special function G_1 may be useful. Spectral decompositions $A^\pm = X_j\, S\, X_j^{-1}$ will need to be used to evaluate $F^\pm$, although errors in X_j and S can be magnified as $y \to 0$. We remark also, that the operators $F^\pm(A^\pm,i\,y)$ are well defined in view of the argument restriction in the above definition of $E_1(z)$, under the assumption that the eigenvalues of the matrices $A^\pm$ are located on the open right half plane. (See the footnote on page 99.)

We can also deduce the Hilbert transform from this result, via use of Sinc convolution. Recall first, the Plemelj identity,

$$\lim_{y \to 0^+} \frac{1}{2\pi i} \int_a^b \frac{f(t)}{t - x - iy} \, dt = \frac{1}{2} \left(f(x) + (\mathcal{S} f)(x) \right). \qquad (1.5.97)$$

It is also known that ([Na, Eq. (5.1.11)])

$$E_1(z) = -\gamma - \log(z) + G_1(z)$$

$$G_1(z) = \sum_{n=1}^{\infty} \frac{(-1)^n z^n}{n \, (n!)}. \qquad (1.5.98)$$

Clearly, G_1 is an entire function, and moreover, $G_1(z) \to 0$ as $z \to 0$.

Consequently, in view of the definitions of V and $\mathbf{w}$ in §1.5.4, $\mathcal{J}^+$, $A^+ = h \, I^{(-1)} \, D(1/\varphi')$, $\mathcal{J}^-$, and $A^- = h \, (I^{(-1)})^T \, D(1/\varphi')$, in §1.5.8, we deduce that

$$\lim_{y \to 0^+} F^+(\mathcal{J}^+, i\,y) + F^-(\mathcal{J}^-, i\,y)$$

$$= \frac{1}{2} + \frac{1}{\pi i} \left(\log(\mathcal{J}^-) - \log(\mathcal{J}^+) \right). \qquad (1.5.99)$$

In view of (1.5.96) and (1.5.97) above, we thus get

Theorem 1.5.24 *If $\mathcal{J}^\pm$ and $A^\pm$ are defined corresponding to an interval $(a, b) \subseteq \mathbb{R}$ as in §1.5.8, the operator V and basis function vector $\mathbf{w}$ as in Definition 1.5.12, then the Hilbert transform of f taken over (a, b) and evaluated at $x \in (a, b)$ is approximated as follows:*

$$(\mathcal{S} f)(x) = \frac{P.V.}{\pi i} \int_a^b \frac{g(t)}{t - x} \, dt$$

$$= \left(\left(\frac{1}{\pi i} \left(\log(\mathcal{J}^-) - \log(\mathcal{J}^+) \right) \right) g \right)(x) \qquad (1.5.100)$$

$$\approx \mathbf{w}(x) \frac{1}{\pi i} \left(\log(A^-) - \log(A^+) \right) V g.$$

If $f \in \mathbf{L}_{\alpha,\beta,d}(\varphi)$, then the error of this approximation is $\mathcal{O}(\varepsilon_N)$, whereas if $f \in \mathbf{M}_{\alpha,\beta,d}(\varphi)$, then $\mathcal{S} f(x)$ may become unbounded as

$x \to a$ or as $x \to b$, and in this case the approximation (1.5.100) is accurate to within a relative error of the order of ε_N.

This last result was originally obtained by T. Yamamoto [Ya1] via a more direct procedure that does not involve the special function E_1.

Remark. It is unfortunate that

$$\log(A^-) - \log(A^+) \neq \log\left(A^- (A^+)^{-1}\right) = \log\left(I^{(-1)} ((I^{(-1)})^T)^{-1}\right).$$

since equality of this result would enable a more efficient procedure for computing Hilbert and Cauchy transforms.

3. *Extension to Analytic Arcs.* We extend here, the results of Theorems 1.5.23 and 1.5.24 to analytic arcs. To this end, let us assume that the arc Γ is defined by

$$\Gamma = \{z \in \mathbb{C} : z = \zeta(t), \ 0 \leq t \leq 1\}, \tag{1.5.101}$$

that we want to approximate the Cauchy transform at z, and that our result should hold uniformly, as $z \to z' \in \Gamma$, where the approach $z \to z' = \zeta(\eta)$ is assumed to be from the left of Γ. Let us also assume that we have for some $i\,y \in \mathbb{C}$, the relation $\zeta(\eta + i\,y) = z$.

Then we get,

$$
\begin{aligned}
(\mathcal{C}f)(z) &= \frac{1}{2\pi i} \int_\Gamma \frac{f(t)}{t - z}\, dt \\
&= \frac{1}{2\pi i} \int_0^1 \frac{\xi - \eta - i\,y}{\zeta(\xi) - z} \frac{f(\zeta(\xi))\,\dot\zeta(\xi)}{\xi - \eta - i\,y}\, d\xi,
\end{aligned}
\tag{1.5.102}
$$

where $\dot\zeta(\xi) = d\zeta(\xi)/d\xi$.

At this point, let $F^\pm$ be defined as in (1.5.92) above, let $A^\pm$ denote the Sinc indefinite integration matrices defined as above, but for the interval $(0,1)$. Let us set

$$B = [b_{j,k}] = \frac{1}{2\pi i} \left(F^+(A^+, i\,y) + F^-(A^-, i\,y)\right). \tag{1.5.103}$$

whose terms need to be evaluated carefully via use of the eigenvalue decompositions of A^+ and A^-, since we may no longer have

$|\arg(iy)| < \pi$ for the arc Γ. Here, too, the above expression (1.5.98) for E_1 in terms of γ, a logarithm, and an entire function G_1 will be useful for purposes of programming the result given in the next equation.

Evidently, the last expression in the above equation for $(\mathcal{C}\,f)(z)$ takes the form of the convolution $r(x)$ in §1.5.11, and we thus appeal to the procedure given in §1.5.11 for approximating $r(x)$. Letting t_j denote the Sinc points of $(0,1)$ for $j = -N \, \ldots , \, N$, we thus get

$$(\mathcal{C}\,f)(\zeta(t_j + iy)) = \sum_{k=-N}^{N} b_{jk} \frac{(t_j - t_k - i\,y)\,\dot{\zeta}(t_k)}{\zeta(t_j) - \zeta(t_k + i\,y)}\, f(\zeta(t_k)). \quad (1.5.104)$$

Remark. *The evaluation of the above entries b_{jk} is non-trivial, and inasmuch as this formula for evaluating $(\mathcal{C}\,f)(z)$ is a formula for analytic continuation, it will be more efficient and just as accurate to use the procedure described below in §1.5.13. The next formula, for approximating the Hilbert transform $\mathcal{S}\,f$ which is derived via the last formula for $\mathcal{C}\,f$, is however, highly recommended.*

For the case of the Hilbert transform, we just take the limit as $y \to 0$ in the above expression for $(\mathcal{C}\,f)(\zeta(t_j + iy))$, and recall the above Plemelj identities (1.5.21). Thus, setting

$$C = [c_{jk}] = \frac{1}{\pi\,i}\left(\log(A^-) - \log(A^+)\right), \quad (1.5.105)$$

we get

$$(\mathcal{S}f)(\zeta(t_j)) = c_{jj}\, f(\zeta(t_j)) + \sum_{k=-N,\,k\neq j}^{N} c_{jk} \frac{(t_j - t_k)\,\dot{\zeta}(t_k)}{\zeta(t_j) - \zeta(t_k)}\, f(\zeta(t_k)).$$
$$(1.5.106)$$

The proofs of the results of these last two equations are left to the problems at the end of §1.5.

Programs. See §5.1 for a listing of wavelet programs for this handbook.

1.5.13   ANALYTIC CONTINUATION

We present here an explicit harmonic extension of the basis defined in Definition 1.5.12, enabling an effective procedure for analytic continuation. The method may be used, for example, to determine a complex valued function that is analytic to the left of an arc Γ, but which reduces to given values on Γ, or to solve Poisson's equation (§4.1.1) on a region whose boundary consists of a finite number of analytic arcs, or for computing conformal maps of bounded or unbounded regions whose boundary is the union of a finite number of analytic arcs, as defined in Definition 1.5.2. Here and henceforth, a function U is called *harmonic* in a region $\mathcal{B}$ of the plane if it satisfies the equation

$$\frac{\partial^2}{\partial x^2} U(x, y) + \frac{\partial^2}{\partial y^2} U(x, y) = 0$$

at all points (x, y) in $\mathcal{B}$.

Let Γ, φ, ρ, N, m and h be defined as in (1.5.24), and let $\mathcal{D}^+$ denote a region to the left of Γ as one traverses Γ from the end–point a to b of Γ. We can give a more precise definition of the region $\mathcal{D}^+$, which depends not only on Γ, but also on φ. In Definition 1.5.2, we defined Γ by the equation $\Gamma = \{z = x + iy \in \mathbb{C} : z = \psi(u), \ u \in \mathbb{R}\}$. Of course, if Γ is a subinterval of $\mathbb{R}$, then $\psi(u) = \varphi^{-1}(u)$ is real valued for all $u \in \mathbb{R}$. On the other hand, if Γ is an arc in the plane, then $\psi(u) = z(u) = x(u) + i y(u)$. In any case, whether Γ is a subinterval of $\mathbb{R}$ or not, we expect $\psi(u + iv)$ to be defined for a set of non–zero values of v. We can therefore define $\mathcal{D}^+$ by $\mathcal{D}^+ = \{\psi(u + iv) : u \in \mathbb{R}, \ v \geq 0, \text{ and } \psi(u + iv) \text{ is well defined}\}$.

We are now in position to define the following functions, which enable analytic continuations to $\mathcal{D}^+$. In particular, $\sigma_j(z)$ is the imaginary part of the analytic continuation of $S(j\, h) \circ \varphi$ and $\tau_a(z)$ and $\tau_b(z)$ are the imaginary parts of the analytic continuation of $1/(1 + \rho(z))$ and $\rho(z)/(1 + \rho(z))$ respectively. See [S1, §3.4] for details.

$$\sigma_j(z) \;=\; \Im\left\{\frac{e^{i\pi[\varphi(z)-jh]/h} - 1}{\pi[\varphi(z) - jh]/h}\right\}, \qquad j \in \mathbb{Z},$$

$$\tau_a(z) \;=\; \left[1 - \frac{\Im\varphi(z)}{\pi}\right]\Re\left\{\frac{1}{1+\rho(z)}\right\} - \frac{\Re\varphi(z)}{\pi}\Im\left\{\frac{1}{1+\rho(z)}\right\},$$

$$\tau_b(z) \;=\; \left[1 - \frac{\Im\varphi(z)}{\pi}\right]\Re\left\{\frac{\rho(z)}{1+\rho(z)}\right\} - \frac{\Re\varphi(z)}{\pi}\Im\left\{\frac{\rho(z)}{1+\rho(z)}\right\},$$

$$\delta_j \;=\; \sigma_j, \qquad -M < j < N,$$

$$\delta_{-M} \;=\; \tau_a - \sum_{j=-M+1}^{N}\frac{1}{1+e^{jh}}\sigma_j,$$

$$\delta_N \;=\; \tau_b - \sum_{j=-M}^{N-1}\frac{e^{jh}}{1+e^{jh}}\sigma_j.$$

$$(1.5.107)$$

The proof of the following theorem is a matter of inspection, based, in part, on [S1, Eq. (3.4.1)], and also, on the following explicit expression for the Cauchy transform of a linear function: If φ is defined as in (1.5.1) and Lg is given in (1.5.6), then for $z = x + iy$, with $y > 0$,

$$\frac{1}{\pi}\int_a^b \frac{Lg(t)}{t - z}\,dt = g(b) - g(a) - \frac{\varphi(z)}{\pi}\,Lg(z). \qquad (1.5.108)$$

This is a complex valued function whose real and imaginary parts are harmonic in the upper half of the complex plane, $\{z = x + iy : \Im z > 0\}$.

Theorem 1.5.25 *Let the Sinc basis $\{\omega_j\}$ be defined as in Definition 1.5.12, and let the functions δ_j be defined as in (1.5.107). Given any m complex numbers $c_{-M},\dots,c_N$, the expression*

$$u_m = \sum_{j=-M}^{N} c_j\,\delta_j \qquad (1.5.109)$$

is harmonic on $\mathcal{D}^+$. Moreover, if $\zeta \in \Gamma$,

$$\lim_{z\to\zeta,\, z\in\mathcal{D}^+} u_m(z) = \sum_{j=-M}^{N} c_j\,\omega_j(\zeta), \qquad (1.5.110)$$

where the basis functions ω_j are defined as in (1.5.24).

These results are used in §2.5.4 and §4.1.1 below to solve Dirichlet's problem over a region bounded by a finite number of arcs Γ_j, $(j = 1, 2, \ldots, n)$.

1.5.14 INITIAL VALUE PROBLEMS

We consider here, the solution of the model differential equation IVP (initial value problem)

$$
\begin{aligned}
\frac{dy}{dx} &= F(x, y), \quad x \in \Gamma \\
y(a) &= y_a.
\end{aligned}
\tag{1.5.111}
$$

At the outset, we transform the problem (1.5.111) into the Volterra integral equation problem

$$
y(x) = (\mathcal{T}y)(x) \equiv \int_a^x F(t, y(t)) \, dt + y_a, \quad x \in \Gamma.
\tag{1.5.112}
$$

Assumption 1.5.26 *Corresponding to a some positive number r and a fixed function $g \in \mathbf{M}_{\alpha,\beta,d}(\varphi)$, let $\mathcal{B}(g; r)$ denote the family of all functions $y \in \mathbf{M}_{\alpha,\beta,d}(\varphi)$ such that $\|y - g\| < r$. We make the following assumptions on $\mathcal{T}y$:*

(a) $\mathcal{T}y \in \mathcal{B}(g; r) \cap \mathbf{M}_{\alpha,\beta,d}(\varphi)$ *for all $y \in \mathcal{B}(g; r)$. (We remark that this condition may be easily checked, i.e., if $F(\cdot, y(\cdot))/\phi'(\cdot) \in \mathbf{L}_{\alpha,\beta,d}(\varphi)$);*

(b) *There exists a positive number $\gamma \in (0, 1)$ such that for all u and v in $\mathcal{B}(g; r)$ we have $\|\mathcal{T}u - \mathcal{T}v\| \leq \gamma \|u - v\|$; and*

(c) $\|\mathcal{T}g - g\| < (1 - \gamma)r$.

Let a sequence of functions $\{y^{(m)}\}$ be defined for $x \in \Gamma$ by

$$
\begin{aligned}
y^{(0)}(x) &= g(x), \\
y^{(k+1)}(x) &= \mathcal{T}y^{(k)}(x), \quad k = 0, 1, \cdots.
\end{aligned}
\tag{1.5.113}
$$

Theorem 1.5.27 *Let Assumptions 1.5.27 be satisfied, and let the sequence of functions $\{y^{(k)}\}$ be defined for $x \in \Gamma$ by Eq. (1.5.113). Then $y^{(k)} \to y^*$, as $k \to \infty$, where y^* is the unique solution in the ball $\mathcal{B}(g; r)$ of the initial value problem (1.5.111). Moreover, for any fixed positive integer k, we have $\|y^{(k)} - y^*\| \leq \gamma^k r$.*

Proof. The proof of this theorem is a straightforward application of Picard's proof of the existence of a solution of (1.5.111), based on successive approximation. We omit the details. ∎

Let us now illustrate getting an approximate solution to (1.5.111) for the case when Γ is a finite interval (a, b). To this end, we first replace (1.5.111) by an equivalent integral equation, i.e., if $\mathcal{J}g(x) = \int_a^x g(t)\, dt$ over $(a, a+T)$ for some $T \in (0, (b-a)]$; and if the operator $\mathcal{J}_m$ is defined as in (1.5.49) (i.e., for the interval $(a, a+T)$), then

$$y(x) = y_a + \mathcal{J}(F(y, \cdot))(x) \approx y_a + \mathcal{J}_m\, F(y, \cdot))(x). \qquad (1.5.114)$$

By applying the operator V (see (1.5.23) on the extreme left and right hand sides of (1.5.114), and using the notation of (1.3.20), (1.3.41) as well as

$$
\begin{aligned}
\mathbf{y} &= (y_{-M}, \ldots, y_N)^T \\
\mathbf{F}(\mathbf{y}) &= (F(z_{-M}, y_{-M}), \ldots, F(z_N, y_N))^T \qquad (1.5.115) \\
\mathbf{c} &= (y_a, \ldots, y_a)^T
\end{aligned}
$$

we get the system of nonlinear equations

$$\mathbf{y} = \mathbf{c} + A\, \mathbf{F}(\mathbf{y}). \qquad (1.5.116)$$

with

$$
\begin{aligned}
A &= h\, I^{(-1)}\, D(1/\varphi') = X\, \Lambda\, X^{-1} \\
&\qquad\qquad\qquad\qquad\qquad\qquad (1.5.117) \\
\Lambda &= \operatorname{diag}(\lambda_{-M}, \ldots, \lambda_N).
\end{aligned}
$$

We should perhaps mention that an algorithm for solving (1.3.67) based on Newton's method was published in [SGKOrPa]. This algorithm may be downloaded from *Netlib*.

Now recall that $A = h\, I^{(-1)}\, D(1/\varphi')$, with $\varphi(z) = \log((z-a)/(a + T - z))$, so that $1/\varphi'(z_j) = T\, e^{jh}/\left(1 + e^{jh}\right)^2$. That is, if λ_j is defined as in (1.5.117), then $\lambda_j = T\, \lambda_j^*$, where λ_j^* is independent of T. Similarly, the matrix X in (1.5.117) is also independent of T. It thus follows, by taking $|\mathbf{y}| = \sqrt{|y_{-M}|^2 + \ldots, + |y_N|^2}$, that the corresponding matrix norm of A is proportional to T, and thus, under our above assumptions on f we have a contraction for all sufficiently small T, i.e., $|A F(\mathbf{y}) - A F(\mathbf{z})| \leq \delta |\mathbf{y} - \mathbf{z}|$, with $0 < \delta < 1$. We can thus solve (1.5.116) by successive approximation, by generating a sequence of vectors $\{\mathbf{y}_k\}_{k=0}^{\infty}$ via the scheme

$$
\begin{aligned}
\mathbf{y}_0 &= \mathbf{c} \\
\mathbf{y}_{k+1} &= \mathbf{c} + A \mathbf{F}(\mathbf{y}_k).
\end{aligned}
$$

(1.5.118)

Moreover, this sequence of vectors will thus converge to a vector $\mathbf{y}^*$. Letting $\mathbf{w}$ denote the vector of basis functions defined for the interval $(a, a + T)$ as in (1.5.24), we will then get an approximate solution $y = \mathbf{w} \mathbf{y}^*$ to (1.5.111) on the interval $(a, a + T)$.

The same procedure can then be repeated to produce a solution on the interval $(a + T, a + 2T)$, starting with an initial value y_a^*, where y_a^* is the last component of the vector gotten from the scheme (1.5.118), and so on.

1.5.15   WIENER–HOPF EQUATIONS

The classical Wiener–Hopf integral equation takes the form

$$
f(x) - \int_0^\infty k(x - t) f(t) \, dt = g(x), \qquad x \in (0, \infty).
$$

(1.5.119)

We are asked to solve for f, given k and g. The classical method of Wiener and Hopf is mathematically very pretty, particularly in the case when both k and g are (Lebesgue integrable) $\mathbf{L}^1$ functions. Suppose, for example, that this is the case, and that (1.5.119) has a unique solution f. In this case, one first extends the definition of both f and g to all of $\mathbb{R}$ by defining these functions to be zero on the negative real axes, to get an equation of the form

$$
f_+(x) - \int_{\mathbb{R}} k(x - t) f_+(t) \, dt = g_+(x) + s_-(x), \qquad x \in \mathbb{R},
$$

(1.5.120)

where s_- is an $\mathbf{L}^1(\mathbb{R})$ function which vanishes on the positive real axis. One then takes a Fourier transform of (1.5.120), to get the equation

$$
F_+ - K F_+ = G_+ + S_-
$$

(1.5.121)

with F_+ denoting the Fourier transform of f_+, and similarly for S_-. Using the theorem of Wiener & Levy (S1, Theorem 1.4.10]) one can prove that there exist Fourier transforms K_+ and K_- of functions in $\mathbf{L}^1(\mathbb{R})$ such that $(1 - K) = (1 + K_+)/(1 + K_-)$. Further details are provided in [S1, §6.8]. This yields the relation

$$(1 + K_+) F_+ - G_+ - \{K_- G_+\}_+ = \{K_- G_+\}_- + (1 + K_-) S_-.$$
$$(1.5.122)$$

It is then a straight forward argument to deduce that both sides of (1.5.122) must vanish identically, which enables us to write

$$F_+ = \{(1 + K_-) G_+\}_+ , \qquad (1.5.123)$$

and which enables us to get $f_+ = f$ by taking the inverse Fourier transform of the right hand side of (1.5.123).

The above procedure has two flaws from the point of view of computing an approximate solution f:

1. The explicit form of the factorization $(1 - K) = (1 + K_+)/(1 + K_-)$ used above is not easy to obtain in practice. And while such an approximate factorization was obtained, in §6.8 of [S1], its accurate computation requires a lot of work.

2. The resulting function F_+ decays to zero slowly on the real line, due to the fact that $f_|$ has a discontinuity at the origin. Many evaluation points are therefore required to get an accurate approximation of F_+, and hence also of f.

An explicit algorithm based on this method is nevertheless derived in [S12].

Fortunately, both of these difficulties are easily amended via Sinc convolution. This involves writing Eq. (1.5.119) in the form

$$f(x) - \int_0^x k(x - t) f(t)\, dt - \int_x^\infty k(x - t) f(t)\, dt = g(x), \quad x > 0.$$
$$(1.5.124)$$

Now discretizing this equation at the Sinc points by application of Sinc convolution, we are able to replace Eq. (1.5.124) by the system of linear algebraic equations

$$\mathbf{f} - \mathcal{F}_+(A)\,\mathbf{f} - \mathcal{F}_-(B)\,\mathbf{f} = \mathbf{g}\,, \qquad (1.5.125)$$

where, $\mathcal{F}_\pm$ denote the Laplace transforms

$$\mathcal{F}_+(s) = \int_0^\infty e^{-t/s}\,k(t)\,dt\,,$$

$$\qquad\qquad\qquad\qquad\qquad\qquad\qquad (1.5.126)$$

$$\mathcal{F}_-(s) = \int_0^\infty e^{-t/s}\,k(-t)\,dt\,,$$

where $\mathbf{g}$ is a vector of function values of g evaluated at a finite number of Sinc points on $(0, \infty)$, and where A and B are integration matrices defined in §1.5.8.

After getting the factorizations $A = X\,S\,X^{-1}$ and $B = Y\,SY^{-1}$, where S is a diagonal matrix, we can get $\mathcal{F}_+(A) = X\,\mathcal{F}_+(S)\,X^{-1}$, and similarly for $\mathcal{F}_-(B)$, provided that the functions $\mathcal{F}_\pm$ are explicitly known. In the cases when these functions are not explicitly known, then we can approximate $\mathcal{F}_\pm(s)$ for every value s of S using Sinc quadrature.

We mention that the method described above can be just as easily applied to linear convolution integral equations over finite intervals or over the real line $\mathbb{R}$.

PROBLEMS FOR SECTION 1.5

1.5.1 (a) Show that (1.5.4) is equivalent to the equation

$$F \circ \varphi^{-1}(w) = \mathcal{O}\left(e^{-\alpha|w|}\right)\,, \qquad w \to \pm\infty \;\; \text{on} \;\; \mathbb{R}.$$

(b) Prove that $x/\sinh(\pi\,x) = \mathcal{O}\left(e^{-d\,|x|}\right)$ as $x \to \pm\infty$ for any $d \in (0, \pi)$.

(c) Let F be defined by

$$F(x) = \int_\mathbb{R} \frac{e^{(\alpha+i\,x)\,t}}{(1 + e^t)^{\alpha+\beta}}\,dt.$$

Prove by setting $t = u/(1 - u)$ in the integral defining F, that

$$F(x) = \frac{\Gamma(\alpha + i\,x)\,\Gamma(\beta - i\,x)}{\Gamma(\alpha + \beta)}.$$

[Hint: Use the special functions identity

$$\int_0^1 u^{a-1}\,(1-u)^{b-1}\,du = \frac{\Gamma(a)\,\Gamma(b)}{\Gamma(a+b)}.]$$

(d) Use Stirling's formula (see Eq. (1.5.2)) to deduce that

$$F(x) = \mathcal{O}\left(\exp(-\pi\,|x|)\right), \quad |x| \to \infty.$$

1.5.2 (a) Let the finite interval $\Gamma = (a, b)$ be given, and let $w = \varphi(x) = \log((x-a)/(b-x))$ $(\Leftrightarrow x = \varphi^{-1}(w) = (a + b\,e^w)/(1 + e^w))$ be the transformation that transforms Γ to $\mathbb{R}$. Prove that if f satisfies the inequalities (1.3.3), then

$$f \circ \varphi^{-1}(w) = \mathcal{O}\left(\exp(-\pi\,|w|)\right), \quad w \to \pm\infty \text{ on } \mathbb{R}.$$

(b) Prove that $F \in \mathbf{L}_{\alpha,d}(\varphi) \Leftrightarrow F \circ \varphi^{-1} \in \mathbf{L}_{\alpha,d}(\mathrm{id})$.

(c) Use the result of Part (b) of this problem to show that $\{F' \in \mathbf{L}_{\alpha,d}(\varphi) \Rightarrow F \in \mathbf{M}_{\alpha,d}(\varphi)\}$ yields the statement of Part iii. of Theorem 1.5.3 .

1.5.3 Prove that the sum of the terms in (1.5.34) is, in fact, equal to the sum of the terms in (1.5.35)–(1.5.40).

1.5.4 Let $\{F \circ \varphi^{-1}\}(x) = \mathcal{O}(\exp(-d|x|))$ $x \to \pm\infty$ on $\mathbb{R}$, and for some positive numbers α and β on $(0, 1]$, let

$$F(z) = \begin{cases} \mathcal{O}\left(\rho(z)^\alpha\right) & z \to a, \text{ and} \\ \mathcal{O}\left(\rho(z)^{-\beta}\right) & z \to b. \end{cases}$$

Prove that if corresponding to an arbitrary integer N we:
(i) Select $h = (\pi\,d/(\beta N))^{1/2}$,
(ii) Select an integer $M = [(\beta/\alpha)N]$, where $[\cdot]$ denotes the greatest integer function, and
(iii) Set $\varepsilon_N = N^{1/2} \exp\left(-(\pi d\beta N)^{1/2}\right)$,
Then with z_k defined by (1.5.8),

$$\left\| F - \sum_{k=-M}^{N} F(z_k)\, S(k,h) \circ \varphi \right\| = \mathcal{O}\left(\varepsilon_N\right), \quad N \to \infty.$$

1.5.5 Let $\mathcal{S}f$ be defined as in (1.5.15). Prove that if Γ is a finite interval or arc, and if $f \in \mathbf{L}_{\alpha,d}(\varphi)$ then $\mathcal{S}f \in \mathbf{M}_{\alpha,d}(\varphi)$.

1.5.6 Consider the following interpolation scheme, defined on $\Gamma = (0,1)$, in which $\varphi(x) = \log(x/(1-x))$, and $z_k = \varphi^{-1}(k\,h)$:

$$
\begin{aligned}
h &= \left(\frac{\pi d}{\alpha N}\right)^{1/2} \\
\gamma_j(x) &= \varphi'(z_j)/\varphi'(x)\, S(j,h) \circ \varphi(x), \quad |j| \le N \\
\omega_j(x) &= \gamma_j \quad |j| < N, \\
\omega_{-N}(x) &= L(x) - \sum_{j=-N+1}^{N} L(z(j))\, \gamma_j(x) \\
\omega_N(x) &= R(x) - \sum_{j=-N}^{N-1} R(z_j)\, \gamma_j(x), \\
\varepsilon_N &= N^{1/2} e^{-(\pi \alpha d N)^{1/2}},
\end{aligned}
$$

where $L(x) = (1-x)^2(1+2x)$ and $R(x) = 3x^2 - 2x^3$.

(a) Prove that if $F \in \mathbf{M}_{\alpha,d}(\varphi)$, then

$$\left\| F - \sum_{k=-N}^{N} f(z_k)\, \omega_k \right\| = \mathcal{O}(\varepsilon_N), \quad N \to \infty.$$

(b) Prove that if $F' \in \mathbf{L}_{\alpha,d}(\varphi)$, then

$$\left\| F' - \sum_{k=-N}^{N} f(z_k)\, \omega_k' \right\| = \mathcal{O}(N^{1/2}\varepsilon_N), \quad N \to \infty.$$

1.5.7 Let $R > 1$, and let $\alpha \in (1/R, R)$.

(a) Set $|x| = g(y) = (1 - y^2)^{1/2}$. The function $g \in \mathbf{L}_{\alpha,\pi}(\varphi)$ with φ defined in Example 1.5.5, with $(a,b) = (-1,1)$. Prove, in the notation of Definition 1.5.12, that by taking $u(x) = (1-x^2)^{1/2}\mathrm{sgn}(x)$, $v(x) = \log\{[1+u(x)]/[1-u(x)]\}$, and $\omega_k(x) = S(k,h) \circ v(x) + S(-k,h) \circ v(x)$, that as $N \to \infty$,

$$\sup_{-1<x<1} \left| |x|^\alpha - \frac{1}{2}w_0(x) - \sum_{k=1}^{N} \frac{(2e^{kh})^\alpha}{(1+e^{kh})^\alpha} w_k(x) \right|$$
$$= \mathcal{O}\left(\varepsilon_N\right).$$

(b) Let G be defined by $G(u) = g(\tanh(u)) = [\cosh(u)]^{-\alpha}$, and note that $\tilde{G}(y) = \mathcal{O}\left(e^{-\pi |y|/2}\right)$ as $y \to \pm\infty$ on $\mathbb{R}$, with $\tilde{G}$ the Fourier transform of G, and also, that

$$G(\sinh(kh)) = \mathcal{O}\left(\exp\left\{-(\alpha/2)\,e^{-|k|h}\right\}\right),$$

as $k \to \pm\infty$. Conclude, by taking $h = \log(N)/N$, and then by proceeding as in the proof of the (a)–Part of this problem, that as $N \to \infty$, and with $v = u + \sqrt{1+u^2}$,

$$\sup_{u\in\mathbb{R}} \left| G(u) - \sum_{k=-N}^{N} G(\sinh(kh))S(k,h) \circ \log(v) \right|$$
$$\leq C \exp\{-\tfrac{N}{\log(N)}\}.$$

(c) Use the result of the (b) – part of this problem to construct an $N+1$ – basis approximation to $|x|^\alpha$ on $[-1,1]$, which is accurate to within an error bounded by $C\exp\{-\tfrac{N}{\log(N)}\}$. (Keinert, [Kn], Sugihara [Su3]).

1.5.8 Prove that if $F/\varphi' \in \mathbf{L}_{\alpha,d}(\varphi)$, then there exists a constant c which is independent of h and k, such that

$$\left| \int_\Gamma F(t)\,S(k,h) \circ \varphi(t)\varphi'(t)dt - hF(z_k) \right| \leq c\,h\,e^{-\pi d/h}.$$

1.5.9 Derive explicit expressions of Equation (1.5.10) for each of Examples 1.5.4 to 1.5.10.

1.5.10 Give an explicit approximation of

$$\frac{h^n}{\{\varphi'(x)\}^n} \left(\frac{d}{dx}\right)^n F(x)$$

for the case when $x = z_k$, $k = -M, \cdots, N$, both in general, and for each of Examples 1.5.4 to 1.5.10. Use the notation

$$\delta_{j-k}^{(n)} = h^n \left(\frac{d}{dx}\right)^n S(k, h)(x)\Bigg|_{x=j\,h}.$$

1.5.11 Prove that if $x \in \mathbb{R}$, and if $\mathcal{J}^+$ is defined as in §1.5.8, then
$|(\mathcal{J}^+ S(k, h))(x)| \leq 1.1\,h$.

1.5.12 Let u be defined and integrable over (a, b), let $\mathcal{J}^+$ be defined as in §1.5.8, and set

$$(u, v) \equiv \int_a^b u(x)\overline{v(x)}\,dx.$$

Prove that

$$\Re(\mathcal{J}^+ u, u) = \frac{1}{2}\left|\int_a^b u(x)\,dx\right|^2 \geq 0.$$

1.5.13 Let $\hat{f}$ be analytic on the right half plane, and assume that $f(t)$ is given for all $t > 0$ by the formula

$$f(t) = \frac{1}{2\pi i}\int_{-i\infty}^{i\infty} e^{st}\,\hat{f}(s)\,ds.$$

Prove that $\hat{f}(s) = \int_0^\infty e^{-st} f(t)\,dt$.

1.5.14 Let $\mathcal{J}^\pm$ be defined as in §1.5.8. Let $\hat{f}(s) = \int_0^\infty e^{-st} f(t)\,dt$ denote the (ordinary) Laplace transform of f. As in §1.5.10, set $\mathcal{G}(s) = (1/s)\,\hat{f}(1/s)$.

(a) Prove that along with the formula $f = \mathcal{G}(\mathcal{J}^+)\,g$ (where $g(x) = 1$ for all x), we also have $f = -\mathcal{G}(\mathcal{J}^-)\,g$.

Write a MATLAB® program for approximating f on (a, ∞), with a an arbitrary positive number, given $\hat{f}(s) = \int_a^\infty f(t)\,e^{-st} f(t)\,dt$.

1.5.15 Use the method of §1.5.14 to construct an approximate solution to the problem

$$y' = \frac{1}{\sqrt{1 - x^2}}\,y, \quad 0 < x < 1;$$

$$y(0) = 1.$$

1.5.16 (a) Prove in the notation of Definition 1.5.12 that $V\,\mathbf{w}$ is the identity matrix.

(b) Given that A^+ can be diagonalized as $X\,\Lambda\,X^{-1}$, use this, and Part (a) to show that $(\mathcal{J}^+)^k = \mathbf{w}\,X\,\Lambda^k\,X^{-1}\,V$.

(c) Use the Part (b) to prove that if $F(s) = \hat{f}(1/s)$, with $\hat{f}$ as in Problem 1.5.13, then $F(\mathcal{J}_m^+) = \mathbf{w}\,X\,F(\Lambda)\,X^{-1}\,V$.

1.5.17 Set $A^+ = I^{(-1)}\,D$, $A^- = \left(I^{(-1)}\right)^T D$, where

$$D = \operatorname{diag}\left(h/\varphi'\left(z_{-M}\right),\ \ldots,\ h/\varphi'(z_N)\right),$$

and set $C = D^{1/2}\,I^{(-1)}\,D^{1/2} = Z\,\Lambda\,Z^{-1}$, with Λ a diagonal matrix. Prove that if $A^+ = X\,\Lambda\,X^{-1}$, then $A^- = Y\,\Lambda\,Y^{-1}$, and that for X and Y we can take $X = D^{-1/2}\,Z$ and $Y = D^{-1/2}\left(Z^{-1}\right)^T$.

1.5.18 Give a detailed proof for the derivation of the expression for $\mathcal{C}\,f(\zeta(t_j + iy))$ given at the end of §1.5.12.

1.5.19 Let φ_1 and φ_2 denote transformations that satisfy the conditions of Definition 1.5.2. Let $f \in \mathbf{M}_{\alpha,\beta,d}(\varphi_1)$ where $0 < \alpha \le 1$, $0 < \beta \le 1$, and $0 < d < \pi$. Prove that $g = f \circ \varphi_1^{-1} \circ \varphi_2 \in \mathbf{M}_{\alpha,\beta,d}(\varphi_2)$.

1.6   Rational Approximation at Sinc Points

In this section we construct a family of rational functions for approximating in the space $\mathbf{M}_{\alpha,\beta,d}(\varphi)$. These rationals behave like Sinc approximations in that they converge at (roughly) the same rate as Sinc methods. Thus they do not have any advantages over Sinc methods for purposes of approximating the operations of calculus which were presented in §1.5. However, they do have an advantages for purposes of extrapolation to the limit. It is a fact that extrapolation to the limit based on polynomial interpolation (e.g. Aitken's by method, see [S1, §2.3]) fails when it is used to approximate f at a point ξ where f is not analytic. In this section we discuss methods of extrapolation to the limit via *Thiele's method* that uses rational functions. Thiele's method employs rational functions to get accurate approximations of f on the closure $\overline{\Gamma}$ of Γ by using values of f on a subinterval

Γ_1. Moreover, we are able to give an *a priori* condition for these methods to work, namely, that f should belong to $\mathbf{M}_{\alpha,\beta,d}(\varphi)$, where $\varphi : \Gamma \to \mathbb{R}$.

1.6.1 Rational Approximation in $\mathbf{M}_{\alpha,\beta,d}(\varphi)$

Let z_j, φ, ρ, and Γ be defined as in Definition 1.5.2, and for some positive integers M and N, set

$$B(x) = \frac{\rho(x)}{1 + \rho(x)} \prod_{j=-M}^{N} \frac{\rho(x) - e^{jh}}{\rho(x) + e^{jh}} \qquad (1.6.1)$$

and

$$\eta_N(x) = f(x) - \sum_{j=-M}^{N} \frac{f(z_j) \, e^{jh} \phi'(z_j) \, B(x)}{[\rho(x) - e^{jh}] \, B'(z_j)}. \qquad (1.6.2)$$

Notice from (1.6.1) that $B(x)$ vanishes at the Sinc points $z_j = \varphi^{-1}(j\,h)$, so that (1.6.2) is a rational function with fixed poles for interpolation of f at these Sinc points.

We omit the proof of the following theorem, which can be found in [S1, §5.5].

Theorem 1.6.1 *Let $0 < d \le \pi/2$, let α and β belong to the interval $(0,1)$, and let $f \in \mathbf{L}_{\alpha,\beta,d}(\varphi)$. Let h be selected by the formula*

$$h = \frac{\pi}{(2\beta N)^{1/2}}. \qquad (1.6.3)$$

Let M and N be related by $N = [\alpha/\beta\,M]$, where $[\cdot]$ denotes the greatest integer function. Let $\eta_N(x)$ be defined by (1.6.2). Then there exists a constant C, independent of N, such that for $x \in \Gamma$,

$$|\eta_N(x)| \le CN^{1/2} \exp\{-d(2\beta N)^{1/2}\}. \qquad (1.6.4)$$

Remark. Suppose we want to obtain an accurate approximation of a function f at a point $\xi \in \Gamma$ using values of f on a sub–arc Γ_1 of Γ that does not contain ξ. In words, the above theorem tell us that if $f \in \mathbf{M}_{\alpha,\beta,d}(\varphi)$, and ξ is an end–point of Γ where f is singular, then ξ, then polynomials converge to f too slowly to the value $f(\xi)$, and we cannot expect polynomial extrapolation to give an accurate estimate of $f(\xi)$. However, Theorem 1.6.1 guarantees that rational approximation converges rapidly to f at all points of Γ, including ξ. This value $f(\xi)$ is given to us by Thiele's algorithm.

1.6.2 THIELE–LIKE ALGORITHMS

In this section we first consider the case of $\Gamma = (0, \infty)$, and we describe the usual Thiele algorithm for approximating $f(\infty)$ using values of f on some closed subinterval. We then also describe a variant of this algorithm to approximate a function f at the end–point b of Γ, with input the values of f on a closed sub–arc of Γ.

Thiele's algorithm on $(0, \infty)$. Let us assume that we are given data $\{(x_j, y_j)\}_{j=0}^m$, with $y_j = f(x_j)$. Using this data, we form a table of numbers R_i^j, which are defined by the following equations:

$$
\begin{aligned}
R_0^j &= y_j, & j &= 0, 1, \cdots, m; \\
R_1^j &= \frac{x_{j+1} - x_j}{R_0^{j+1} - R_0^j}, & j &= 0, 1, \cdots, m - 1; \\
R_i^j &= R_{i-2}^{j+1} + \frac{x_{j+i} - x_j}{R_{i-1}^{j+1} - R_{i-1}^j}, & \left\{ \begin{array}{l} j = 0, 1, \cdots, m - i; \\ i = 2, 3, \cdots, m. \end{array} \right.
\end{aligned}
$$
$$\tag{1.6.5}$$

Using these numbers, we can define the partial fraction decomposition rational function

$$
r(x) = R_0^0 + \frac{x - x_0}{R_1^0 +} \ \frac{x - x_1}{R_2^0 - R_0^0 +} \ \cdots \ \frac{x - x_{m-1}}{R_m^0 - R_{m-2}^0}. \tag{1.6.6}
$$

It can then be shown that $r(x_j) = y_j$, and if $m = 2n$, then the function $r(x)$ has the form

$$
r(x) = \frac{p_n(x)}{q_n(x)} \tag{1.6.7}
$$

with both p_n and q_n polynomials of degree at most n in x; while if $m = 2n + 1$,

$$
r(x) = \frac{p_{n+1}^*(x)}{q_n^*(x)} \tag{1.6.8}
$$

with p_{n+1}^* a polynomial of degree at most $n + 1$ in x and q_n^* a polynomial of degree at most n in x.

The case of $m = 2n$ is useful for approximating the limit $f(\infty)$, since in this case the ratio of the two polynomials given in (1.5.7) reduces to the expression

$$r(x) = \frac{R_{2n}^0 x^n + c_1 x^{n-1} + \cdots + c_n}{x^n + d_1 x^{n-1} + \cdots + d_n}, \qquad (1.6.9)$$

with R_{2n}^0 the single entry in the last row of the Thiele table (1.6.5), and c_j, d_j constants. We note, in this case, that

$$R_{2n}^0 = \lim_{x \to \infty} r(x). \qquad (1.6.10)$$

Suppose that f also has a finite limit at ∞. Then, by our assumptions already discussed above, we may expect the approximation

$$R_{2n}^0 \approx f(\infty) \qquad (1.6.11)$$

to be accurate.

A simple variant of the above Thiele algorithm can be used to approximate a function f at a finite point. Suppose that $f \in \mathbf{M}_{\alpha,\beta,d}(\varphi)$, with $\varphi(x) = \log(x/(1-x))$, and that we wish to approximate $f(1)$ using values $\{(\xi_j, f(\xi_j))\}_0^{2n}$ of ξ_j on $(1/4, 3/4)$. To this end, we set $(x_j, y_j) = (1/(1-\xi_j), f(\xi_j))$, proceed to set up the above Thiele table as defined in (1.6.5) to get the approximation $f(1) \approx R_0^{2n}$.

A Sinc Transformed Thiele Algorithm. Thiele's method readily extends from using a variable x to using a variable $\rho(x) = e^{\varphi(x)}$. Suppose that $f \in \mathbf{M}_{\alpha,d}(\varphi)$, for general $\varphi : \Gamma \to \mathbb{R}$, and that we wish to use values $\{(\xi_j, f(\xi_j))\}_0^{2n}$ with the distinct points $\{\xi_j\}_0^{2n}$ located on a closed sub–arc of Γ to get an accurate approximation to f at the end–point b of Γ. Then we can again use the above Thiele table, by first setting $(x_j, y_j) = (\rho(\xi_j), f(\xi_j))$, $j = 0, 1, \ldots, 2n$, and then proceeding as above. We are thus assured of obtaining a rapidly convergent approximation.

PROBLEMS FOR SECTION 1.6

1.6.1 Prove that if the numbers R_i^j are determined as in (1.6.5), and if $r(x)$ is defined by (1.6.7), then $r(x_j) = y_j$.

1.6.2 (a) Prove that if $m = 2n$ in (1.6.5), then $r(x)$ as given in (1.6.6) has the form (1.6.7).

(b) Prove that under the conditions of the (a)–part of this problem, the rational $r(x)$ defined in (1.5.7) has the form (1.6.9).

1.6.3 Let $f \neq 0$ on Γ, let $f \in \mathbf{L}_{\alpha,d}(\varphi)$, let $z_j = \varphi^{-1}(jh)$, and let $4N + 1$ points (x_j, R_0^j) be defined by the formulas

$$(x_{2j}, R_0^{2j}) =$$
$$\left(\rho(z_{-N+j}), \frac{\rho(z_{-N+j})}{(1 + \rho(z_{-N+j})^2 f(z_{-N+j})} \right),$$
$$j = 0, 1, \cdots, 2N$$
$$(x_{2j-1}, R_0^{2j-1}) = (1/z_{-N+j-1}, 0), \quad j = 1, \cdots, N,$$
$$(x_{2j-1}, R_0^{2j-1}) = (1/z_{-N+j}, 0), \quad j = N+1, \cdots, 2N.$$

(a) Use these $4N + 1$ values to determine a rational function $r(x)$, via the Thiele algorithm.

(b) Show that if $r(x)$ is determined as in (a), then

$$R(x) \equiv \frac{\rho(x)}{(1 + \rho(x))^2 \, r(x)} = \sum_{j=-N}^{N} \frac{f(z_j) \, B(x)}{(x - z_j) \, B'(z_j)},$$

with $B(x)$ is defined as in (1.6.1).

1.7   Polynomial Methods at Sinc Points

It was discussed in §1.5.5 that the derivative of the Sinc interpolant function (1.5.27) as an approximation for $f'(x)$ is inaccurate since bases functions $\omega_k(x)$ have infinite derivatives at finite end–points of Γ. Approximating $f'(x)$ however, is necessary for many problems. Here we derive a novel polynomial–like interpolation function which interpolates a function f at the Sinc points and whose derivative approximates the derivative of f at these Sinc points.

It is well known that polynomials as well as their derivatives converge rapidly for functions that are analytic in a region containing the closed interval of approximation. Hence we shall consider this case first.

1.7.1 SINC POLYNOMIAL APPROXIMATION ON $(0, 1)$

We initially form the Lagrange polynomial which interpolates the given data $\{(x_k, f(x_k))\}^N_{-M}$ on $(0, 1)$, where the the x_k denote Sinc points of $(0, 1)$, where for some positive number h, $x_k = \varphi_2^{-1}(k\,h)$, and where $v_2(x) \log(x/(1-x))$

The Lagrange Polynomial. Setting

$$g(x) \quad = \quad \prod_{\ell=-M}^{N} (x - x_\ell)$$

$$(1.7.1)$$

$$b_k(x) \quad = \quad \frac{g(x)}{(x - x_k)\, g'(x_k)},$$

we immediately get the Lagrange polynomial

$$p(x) = \sum_{k=-M}^{N} b_k(x)\, f(x_k). \qquad (1.7.2)$$

This polynomial p is of degree at most $m-1$ in x, and it interpolates the function f at the m Sinc points $\{x_k\}^N_{k=-M}$.

The Derivative of $p(x)$. Let us differentiate the formula (1.7.2) with respect to x, and let us then form an $m \times m$ matrix $A = [a_{j,k}]$, $j, k = -M, \ldots, N$ with the property that

$$f'(x_j) \approx p'(x_j) = \sum_{k=1}^{m} a_{j,k}\, f(x_k). \qquad (1.7.3)$$

Then (1.7.1) immediately yields for $k \neq j$, that $a_{j,k} = b_k'(x_j) = g'(x_j)/((x_j - x_k)\, g'(x_k))$, whereas for $k = j$, $a_{j,j} = \sum_{\ell \neq j} 1/(x_j - x_\ell)$. That is,

$$a_{j,k} = \begin{cases} \dfrac{g'(x_j)}{(x_j - x_k)\, g'(x_k)} & \text{if } k \neq j \\[2em] \displaystyle\sum_{\ell \in [-M,N],\, \ell \neq j} \dfrac{1}{x_j - x_\ell} & \text{if } k = j. \end{cases} \qquad (1.7.4)$$

In the theorem which follows, we obtain bounds on the errors of these approximations.

Errors of the Approximations. We shall obtain bounds on the errors of approximation of $f(x)$ by $p(x)$ on $[0, 1]$, and of $f'(x_j)$ by the sum on the right hand side of (1.7.3). We take $M = N$ for sake of simplicity in our proofs.

Theorem 1.7.1 *Let $r > 0$, let $\mathcal{D} = \varphi_2^{-1}(\mathcal{D}_d)$, let $B(t\,r) \subseteq \mathbb{C}$ denote the disc with center t and radius r, and take $\mathcal{C} = \cup_{t \in [-1,1]} B(t\,r)$. Set $\mathcal{D}_2 = \mathcal{D} \cup \mathcal{C}$, and assume that f is analytic and uniformly bounded by $\mathcal{M}(f)$ in $\mathcal{D}_2$. Let $M = N$, take $h = c/\sqrt{N}$ with c a positive constant independent of N, and let $\{x_j\}_{-N}^N$ denote the Sinc points, as above. Then there exist constants C and C', independent of N, such that:*

(i) An estimate of a bound on the error, $f(x) - p(x)$, for $x \in [0, 1]$ is given by

$$|f(x) - p(x)| \leq \frac{C\,N^{1/2}}{(2\,r)^{2\,N+2}} \exp\left(-\frac{\pi^2\,N^{1/2}}{2\,c}\right), \qquad (1.7.5)$$

and

(ii) A bound on the difference $|f'(x_j) - p'(x_j)|$ is given by

$$\max_{j=-N,\ldots,N} |f'(z_j) - p'(z_j)| \leq \frac{B\,N}{(2\,r)^{2\,N}} \exp\left(-\frac{\pi^2\,N^{1/2}}{2\,c}\right). \qquad (1.7.6)$$

Proof of (1.7.5). The error of f via p on $[-1, 1]$ can be expressed as a contour integral,

$$E_m(f, x) = \frac{g(x)}{2\,\pi\,i} \int_{\partial D_2} \frac{f(\xi)}{(\xi - x)\,g(\xi)}\,d\xi, \qquad (1.7.7)$$

with g given in (1.7.1). Now, by our definition of D_2, we have $|\xi - x| \geq r$ and also, $|\xi - x_j| \geq r$ for x and x_j belonging to $[0, 1]$, and $\xi \in \partial D_2$, $j = -N$, $\ldots$, N. Hence the denominator of the integrand in (1.7.7) is at least as large as r^{m+1}. By assumption, f is uniformly bounded by $\mathcal{M}(f)$ in D_2. Hence

$$|E_m(f, x)| \leq \frac{\mathcal{M}(f)}{r^{m+1}} \max_{x \in [-1,1]} |g(x)| \frac{L(\partial D_2)}{2\,\pi}, \qquad (1.7.8)$$

where $L(\partial D_2) = 2 + 2\,\pi\,r$ is the length of the boundary of D_2.

Setting $x = z/(1 + z)$, which transforms $z \in (0, \infty)$ to $x \in (0, 1)$, we get

$$\left| g\left(\frac{z}{z+1} \right) \right| \;=\; \prod_{j=-N}^{N} \left| \frac{z}{z+1} - \frac{e^{jh}}{1+e^{jh}} \right|$$

$$= \; \frac{z^{N+1}}{2(1+z))^{2\,N+1}} \left(\prod_{j=1}^{N} \frac{1 - z\,e^{-jh}}{1+e^{jh}} \right)^{2} \; w(N,h)$$

$$\leq \; \frac{z^{N+1}}{2\,(1+z)^{2\,N+1}} \left(\prod_{j=1}^{N} \frac{1 - z\,e^{-jh}}{1+e^{jh}} \right)^{2} ,$$

$$(1.7.9)$$

where $w(N,h) = 1 - z\,\exp(-(N+1)\,h)$.

It can be deduced at this point that the maximum value of $|g(x)|$ occurs between the two zeros, $x = x_{-1}$ and $x = x_0 = 0$, and also between $x = x_0 = 0$ and $x = x_1$. We thus pick the approximate mid–point between x_0 and x_1 with the choice $x = z/(z+1)$, where $z = e^{h/2}$. The exact maximum of $g(x)$ for $x \in [0,1]$ does not occur exactly at $x = z/(1+z)$, but it is close to this point, and this is why (1.7.5) is only an estimate of this maximum. (Although complicated to prove, it could, however, be shown that the estimate in the (ii)– Part of Theorem 1.7.1 is, in fact, a bound. One can use the proof of Lemma 5.5.1 of [S1] to establish this).

Now, for $z = e^{h/2}$,

$$\frac{2\,z}{1+z} \; \leq \; 1 ,$$

$$(1.7.10)$$

$$1 - z\,\exp(-(N+1)\,h) \; \leq \; 1 ,$$

which gives (1.7.5) upon applying the results of Problem 1.7.1.

Proof of (1.7.6). Similarly, we have

$$|f'(z_j) - p'(z_j)| \;=\; \left| \frac{1}{2\,\pi\,i}\,g'(z_j) \int_{\partial D_2} \frac{f(z)}{(z-x)\,g(z)}\,dz \right| ,$$

$$(1.7.11)$$

$$\leq \; |g'(z_j)|\,\frac{M(f)\,L(\partial D_2)}{2\,\pi\,r^{m+1}} .$$

It is easy to deduce that the maximum of $|g'(z_j)|$ occurs when $x_j = x_0 = 0$. We thus get

$$\max_{j=-N,\ldots,N} |g'(z_j)| \le |g'(1/2)| = \frac{1}{2^{2N}} \left(\prod_{j=1}^{N} \frac{1 - e^{-jh}}{1 + e^{-jh}} \right)^2. \quad (1.7.12)$$

By again using the results of Problem 1.7.1, we get (1.7.6). ∎

Remarks 1. We may note that convergence in the bounds of (1.7.5) and (1.7.6) does not occur as $N \to \infty$ unless $r \ge 1/2$. If however, we replace x by Ty and then perform Sinc approximation of the same function f but now on the interval $y \in [0,1]$ rather than as before, on the interval $x \in [0,1]$ (in effect, doing Sinc approximation on $[0,T]$), we get an additional factor T^{4N+4} in the errors of the above approximations. Thus, convergence of the above approximations as $N \to \infty$ occurs whenever $2r/T^2 > 1$. On the other hand, shortening the length of the interval from $(0,1)$ to $(0,T)$ does not enable accurate uniform approximation of the derivative when we use the derivative of a Sinc approximation (1.5.27), or the inverse of the Sinc indefinite integration matrix of §1.5.8 for the interval $(0,T)$. This is due to the fact that the derivative of the map $\varphi_2(x) = \log(x/(T-x))$ is unbounded at the end-points of $(0,T)$.

2. The bound of order $1/r^{m+1}$ in (1.7.9) yields a gross overestimate of the error. We proceed with it here, though we are confident better bounds are attainable.

1.7.2 POLYNOMIAL APPROXIMATION ON Γ

Examination of the transformation #2 in §1.5.3 shows that $x = \rho_2/(\rho_2+1)$, where $\rho_2(x) = \exp(\varphi_2(x))$, and where $\varphi_2(x) = \log(x/(1-x))$. It follows by Problem 1.5.19, that if $f \in \mathbf{M}_{\alpha,\beta,d}(\varphi_2)$, and if φ is any other transformation that satisfies the conditions of Definition 1.5.2, then $F = f \circ \varphi_2^{-1} \circ \varphi \in \mathbf{M}_{\alpha,\beta,d}(\varphi)$. Let g be defined as in the above subsection, and define G by $G(t) = g(\rho(t)/(1 + \rho(t)))$, where $\rho = e^{\varphi}$. Letting $t_k = \varphi^{-1}(k\,h)$ denote the Sinc points of this map on the arc $\Gamma = \varphi^{-1}(\mathbb{R})$, we form the the composite polynomial

$$P(t) = \sum_{k=1}^{m} \frac{G(t)}{(t - t_k)\,G'(t_k)}\, F(t_k). \quad (1.7.13)$$

It then follows that we can achieve the exact same accuracy in approximating F via this polynomial $P(t)$ as we did above, in approximating f via use of the above polynomial p. The above formula for $p(x)$ is thus readily extended to any arc Γ.

Note: Note: if we set $x = \rho(t)/(1 + \rho(t))$ in (1.7.1), then

$$P(t) = p(x) = \sum_{k=1}^{m} b_k(x) F(t_k), \tag{1.7.14}$$

where the $b_k(x)$ are exactly those basis functions defined as in (1.7.2). These functions can be evaluated using programs of this handbook.

1.7.3 Approximation of the Derivative on Γ

Upon differentiating the formula (1.7.14) with respect to t, we can determine a matrix $B = [b_{j,k}]$ such that

$$P'(t_j) = \sum_{k=1}^{m} b_{j,k} F(t_k). \tag{1.7.15}$$

Let us now derive the expressions for $b_{j,k}$. To this end, we readily arrive at the formula

$$b_{j,k} = \begin{cases} \dfrac{G'(t_j)}{(t_j - t_k) G'(t_k)} & \text{if } k \neq j, \\ \dfrac{G''(t_j)}{2\, G'(t_j)} & \text{if } k = j. \end{cases} \tag{1.7.16}$$

Since

$$\begin{aligned} G'(t) &= g'(x)\,\frac{dx}{dt} \\[2mm] G''(t) &= g''(x) \left(\frac{dx}{dt}\right)^2 + g'(x)\,\frac{d^2 x}{dt^2}\,, \end{aligned} \tag{1.7.17}$$

we find that

$$\begin{aligned} \frac{dx}{dt} &= \frac{\rho(t)}{(1 + \rho(t))^2} \\[2mm] \frac{d^2 x}{dt^2} &= \frac{\rho(t)\,\varphi'(t)}{(1 + \rho(t))^2} \left(\varphi''(t) - \frac{\rho(t) - 1}{\rho(t) + 1}\varphi'(t)^2\right), \end{aligned} \tag{1.7.18}$$

and these expressions readily enable us to compute values for $b_{k,j}$.

This handbook contains programs for evaluating derivative formulas.

Near Uniform Approximation

It is more convenient to discuss uniform approximation over $(-1, 1)$ in place of approximation over $(0, 1)$. If we wish to approximate a function f and if we are free to choose M, N and h, we could proceed as described at the end of §3.2. Let us assume, on the other hand, that we given the data $\{(x_k, f_k)\}_{-M}^{N}$, with Sinc–points x_k of $\Gamma = (-1, 1)$, and that we have constructed a polynomial $p(x)$ for interpolating this data. We discuss here the replacement of $p(x)$ with $p(w(x))$, where $p(w(x))$ interpolates the same data as $p(x)$, but where $p(w(x))$ provides a uniformly better approximation than $p(x)$ on $[-1, 1]$. To this end, we consider, in turn, each of the following situations:

(i) The errors of interpolation near the ends of $\Gamma = [-1, 1]$ are appreciably different; or else,

(ii) The error of interpolation at the two ends of $\Gamma = [-1, 1]$ are nearly the same, but the errors near the ends of Γ are appreciably different from the error near the center of Γ.

It is convenient to first make an adjustment to correct (i), if necessary. We then proceed to make an adjustment to correct (ii).

The adjustments derived here are based on our knowledge of the error of Sinc approximation, and are therefore for Sinc approximation. However, the analysis of §1.7.1 shows that these adjustments also apply to the above derived polynomial methods, since errors of these polynomial methods increase and decrease with Sinc errors.

(i) We test this case first, setting $u_{-1} = (x_{-M+1} + x_{-M})/2$, $u_1 = (x_{N-1} + x_N)/2$, and then evaluating the "Lipschitz differences"

$$
\begin{aligned}
A_{-1} &= \left| \frac{f(x_{-M+1}) + f(x_{-M})}{2} - p(u_{-1}) \right|, \\
A_1 &= \left| \frac{f(x_{N-1}) + f(x_N)}{2} - p(u_1) \right|,
\end{aligned}
\tag{1.7.19}
$$

Suppose that these two numbers $A_{\pm 1}$ are appreciably different.

Let $c \in (-1, 1)$, and consider making the substitution

$$
x = \frac{\xi + c}{1 + c\xi} \longleftrightarrow \xi = \frac{x - c}{1 - cx},
\tag{1.7.20}
$$

which is a one–to–one transformation of $[-1, 1]$ to $[-1, 1]$ (indeed, also a conformal transformation of the unit disc in the complex plane $\mathbb{C}$ to itself).

Suppose now, that A_1 is greater than A_{-1}. Under the transformation (1.7.20), $f(x) \to F(\xi)$, and the Lipschitz conditions at ± 1

$$
\begin{aligned}
|f(x) - f(-1)| &\leq C_{-1} |1 + x|^\alpha \\
|f(x) - f(1)| &\leq C_1 |1 - x|^\beta
\end{aligned}
\tag{1.7.21}
$$

become

$$
\begin{aligned}
|F(\xi) - F(-1)| &\leq C'_{-1} |1 + c|^\alpha |1 + \xi|^\alpha \\
|F(\xi) - F(1)| &\leq C'_1 |1 - c|^\beta |1 - \xi|^\beta.
\end{aligned}
\tag{1.7.22}
$$

where C'_{-1} (resp., C'_1) differs only slightly from C_{-1} (resp., C_1). On the other hand, by taking c in $(0, 1)$ we can get a considerable reduction in the resulting Lipschitz constant of F from that of f in the neighborhood of 1.

Hence, if $A_1 > A_{-1}$ (resp., if $A_1 < A_{-1}$) we can take c to be the Sinc point x_k with $k > 0$ (resp., $k < 0$) selected such that $1 - c|^\beta \approx |1 - x_m|^\beta \approx A_{-1}/A_1$.

However, rather than change f to F, we opt to replace $p(x)$ with $p(w(\xi))$ with

$$
w(\xi) = \frac{x - x_k}{1 - x\, x_k}.
\tag{1.7.23}
$$

Note: By taking $c < 0$, we increase the number of Sinc points to the right of the origin and decrease the number to the left of the origin, and vice–versa, for $c > 0$. That is, taking $c < 0$ causes the error near the right hand end–point of $[-1, 1]$ to decrease, and at the same time, it causes the error near the left hand end–point to increase.

(ii) We assume in this case that we already have a polynomial $p(x)$ for which the above defined numbers $A_{\pm 1}$ are roughly equal. We also test the mid–range error e.g., by assuming that $x_0 = 0$,

$$v_{1/2} = (x_1 + x_0)/2,$$

$$A_0 = \left| \frac{f(0) + f(x_1)}{2} - p(v_{1/2}) \right|.$$

(1.7.24)

Suppose then, that A_0 differs appreciably from, e.g., A_1, with A_1 defined as above, and with $A_1 \approx A_{-1}$. Let us now replace the original definition of $x = (\rho - 1)/(\rho + 1)$, with $\xi = (\rho^a - 1)/(\rho^a + 1)$, where $\rho(x) = \exp(\varphi_2(x)) = (1 + x)/(1 - x)$ as before, and where a is a positive number to be determined. Since $\rho(x_j) = e^{jh}$, we have $\rho^a(x_j) = e^{ajh}$, i.e., the transformation $x' = (\rho^a - 1)/(\rho^a + 1)$ which is also a one–to–one transformation of the interval $(-1, 1)$ onto itself has the effect of replacing h by $a\,h$.

Now, for the case of Sinc interpolation, the error in the mid–range of the interval is proportional to $A_0 \approx \exp(-\pi\, d/h)$, while the error at the ends is proportional to $A_1 \approx \exp(-\beta\, N\, h)$. Upon replacing h with $a\,h$, we get new errors, A_0' and A_1'. By equating these errors we arrive at the result that

$$a = \left(\frac{|\log(A_0)|}{|\log(A_1)|} \right)^{1/2}$$

(1.7.25)

We can therefore conclude that by increasing a we increase the error in the mid–range of the interval, and decrease it at the ends. Indeed, choosing a in this manner yields the right improvement for both Sinc and polynomial approximation.

PROBLEMS FOR SECTION 1.7

1.7.1 Let $h = c/\sqrt{N}$, with c a positive constant and with N a positive integer and prove the following results:

(a) If $z = e^{h/2}$, then

$$\frac{z^N}{(1 + z)^{2N}} \le 2^{-2N};$$

(b) If $z = e^{h/2}$, then there exists a constant A_1, that is independent of N, such that

$$\sum_{j=1}^{N} \log \left(1 - \exp \left(-z\, e^{-jh}\right)\right) \leq -\frac{\pi^2 \, N^{1/2}}{6\,c} + \log \left(N^{1/4}\right) + A_1 \, ;$$

(c) There exists a positive constant A_2 that is independent of N, such that

$$\sum_{j=1}^{N} \log \left(1 - e^{-jh}\right) \leq -\frac{\pi^2 \, N^{1/2}}{6\,c} + \log \left(N^{1/2}\right) + A_2 \, ;$$

(d) There exists a positive constant A_3 that is independent of N, such that

$$\sum_{j=1}^{N} \log \left(1 + e^{-jh}\right) \geq \frac{\pi^2 \, N^{1/2}}{12\,c} - A_3.$$

1.7.2 Let Γ, ρ, t_j and $b_{j,k}$ be as above. Obtain an accurate bound on the error of the approximation

$$F'(t_j) \approx \sum_{k=-N}^{N} b_{j,k}\, F(t_j)$$

for each of the transformations #1, #3, $\ldots$,, #7 discussed in §1.5.3.

2

Sinc Convolution–BIE Methods for PDE & IE

ABSTRACT   This chapter applies a separation of variables method to finding approximate solutions to virtually any linear PDE in two or more dimensions, with given boundary and/or initial conditions. The method also applies to a wide array of (possibly non–linear) differential and integral equations whose solutions are known to exist through theoretical considerations. Far from being complete in itself, this chapter will serve as a springboard for numerically conquering unsolved problems of importance. The heart of the method is the evaluation of multidimensional Green's function integrals into one–dimensional integrals by iterated application of the one-dimensional convolution procedure of §1.5.9, but now using the known multidimensional Laplace transform of the Green's function.

We contrast this with Morse and Feshbach [MrFe] who proved in 1953 that when solving Laplace or Helmholtz problems in 3 dimensions it is possible to achieve a separation of variables solution for only 13 coordinate systems.

2.1   Introduction and Summary

This chapter contains both known and new results for obtaining solution of partial differential equations (PDE) via combination of boundary integral equation (BIE) and Sinc methods. The results derived here are theoretical. Explicit programs illustrating solutions of PDE using the results of this chapter are presented in Chapter 4. Features of this chapter with reference to solving elliptic, parabolic, and hyperbolic PDE are:

- Using one dimensional Sinc methods of Chapter 1 for solving linear PDE;

- Elimination the use of large matrices required by other methods for solving linear PDE;

- Achieving "separation of variables" in solving elliptic, parabolic and hyperbolic PDE, even over curvilinear regions.

139

- Identifying the properties possessed by solutions of linear PDE which enable us to achieve each of the above, and in addition, to achieving the exponential convergence to a uniformly accurate solution.

In §2.2, we present well known properties of Green's functions, such as their defining differential equations, and how to represent solutions of PDE using these Green's functions.

In §2.3 we list standard free space Green's functions in two and three space dimensions for linear elliptic problems, for linear parabolic and hyperbolic equations in up to three dimensions, for Helmholtz equations, for biharmonic equations, and Green's functions for Navier–Stokes problems. Finally, we illustrate in this section classical Green's function methods for representing solutions of PDE[1].

In this section we derive multidimensional Laplace transforms of all of the Green's functions discussed in §2.3. These multidimensional Laplace transforms allow accurate evaluation of the multidimensional convolution integrals in the IE formulation of the PDE by moving the singularity of the Green's function to end–points of a finite sequence of one–dimensional indefinite convolution integrals.

Nearly all of the results of this section are new, and are required in the Sinc convolution procedures for solving PDE which follow.

In §2.5 we study the procedures for solving linear PDE whose solutions can be expressed by means of Green's functions. These methods are based on Sinc evaluation of one-dimensional convolution integrals. It has been known since 1993 ([S7, S1, §4.6]) that the one–dimensional Sinc convolution procedure of §1.5.9 can be used to reduce multidimensional convolution integrals over rectangular regions to one one dimensional convolutions. Separation can also occur over curvilinear regions, made possible by the formulas of §1.5.11. Classical methods for solving PDE are based on approximation of the highest derivative, and lead to inverting very large matrices. Contrast this to the approach of this chapter which provides algorithms

---

[1] Omitted are Sinc methods applied to Green's functions for three–dimensional stress–strain problems and for Christoffel's equation. Publication of this work will appear in a future monograph.

that require but a relatively small number of one–dimensional matrix multiplications for the solution.

§2.6 studies PDE of interest to scientists or engineers. We assume that the PDE has piecewise smooth components. This is nearly always the case, based on our experience. Therefore the coefficients of the PDE, as well as the boundary of region on which they are defined, and the initial and/or boundary conditions satisfy certain smoothness assumptions. Based on known theorems and some of our own results, we prove that the solutions to the PDE have similar smoothness properties. We are then able to achieve exponential convergence of solution of the PDE, the same as for the one–dimensional Sinc methods of §1.5. Furthermore our solutions remain uniformly accurate over the entire domain of the PDE.

2.2 Some Properties of Green's Functions

In this chapter we discuss *Green's functions*, and we illustrate the use of these functions for converting partial differential equations into integral equations. To this end, we illustrate in detail the procedure of dealing with the singularities of potentials, which give the basic idea of how Green's functions work. At the outset we review some known concepts of calculus operations which are required for the derivation of the Green's functions.

We first define directional derivatives, for integrals over arcs and surface patches.

2.2.1 DIRECTIONAL DERIVATIVES

Let us take $\bar{r} = (x, y, z) \in \mathbb{R}^3$, and let $\mathbf{n} = (n_1, n_2, n_3)$ denote a unit vector in $\mathbb{R}^3$. Given a differentiable function $u(\bar{r})$, we denote the *gradient* of u at $\bar{r}$ by $\nabla u(\bar{r}) = (u_x(\bar{r}), u_y(\bar{r}), u_z(\bar{r}))$. The directional derivative in the direction of the unit vector $\mathbf{n}$ is given by

$$\frac{\partial u}{\partial n}(\bar{r}) = \mathbf{n} \cdot \nabla u(\bar{r}). \tag{2.2.1}$$

More generally, if $\mathbf{v} = (v_1, v_2, v_3)$ is an arbitrary vector that is independent of s, then

$$\frac{d}{ds} u \left(\bar{r} + s\,\mathbf{v} \right) \Big|_{s=0} = \mathbf{v} \cdot \nabla u(\bar{r}). \qquad (2.2.2)$$

2.2.2   INTEGRALS ALONG ARCS

We define an arc $\mathcal{P}$ in the plane by

$$\mathcal{P} = \{ \bar{\rho} = (x, y) = (\xi(t), \eta(t)), \;\; 0 \le t \le 1 \}, \qquad (2.2.3)$$

where ξ and η belong to $\mathbf{C}^1[0, 1]$, and where $(\dot{\xi}(t))^2 + (\dot{\eta}(t))^2 > 0$ for all $t \in (0, 1)$ (with the "dot" denoting differentiation).

Let $ds = \sqrt{(dx)^2 + (dy)^2}$ denote the element of arc length on $\mathcal{P}$. Given a function f defined on $\mathcal{P}$, we define the integral of f along $\mathcal{P}$ by

$$\int_{\mathcal{P}} f(x, y)\, ds = \int_0^1 f(\xi(t), \eta(t)) \frac{ds}{dt}\, dt, \qquad (2.2.4)$$

where

$$\frac{ds}{dt} = \sqrt{\dot{\xi}^2(t) + \dot{\eta}^2(t)}. \qquad (2.2.5)$$

2.2.3   SURFACE INTEGRALS

Setting $\bar{\rho} = (\xi, \eta)$, we may define a patch S of surface by an expression of the form

$$S = \left\{ \bar{r} = (x(\bar{\rho}), y(\bar{\rho}), z(\bar{\rho})), \;\; \bar{\rho} \in (0, 1)^2 \right\}. \qquad (2.2.6)$$

An element of surface area is then given in this notation by means of the vector product expression

$$dA = |\bar{r}_\xi(\bar{\rho}) \times \bar{r}_\eta(\bar{\rho})| \; d\xi\, d\eta. \qquad (2.2.7)$$

Here $\bar{r}_\xi \times \bar{r}_\eta$ is defined by the (usual) determinant

$$\bar{r}_\xi \times \bar{r}_\eta = \begin{vmatrix} i & j & k \\ x_\xi & y_\xi & z_\xi \\ x_\eta & y_\eta & z_\eta \end{vmatrix}, \qquad (2.2.8)$$

with $i = (1, 0, 0)$, $j = (0, 1, 0)$, and with $k = (0, 0, 1)$.

2.2.4 SOME GREEN'S IDENTITIES

Green's theorem can be found in standard advanced calculus textbooks.

– In Two Dimensions.

> **Theorem 2.2.1** *(Green's Theorem) Let $\mathcal{B} \subset \mathbb{R}^2$ be simply connected region with piecewise smooth boundary $\partial \mathcal{B}$, and let p and q belong to $\mathbf{C}^2(\bar{\mathcal{B}})$, where $\bar{\mathcal{B}}$ denotes the closure of $\mathcal{B}$. If $\mathbf{n}' = \mathbf{n}(\bar{\rho}')$ denotes the unit outward normal to $\partial \mathcal{B}$ at $\bar{\rho}' \in \partial \mathcal{B}$, where $\bar{\rho} = (x, y)$, $\bar{\rho}' = (x', y')$, $d\bar{\rho}' = dx'\,dy'$, and $ds(\bar{\rho}')$ is the element of arc length at $\bar{\rho}' \in \partial \mathcal{B}$, then*

$$\int\int_{\mathcal{B}} \left(p(\bar{\rho})\, \nabla^2 q(\bar{\rho}) - q(\bar{\rho})\, \nabla^2 p(\bar{\rho}) \right) d\bar{\rho}$$
$$= \int_{\partial \mathcal{B}} \left(p(\bar{\rho}')\, \frac{\partial q(\bar{\rho}')}{\partial n(\bar{\rho}')} - q(\bar{\rho}')\, \frac{\partial p(\bar{\rho}')}{\partial n(\bar{\rho}')} \right) ds(\bar{\rho}'). \tag{2.2.9}$$

This theorem enables us to derive a Green's function integral expression which is useful for our purposes. We first set $R = |\bar{\rho} - \bar{\rho}'|$, and then define the Green's function

$$\mathcal{G}(\bar{\rho} - \bar{\rho}') = \frac{1}{2\pi} \log \frac{1}{R}. \tag{2.2.10}$$

If in Theorem 2.2.1 we replace $p = p(\bar{\rho}')$ by $\mathcal{G}(\bar{\rho} - \bar{\rho}')$ and ignore the fact that this function does not belong to $\mathbf{C}^2(\bar{\mathcal{B}})$ (it is not differentiable at $\bar{\rho}' = \bar{\rho}$), we formally get the result

$$\int\int_{\mathcal{B}} \left(\mathcal{G}\, \nabla^2_{\bar{\rho}'} q - q\, \nabla^2_{\bar{\rho}'} \mathcal{G} \right) d\bar{\rho}' = \int_{\partial \mathcal{B}} \left(\mathcal{G}\, \frac{\partial q}{\partial n'} - q\, \frac{\partial \mathcal{G}}{\partial n'} \right) ds(\bar{\rho}'). \tag{2.2.11}$$

We shall deduce that this result is still correct, by assigning a proper "meaning" to the integral

$$\int\int_{\mathcal{B}} q(\bar{\rho}')\, \nabla^2_{\bar{\rho}'} \mathcal{G}(\bar{\rho} - \bar{\rho}')\, d\bar{\rho}'.$$

To this end, let $\bar{\rho} \in \mathbb{R}^2$, let us define the disk with center $\bar{\rho}$ and radius ε by

$$\mathcal{D}(\bar{\rho}, \varepsilon) = \left\{ \bar{\rho}' \in \mathbb{R}^2 : |\bar{\rho}' - \bar{\rho}| < \varepsilon \right\}.$$

The length of $\partial \mathcal{D}(\bar{\rho}, \varepsilon)$, i.e., the length of the boundary of this disk is, of course $2\pi\varepsilon$. Let $\Gamma(\bar{\rho}, \varepsilon)$ denote the part of $\partial \mathcal{D}(\bar{\rho}, \varepsilon)$ lying in $\mathcal{B}$, let $\beta(\bar{\rho}, \varepsilon)$ denote the length of $\Gamma(\bar{\rho}, \varepsilon)$ divided by $(2\pi\varepsilon)$ and assume that the limit

$$\beta(\bar{\rho}) = \lim_{\varepsilon \to 0} \beta(\bar{\rho}, \varepsilon) \tag{2.2.12}$$

exists. The value of $\beta(\bar{\rho})$ is clearly 1 if $\bar{\rho} \in \mathcal{B}$, 0 if $\bar{\rho} \in \mathbb{R}^2 \setminus \bar{\mathcal{B}}$, and $1/2$ if $\bar{\rho}$ is a point of smoothness of $\partial \mathcal{B}$. It may have values other than these at corner points and at cusps.

Theorem 2.2.2 *In the notation of Theorem 2.2.1, let $\bar{\rho} \in \mathbb{R}^2$, let us set $R = |\bar{\rho} - \bar{\rho}'|$. If $g \in \mathbf{C}(\bar{\mathcal{B}})$, then*

$$\int\int_{\mathcal{B}} g(\bar{\rho}') \, \nabla_{\bar{\rho}'}^2 \, \mathcal{G}(\bar{\rho} - \bar{\rho}') \, d\bar{\rho}' = -\beta(\bar{\rho}) \, g(\bar{\rho}). \tag{2.2.13}$$

Proof. Let us use the notation of Theorem 2.2.1 above, but let us replace $\mathcal{B}$ by $\mathcal{B}_\varepsilon$, with $\mathcal{B}_\varepsilon = \mathcal{B} \setminus \mathcal{D}(\bar{\rho}, \varepsilon)$. It is known there exists $q \in \mathbf{C}^2(\bar{\mathcal{B}})$ such that $\nabla^2 q = g$, and let us define p by

$$p(\bar{\rho}') = \mathcal{G}(\bar{\rho} - \bar{\rho}') = \frac{1}{2\pi} \log \frac{1}{|\bar{\rho} - \bar{\rho}'|}. \tag{2.2.14}$$

We may then readily verify, that $\nabla_{\bar{\rho}'}^2 \, p = 0$ in $\mathcal{B}_\varepsilon$, since $\bar{\rho}' \neq \bar{\rho}$ in this region.

All assumptions of Theorem 2.2.1 with $\mathcal{B}$ replaced by $\mathcal{B}_\varepsilon$ are then satisfied, so that the result of this theorem with $\mathcal{B}$ replaced by $\mathcal{B}_\varepsilon$ are then valid.

Note now, that $\partial \mathcal{B}_\varepsilon$ includes the arc $\Gamma(\bar{\rho}, \varepsilon)$, and that the outward normal $\mathbf{n}'$ at the point $\bar{\rho}'$ points into the disk $\bar{\mathcal{D}}(\bar{\rho}, \varepsilon)$ directly toward $\bar{\rho}$, i.e. that

$$\mathbf{n}' = \frac{\bar{\rho} - \bar{\rho}'}{R}.$$

Note also, that $R = |\bar{\rho}' - \bar{\rho}| = \varepsilon$. Hence

$$\begin{aligned}
\frac{\partial}{\partial n'}\mathcal{G}(\bar{\rho} - \bar{\rho}') &= \nabla_{\bar{\rho}'}\mathcal{G}(\bar{\rho} - \bar{\rho}') \cdot \mathbf{n}' \\
&= \frac{-1}{2\pi}\frac{\bar{\rho}' - \bar{\rho}}{\varepsilon^2} \cdot \frac{\bar{\rho} - \bar{\rho}'}{\varepsilon} \\
&= \frac{1}{2\pi\varepsilon}.
\end{aligned}$$

Thus, taking $C = \Gamma(\bar{\rho}, \varepsilon)$, and noting that along C, $ds(\bar{\rho}') = \varepsilon\, d\theta$ for integration along C, we have

$$\begin{aligned}
&\int_{\partial\mathcal{B}}\left(\mathcal{G}(\bar{\rho} - \bar{\rho}')\frac{\partial q(\bar{\rho}')}{\partial n'} - q(\bar{\rho}')\frac{\partial}{\partial n'}\mathcal{G}(\bar{\rho} - \bar{\rho}')\right)ds(\bar{\rho}') \\
&+ \int_C\left(\frac{\log(1/\varepsilon)}{2\pi}\frac{\partial q(\bar{\rho}')}{\partial n'} - q(\bar{\rho}')\frac{1}{2\pi\varepsilon}\right)\varepsilon\, d\theta \\
&= \int\int_{\mathcal{B}\backslash\bar{D}(\bar{\rho},\varepsilon)}\mathcal{G}(\bar{\rho} - \bar{\rho}')\nabla_{\bar{\rho}'}^2\, q(\rho')\, d\bar{\rho}'.
\end{aligned}$$

Also,

$$\int_C \mathcal{G}\frac{\partial q}{\partial n'}\varepsilon\, d\theta \to 0$$

$$\int_C q\frac{\partial\mathcal{G}}{\partial n'}\varepsilon\, d\theta \to -\beta(\bar{\rho})\, g(\bar{\rho}),$$

(2.2.15)

as $\varepsilon \to 0$.

Recalling that $\nabla^2 q = g$, letting $\varepsilon \to 0$, we arrive at the equation

$$\int\int_{\mathcal{B}}\mathcal{G}(\bar{\rho}-\bar{\rho}')\, q(\bar{\rho}')\, d\bar{\rho}' + \beta\, g(\bar{\rho}) = \int_{\partial\mathcal{B}}\left(\mathcal{G}\frac{\partial q}{\partial n'} - q\frac{\partial}{\partial n'}\mathcal{G}\right)ds(\bar{\rho}').$$

(2.2.16)

Upon comparing this equation with (2.2.11) above, we conclude that:

Equation (2.2.9) holds for $q \in \mathbf{C}^2(\bar{\mathcal{B}})$ and $p(\bar{\rho}') = \mathcal{G}(\bar{\rho} - \bar{\rho}')$, provided that by $\nabla_{\bar{\rho}'}^2 \mathcal{G}(\bar{\rho} - \bar{\rho}')$ we mean

$$\nabla_{\bar{\rho}'}^2 \mathcal{G}(\bar{\rho} - \bar{\rho}') = -\delta^2(\bar{\rho} - \bar{\rho}'). \qquad (2.2.17)$$

where $\delta^2(\bar{\rho} - \bar{\rho}')$ denotes the two dimensional delta function. ∎

– Three Dimensional Version.

The proof of the following three dimensional Green's theorem can also be found in most advanced calculus texts.

Theorem 2.2.3 *Let $\mathcal{B} \in \mathbb{R}^3$ be a simply connected region with piecewise smooth boundary, $\partial \mathcal{B}$, and let p and q belong to $\mathbf{C}^2(\bar{\mathcal{B}})$, where $\bar{\mathcal{B}}$ denotes the closure of $\mathcal{B}$, and let $dA = dA(\bar{r})$ denote the element of surface area at $\bar{r}$ of $\partial \mathcal{B}$. Then*

$$\int \int \int_{\mathcal{B}} \left(p \nabla^2 q - q \nabla^2 p \right) d\bar{r} = \int \int_{\partial \mathcal{B}} \left(p \frac{\partial q}{\partial n} - q \frac{\partial p}{\partial n} \right) dA.$$
$$(2.2.18)$$

Just as for the two dimensional case, we can also deduce a useful result for the three dimensional case. This will be done via the Green's function

$$\mathcal{G}(\bar{r} - \bar{r}') = \frac{1}{4\pi |\bar{r} - \bar{r}'|}. \qquad (2.2.19)$$

It is easily deduced that $\nabla_{\bar{r}'}^2 \mathcal{G}(\bar{r} - \bar{r}') = 0$ if $\bar{r}' \neq \bar{r}$. Analogously to the two dimensional case, we define the ball with center $\bar{r}$ and radius ε by

$$\mathcal{D}(\bar{r}, \varepsilon) = \left\{ \bar{r}' \in \mathbb{R}^3 : |\bar{r}' - \bar{r}| < \varepsilon \right\}.$$

The surface area of $\partial \mathcal{D}(\bar{\rho}, \varepsilon)$, i.e., the area of the boundary of this ball is, of course $4\pi \varepsilon^2$. Let $\Gamma(\bar{r}, \varepsilon)$ denote the part of

$\partial \mathcal{D}(\bar{r}, \varepsilon)$ lying in $\mathcal{B}$, let $\beta(\bar{r}, \varepsilon)$ denote the area of $\Gamma(\bar{r}, \varepsilon)$ divided by $(4\pi \varepsilon^2)$ and assume that the limit

$$\beta(\bar{r}) = \lim_{\varepsilon \to 0} \beta(\bar{r}, \varepsilon) \tag{2.2.20}$$

exists. The value of $\beta(\bar{r})$ is clearly 1 if $\bar{r} \in \mathcal{B}$, 0 if $\bar{r} \in \mathbb{R}^3 \setminus \bar{\mathcal{B}}$, and $1/2$ if $\bar{r}$ is a point of smoothness of $\partial \mathcal{B}$. It may have values other than these at corner points or at edges.

Set $\mathcal{B}_\varepsilon = \mathcal{B} \setminus \mathcal{D}(\bar{r}, \varepsilon)$. The result of Theorem 2.2.3 with $\mathcal{B}$ replaced by $\mathcal{B}_\varepsilon$ and $p(\bar{r}')$ by $\mathcal{G}(\bar{r} - \bar{r}')$ is then valid, under the additional assumption that the derivatives and integrals in this expression are taken with respect to $\bar{r}'$.

Setting $C = \partial \mathcal{B}_\varepsilon$, and noting that on C, $dA(\bar{r}') = \varepsilon^2 \, d\Omega'$, where $d\Omega'$ denotes an element of solid angle, it is readily shown by proceeding as in the two dimensional case above, that

$$\int\int_C \mathcal{G} \frac{\partial q}{\partial n'} \varepsilon^2 d\Omega' \to 0$$

$$\int\int_C q \frac{\partial \mathcal{G}}{\partial n'} \varepsilon^2 d\Omega' \to -\beta(\bar{r}) \, q(\bar{r}), \tag{2.2.21}$$

as $\varepsilon \to 0$.

Formally writing the result of Green's identity (2.2.18), i.e., ignoring the fact that $p(\bar{\rho}') = \mathcal{G}(\bar{\rho} - \bar{\rho}') \notin \mathbf{C}^2(\bar{B})$, we thus arrive at the equation

$$\int\int\int_B p \nabla_{\bar{r}'}^2 q \, d\bar{r}' + \beta(\bar{r}) \, q(\bar{r})$$

$$= \int\int_{\partial B} \left(p \frac{\partial q(\bar{r}')}{\partial n'} - q \frac{\partial p}{\partial n'} \right) dA(\bar{r}'). \tag{2.2.22}$$

Comparing this equation with (2.2.18) above, we conclude:

Theorem 2.2.4 *If $q \in \mathbf{C}(\bar{B})$, then*

$$\int\int\int_B q(\bar{r}') \, \nabla_{\bar{r}'}^2 \, \mathcal{G}(\bar{r} - \bar{r}') \, d\bar{r}' = -\beta(\bar{r}) \, q(\bar{r}). \tag{2.2.23}$$

That is, we have shown that

$$\nabla_{\bar{r}'}^2 \mathcal{G}(\bar{r} - \bar{r}') = -\delta^3(\bar{r} - \bar{r}'),\qquad(2.2.24)$$

where $\delta^3(\bar{r} - \bar{r}')$ denotes the three dimensional delta function.

Example 2.2.5 Let $\mathcal{B}$ satisfy the conditions of Theorem 2.2.5. We illustrate the conversion of the following PDE problem to an IE (integral equation) problem:

$$\nabla^2 u(\bar{r}) + f(\bar{r}, u(\bar{r}), \nabla u(\bar{r})) = 0,\quad \bar{r} \in \mathcal{B},$$

$$u(\bar{r}) = g(\bar{r}),\quad \frac{\partial u}{\partial n}(\bar{r}) = h(\bar{r}),\quad \bar{r} \in \partial\mathcal{B}.$$

(2.2.25)

Solution: Let $\mathcal{G}$ be defined as in (2.2.19). Then, using Theorem 2.2.4, we have

$$\int\int\int_{\mathcal{B}} \left(\mathcal{G}\nabla^2 u - u\nabla^2 \mathcal{G}\right) d\bar{r}' = \int\int_{\partial\mathcal{B}} \left(\mathcal{G}\frac{\partial u}{\partial n'} - u\frac{\partial\mathcal{G}}{\partial n'}\right) dA(\bar{r}').$$

(2.2.26)

Now using (2.2.24) and (2.2.25), we immediately arrive at the integral equation

$$u(\bar{r}) - \int\int\int_{\mathcal{B}} \mathcal{G} f(\bar{r}', u(\bar{r}'), \nabla u(\bar{r}')) \, d\bar{r}'$$

$$= \int\int_{\partial\mathcal{B}} \left(\mathcal{G} h - g\frac{\partial\mathcal{G}}{\partial n'}\right) dA(\bar{r}').$$

(2.2.27)

Example 2.2.6 Solving Laplace's equation with Neumann Boundary Conditions, via IE. Let $\mathcal{B}$ be a simply connected bounded region in $\mathbb{R}^3$, with $\partial\mathcal{B}$ smooth, and consider the solution to the Neumann problem:

$$\nabla^2 u = 0,\quad \bar{r} \in \mathcal{B},$$

$$\frac{\partial}{\partial n} u(\bar{r}) = g(\bar{r}),\quad \bar{r} \in \partial\mathcal{B}.$$

(2.2.28)

Solution. We assume a solution of the for

$$u(\bar{r}) = \int\int_{\partial\mathcal{B}} \frac{\sigma(\bar{r}')}{4\pi |\bar{r}' - \bar{r}|} dA(\bar{r}'),\quad \bar{r} \in B,\qquad(2.2.29)$$

where dA is the element of surface area of $\partial \mathcal{B}$, and where σ is a potential of a single layer, to be determined. By letting $\bar{r}$ approach $\partial \mathcal{B}$, noting that when $\bar{r} \in \partial \mathcal{B}$ then[2]

$$\int\int_{\partial\mathcal{B}} \left\{ \frac{\partial}{\partial n(\bar{r})} \frac{1}{4\pi|\bar{r}' - \bar{r}|} \right\} \sigma(\bar{r}') \, dA(\bar{r}') = \frac{\sigma(\bar{r})}{2}, \qquad (2.2.30)$$

and then applying the boundary condition of (2.2.28), we get an integral equation for σ, namely,

$$\sigma(\bar{r}) + \int\int_{\partial\mathcal{B}} \left\{ \frac{\partial}{\partial n(\bar{r})} \frac{1}{2\pi|\bar{r} - \bar{r}'|} \right\} \sigma(\bar{r}') \, dA(\bar{r}') = 2\,g(\bar{r}), \qquad \bar{r} \in \partial \mathcal{B}.$$
$$(2.2.31)$$

It may be shown that this equation has a solution if and only if

$$\int\int_{\partial\mathcal{B}} g(\bar{r}') \, dA(\bar{r}') = 0, \qquad (2.2.32)$$

and moreover, this solution, when it exists, is not unique.

After solving (2.2.31) for σ (see [S1 , §2.6] , and §2.6 of this handbook) we can use (2.2.29) to obtain a solution u at any point in $\mathcal{B}$.

Example 2.2.7 *Solving Laplace's Equation with Dirichlet Boundary Conditions via IE.* Let $\mathcal{B}$ be a simply connected region in $\mathbb{R}^3$, with $\partial \mathcal{B}$ smooth, and we consider the the *Dirichlet problem*:

$$\nabla^2 u = 0, \quad \bar{r} \in \mathcal{B},$$
$$(2.2.33)$$
$$u(\bar{r}) = g(\bar{r}), \quad \bar{r} \in \partial \mathcal{B}.$$

Solution. We assume a solution of the form

$$u(\bar{r}) = \int\int_{\partial\mathcal{B}} \left\{ \frac{\partial}{\partial n(\bar{r}')} \frac{1}{4\pi|\bar{r}' - \bar{r}|} \right\} \sigma(\bar{r}') \, dA(\bar{r}'), \quad \bar{r} \in \mathcal{B}, \quad (2.2.34)$$

where the double layer potential σ is to be determined. By letting $\bar{r}$ approach $\partial \mathcal{B}$, noting that when $\bar{r} \in \partial \mathcal{B}$ then

$$\int\int_{\partial\mathcal{B}} \left\{ \frac{\partial}{\partial n(\bar{r}')} \frac{1}{4\pi|\bar{r}' - \bar{r}|} \right\} \sigma(\bar{r}') \, dA(\bar{r}') = -\frac{\sigma(\bar{r})}{2}, \qquad (2.2.35)$$

---

[2]The proof of (2.2.30) is similar to the proof of Theorem 2.2.2.

and then applying the boundary condition of (2.2.33), we get an integral equation for σ, namely,

$$\sigma(\bar{r}) - \int \int_{\partial B} \left\{ \frac{\partial}{\partial n(\bar{r}')} \frac{1}{2\pi|\bar{r}' - \bar{r}|} \right\} \sigma(\bar{r}') \, dA(\bar{r}') = -2\,g(\bar{r}), \quad \bar{r} \in \partial B.$$

(2.2.36)

This equation always has a unique solution.

After solving this equation for σ (as in Example 2.2.6) we can use (2.2.34) to obtain the solution u at any point in B.

PROBLEMS FOR SECTION 2.2

2.2.1 Establish the identities (2.2.15).

2.2.2* *Fredholm Alternative* .

    (a) Prove, in the notation of (2.2.30), that

$$\int \int_{\partial B} \left\{ \frac{\partial}{\partial n(\bar{r})} \frac{1}{4\pi|\bar{r} - \bar{r}'|} \right\} dA(\bar{r}') = -1,$$

    i.e., that $\sigma(\bar{r}) = $ const. is a solution of the integral equation

$$\sigma(\bar{r}) + \int \int_{\partial B} \left\{ \frac{\partial}{\partial n(\bar{r})} \frac{1}{4\pi|\bar{r} - \bar{r}'|} \right\} \sigma(\bar{r}') \, dA(\bar{r}') = 0.$$

    (b) Why can we conclude from (a) that if $g(r)$ satisfies (2.2.32), then (2.2.31) has a non–unique solution.

2.3   Free–Space Green's Functions for PDE

In this section we illustrate Green's functions for the *heat equation*, the *wave equation* and related equations. The derivation of these is based, essentially, on the above obtained results (for elliptic equations) of §2.2. Throughout this section we use the notation $\bar{r} = (x_1, x_2, \ldots, x_d)$, and $r = |\bar{r}| = \sqrt{x_1^2 + x_2^2 + \ldots + x_d^2}$, for $d = 1, 2, 3$, $\mathbb{R} = (-\infty, \infty)$, and $\mathbb{R}_+ = (0, \infty)$.

2.3.1 HEAT PROBLEMS

The Green's function $\mathcal{G}^{(d)} = \mathcal{G}^{(d)}(\bar{r}, t)$ of the heat equation is the bounded solution which solves on $\mathbb{R}^d \times (0, \infty)$ the differential equation problem

$$\mathcal{G}_t^{(d)} - \varepsilon \Delta \mathcal{G}^{(d)} = \delta(t)\, \delta^d(\bar{r} - \bar{r}'), \quad \varepsilon > 0,$$

$$\mathcal{G}(\bar{r}, 0^+) = 0,$$
(2.3.1)

with d denoting the space dimension, with $\delta^d(\bar{r})$ denoting the d – dimensional delta function, with $\Delta = \nabla \cdot \nabla = \nabla^2$ denoting the Laplacian on $\mathbb{R}^d$. The solution to this equation is given by

$$\mathcal{G}^{(d)}(\bar{r}, t) = \frac{1}{(4\pi\varepsilon t)^{d/2}} \exp\left(-\frac{r^2}{4\varepsilon t}\right).$$
(2.3.2)

If $\mathcal{B}$ is an open region in $\mathbb{R}^d$, solution to the problem

$$u_t - \varepsilon \nabla^2 u = f(\bar{r}, t), \quad u = u(\bar{r}, t), \quad (\bar{r}, t) \in \mathcal{B} \times (0, T),$$
(2.3.3)

$$u(\bar{r}, 0^+) = u_0(\bar{r})$$

can be readily expressed in integral form via the above Green's function, i.e.,

$$u(\bar{r}, t) = \int_0^t \int_{\mathcal{B}} \mathcal{G}^{(d)}(\bar{r} - \bar{r}', t - t')\, f(\bar{r}', t')\, d\bar{r}'\, dt'$$

$$+ \int_{\mathcal{B}} \mathcal{G}^{(d)}(\bar{r} - \bar{r}', t)\, u_0(\bar{r}')\, d\bar{r}'$$
(2.3.4)

Clearly, f could also depend (linearly, or nonlinearly) on u and ∇u, in which case we would get an integral equation for u.

2.3.2 WAVE PROBLEMS

Consider the wave equation problem

$$\frac{1}{c^2} \frac{\partial^2 u(\bar{r}, t)}{\partial t^2} - \nabla^2 u(\bar{r}, t) = f(\bar{r}, t), \quad (\bar{r}, t) \in \mathcal{B} \times (0, T),$$
(2.3.5)

$$u(\bar{r}, 0^+) = g(\bar{r}), \quad \frac{\partial u}{\partial t}(\bar{r}, 0^+) = h(\bar{r}),$$

with $\mathcal{B}$ an open region in $\mathbb{R}^d$. The Green's function $\mathcal{G}^{(d)}(\bar{r}, t)$ which we require for the solution to this problem satisfies the equations

$$\frac{1}{c^2} \frac{\partial^2 \mathcal{G}^{(d)}(\bar{r}, t)}{\partial t^2} - \nabla^2 \mathcal{G}^{(d)}(\bar{r}, t) = \delta(t)\, \delta^d(\bar{r}), \quad (\bar{r}, t) \in \mathbb{R}^d \times (0, \infty)$$

$$\mathcal{G}^{(d)}(\bar{r}, 0^+) = 0, \quad \frac{\partial \mathcal{G}^{(d)}}{\partial t}(\bar{r}, 0^+) = 0,$$

$$(2.3.6)$$

where $\delta(\cdot)$ and $\delta^d(\cdot)$ represent the usual one and d–dimensional delta functions respectively. For example, with $r = |\bar{r}|$, and H the Heaviside function, we have

$$\mathcal{G}^{(d)}(\bar{r}, t) = \begin{cases} \dfrac{c}{2} H(ct - r) & \text{if } d = 1, \\[3mm] \dfrac{c}{2\pi \sqrt{c^2 t^2 - r^2}} H(ct - r) & \text{if } d = 2, \\[3mm] \dfrac{c}{2\pi} \delta\left(c^2 t^2 - r^2\right) H(ct - r) & \text{if } d = 3. \end{cases} \quad (2.3.7)$$

If $\mathcal{B} = \mathbb{R}^d$, then the solution on $\mathbb{R}^d \times (0, \infty)$ to the problem (2.3.5) is given in terms of $\mathcal{G}^{(d)}$ by

$$\begin{aligned} u(\bar{r}, t) &= \int_0^t \int_{\mathbb{R}^d} \mathcal{G}^{(d)}(\bar{r} - \bar{r}', t - t')\, f(\bar{r}', t')\, d\bar{r}'\, dt' \\[3mm] &+ \frac{\partial}{\partial t} \int_{\mathbb{R}^d} \mathcal{G}^{(d)}(\bar{r} - \bar{r}', t)\, g(\bar{r}')\, d\bar{r}' \\[3mm] &+ \int_{\mathbb{R}^d} \mathcal{G}^{(d)}(\bar{r} - \bar{r}', t)\, h(\bar{r}')\, d\bar{r}'. \end{aligned} \quad (2.3.8)$$

2.3.3 HELMHOLTZ EQUATIONS

Let us allow $c = c(\bar{r})$ in (2.3.5), and let us assume that the solution to the resulting equation is of the form

$$u(\bar{r}, t) = e^{-i\omega t}\, v(\bar{r}). \quad (2.3.9)$$

Then, upon setting

$$k = \frac{\omega}{c_0}, \quad f(\bar{r}) = \frac{c_0^2}{c^2(\bar{r})} - 1 \tag{2.3.10}$$

we get the Helmholtz equation,

$$\Delta v + k^2 v = -k^2 f v. \tag{2.3.11}$$

We shall here discuss only the equation $\Delta v + k^2 v = - k^2 f v$; below, in §2.4.2, we shall discuss the same equation, but where $k^2 v$ is replaced with $- k^2 v$. It is convenient to take as Green's function the function $\mathcal{G}^{(d)} = \mathcal{G}^{(d)}(\bar{r} - \bar{r}')$ which may be shown by proceeding as for the case of the Laplace operator above, to satisfy the equation

$$\nabla^2 \mathcal{G}^{(d)}(\bar{r} - \bar{r}') + k^2 \mathcal{G}^{(d)}(\bar{r} - \bar{r}') = -\delta^d(\bar{r} - \bar{r}') \tag{2.3.12}$$

Suppose for example, that $c = c(\bar{r})$ in some open, bounded region $\mathcal{B}$, but that $c = c_0$ on $\mathbb{R}^d \setminus \mathcal{B}$. An integral equation whose solution satisfies equation (2.3.11) is then given by

$$v(\bar{r}) - k^2 \int_{\mathbb{R}^d} \mathcal{G}^{(d)}(\bar{r} - \bar{r}') f(\bar{r}') v(\bar{r}') d\bar{r}' = v^i(\bar{r}), \tag{2.3.13}$$

where v^i is the "source", i.e., any function v^i which in $\mathcal{B}$ satisfies the equation

$$\nabla^2 v^i + k^2 v^i = 0. \tag{2.3.14}$$

The integral in (2.3.13) taken over $\mathbb{R}^d$ may, in fact be replaced by an integral over $\mathcal{B}$, since $f = 0$ on $\mathbb{R}^d \setminus \mathcal{B}$. Equation (2.3.13) is called the *Lippmann–Schwinger equation*, and (2.2.11) is called the *reduced wave equation*.

Now, by Green's theorem, and (2.3.12), we have

$$\int_{\mathcal{B}} \left(\mathcal{G}^{(d)}(\bar{r} - \bar{r}') \nabla^2 v(\bar{r}') - v(\bar{r}') \nabla^2_{\bar{r}'} \mathcal{G}^{(d)}(\bar{r} - \bar{r}') \right) d\bar{r}'$$

$$= v(\bar{r}) - k^2 \int_{\mathcal{B}} \mathcal{G}^{(d)}(\bar{r} - \bar{r}') f(\bar{r}') v(\bar{r}') d\bar{r}'$$

$$= \int_{\partial \mathcal{B}} \left(\mathcal{G}^{(d)}(\bar{r} - \bar{r}') \frac{\partial}{\partial n'} v(\bar{r}') - v(\bar{r}') \frac{\partial}{\partial n'} \mathcal{G}^{(d)}(\bar{r} - \bar{r}') \right) dA(\bar{r}'),$$

$$\tag{2.3.15}$$

where $\mathbf{n}' = \mathbf{n}(\bar{r}')$ denotes the outward normal at $\mathbf{r}' \in \partial\mathcal{B}$. Comparing equation (2.3.15) with (2.3.13), we conclude that

$$v^i(\bar{r}) = \int_{\partial\mathcal{B}} \left(\mathcal{G}^{(d)}(\bar{r} - \bar{r}') \frac{\partial}{\partial n'} v(\bar{r}') - v(\bar{r}') \frac{\partial}{\partial n'} \mathcal{G}^{(d)}(\bar{r} - \bar{r}') \right) dA(\bar{r}').$$
(2.3.16)

In practice, v^i is often known, and in this case, there is no need to be concerned with the boundary values of v and $\partial v/\partial n'$.

It may be shown that the Green's functions of the Helmholtz equation in 2 and 3 dimensions are given by

$$\mathcal{G}^{(d)}(\bar{r} - \bar{r}') = \begin{cases} \dfrac{i}{4} H_0^{(1)}(k\,|\bar{r} - \bar{r}'|)\,, & d = 2\,, \\[3mm] \dfrac{\exp\{i\,k|\bar{r} - \bar{r}'|\}}{4\,\pi|\bar{r} - \bar{r}'|}\,, & d = 3\,, \end{cases}$$
(2.3.17)

where $H_0^{(1)}$ denotes the *Hankel function* [Na].

2.3.4   BIHARMONIC GREEN'S FUNCTIONS

The steady–state biharmonic PDE involve problems of the form

$$\nabla^4 W(\bar{\rho}) = f(\bar{\rho})\,, \quad \bar{\rho} = (x, y) \in \mathcal{B}$$

$$W(\bar{\rho}) = g(\bar{\rho})\,, \quad \text{and} \quad \frac{\partial W(\bar{\rho})}{\partial n} = h(\bar{\rho})\,, \quad \bar{\rho} \in \partial\mathcal{B}\,,$$
(2.3.18)

with $\mathcal{B}$ a region in $\mathbb{R}^2$. The Green's function for biharmonic equations is given by

$$\mathcal{G}(\bar{\rho}) = \frac{1}{8\pi} \left(\rho^2\right) \log(\rho).$$
(2.3.19)

This function $\mathcal{G}$ satisfies on $\mathbb{R}^2$ the equation

$$\nabla^4 \mathcal{G}(\bar{\rho}) = \delta^2(\bar{\rho}).$$
(2.3.20)

A *particular solution to the equation* $\nabla^4 V(x, y) = f(x, y)$ is given in terms of this function $\mathcal{G}$ by

$$V(\bar{\rho}) = \int \int_{\mathcal{B}} \mathcal{G}(\bar{\rho} - \bar{\rho}') f(\bar{\rho}') d\bar{\rho}'. \qquad (2.3.21)$$

The time dependent biharmonic PDE involves problems that satisfy the PDE

$$\frac{1}{c^2} \frac{\partial^2 U(\bar{\rho}, t)}{\partial t^2} + \nabla^4 U(\bar{\rho}, t) = f(\bar{\rho}, t), \qquad (\bar{\rho}, t) \in \mathcal{B} \times (0, T), \quad (2.3.22)$$

as well as some initial and boundary conditions, the latter of the form in (2.3.18).

For this case, the Green's function $\mathcal{G}(x, y, t)$ satisfies on $\mathbb{R}^2 \times (0, \infty)$ the equation

$$\frac{1}{c^2} \frac{\partial^2 \mathcal{G}(\bar{\rho}, t)}{\partial t^2} + \nabla^4 \mathcal{G}(\bar{\rho}, t) = \delta^2(\bar{\rho}) \delta(t), \qquad (2.3.23)$$

and is given by [Bk]

$$\mathcal{G}(\bar{\rho}, t) = \frac{1}{4\pi c t} \sin\left(\frac{\rho^2}{4 c t}\right). \qquad (2.3.24)$$

A particular solution to (2.3.22) is given by

$$W(\bar{\rho}, t) = \int_0^t \int \int_{\mathcal{B}} \mathcal{G}(\bar{\rho} - \bar{\rho}', t - t') f(\bar{\rho}', t') d\bar{\rho} dt'. \qquad (2.3.25)$$

2.4 Laplace Transforms of Green's Functions

Our favorite Sinc method for solving the linear convolution–type IE (integral equations) based on the BIE (boundary integral equation) procedure uses Sinc convolution. Effective implementation of this procedure results when the multidimensional Laplace transforms of requisite free space Green's functions are explicitly known. To this end, luck was with us, in that we were able to obtain in this section explicit expressions for the Laplace transforms of all of the free space Green's functions known to us. They enable highly efficient procedures for the simultaneous evaluation at all of the Sinc points of Green's function convolution integrals over rectangular regions (see

§2.5). And although the transformation (2.5.17) which is a transformation of the square $(0, 1) \times (0, 1)$ to the region in defined in (2.5.14) destroys a convolution, (and similarly for the case of the transformation (2.5.37), which transforms the cube $(0, 1) \times (0, 1) \times (0, 1)$ to the region defined in (2.5.35)) we were again lucky, in that we were able to use the methods of §1.5.11 for approximating the more general one dimensional convolutions in order to be able to approximate the transformed multidimensional integral over the rectangular region. That is, we were able to approximate this new integral by *using the Laplace transform of the original free space Green's function.*

We carry out in this section the derivation of the Laplace transforms of standard free–space Green's functions for Laplace, heat, wave, Helmholtz, Navier–Stokes, and biharmonic equations. More specifically, we derive here the multidimensional Laplace transforms of Green's functions for solving Poisson problems in two and three spacial dimensions, biharmonic problems in two spacial dimensions, and also possibly one time dimension, wave equation problems in one, two and possibly also three spacial and one time dimension, Helmholtz equation problems in two and three spacial dimensions, heat equation problems in one, two and three spacial and one time dimension, and Navier–Stokes equations in three spacial and one time dimension.

The procedure used to carry out these derivations is essentially that developed first in [SBaVa, N], and is illustrated below in this section. These transforms enable essentially the same explicit algorithmic "separation of variables" solution of all three types of equations: elliptic, parabolic, and hyperbolic, indeed, even of some nonlinear PDE. The explicit dependence of these algorithms on the eigenvalues of one dimensional indefinite integration matrices suggests the possibility of further work along the lines of [Va1], using the methods of [Va4].

We start by proving a general result which is applicable for evaluation of the Laplace transforms of the Green's functions for all three types of PDEs, elliptic, hyperbolic, and parabolic. The theorem involves evaluation of the integral

$$Q(a) = \int_C \frac{dz}{z - a} \tag{2.4.1}$$

over the circular arc

$$C = \left\{ z = e^{i\theta} \in \mathbb{C} : 0 < \theta < \frac{\pi}{2} \right\} \tag{2.4.2}$$

for any given a in the complex plane $\mathbb{C}$. The purpose of this theorem is for supplementing results of standard texts that tabulate integrals. These texts largely concentrate on real values and consequently, they do not always give us the results we need for expressing Laplace transforms at arbitrary complex values.

As a convention, we let $\log(a)$ denote the principal value of the logarithm, i.e., if $a = \xi + i\eta \in \mathbb{C}$, then $\log(a) = \log|a| + i \arg(a)$, with $\arg(a)$ taking its principal value in the range $-\pi < \arg(a) \leq \pi$.

Theorem 2.4.1 *Let $\xi \in \mathbb{R}$, $\eta \in \mathbb{R}$, let $a = \xi + i\eta \in \mathbb{C}$, set $L(a) = \xi + \eta - 1$ and let A denote the region*

$$A = \{ a = \xi + i\eta \in \mathbb{C} : |a| < 1, \ L(a) > 0 \}. \tag{2.4.3}$$

If $a \in \mathbb{C} \setminus \overline{A}$, then

$$Q(a) = \log\left(\frac{i - a}{1 - a} \right); \tag{2.4.4}$$

If $a = 1$ or $a = i$, then $Q(a)$ is infinite;

If $a \in A$, then

$$Q(a) = \log\left| \frac{i - a}{1 - a} \right| + i \left(2\pi - \arg\left(\frac{i - a}{1 - a} \right) \right); \tag{2.4.5}$$

If $L(a) = 0$ with $|a| < 1$, then

$$Q(a) = \log\left| \frac{i - a}{1 - a} \right| + i\pi; \quad \text{and} \tag{2.4.6}$$

If $L(a) > 0$, and $|a| = 1$, then

$$Q(a) = \log\left| \frac{i - a}{1 - a} \right| + i \frac{\pi}{2}. \tag{2.4.7}$$

Proof. The proof is straightforward, the main difficulty stemming from the fact that if $a \in A$, then the principal value of $\log \left(\frac{i-a}{1-a} \right)$ does not yield the correct value of $Q(a)$ as defined by the integral definition of $Q(a)$, since in that case the imaginary part of $Q(a)$ is larger than π. Perhaps the most difficult part of the proof is the verification of (2.4.7), which results from the fact that if the conditions on a are satisfied, then the integral (2.4.1) defining $Q(a)$ is a principal value integral along the arc C, and this value of $Q(a)$ is thus the average of the two limiting values of $Q(\zeta)$ as ζ approaches the point a on the circular arc from interior and from the exterior of A. This value is easily seen to be

$$\log \left| \frac{i-a}{1-a} \right| + i\pi/2 \qquad (2.4.8)$$

by use of the Plemelj formulas (1.5.21). ∎

2.4.1   TRANSFORMS FOR POISSON PROBLEMS

We found in the previous section the Green's functions of the Laplacian in $\mathbb{R}^d$ for $n = 2$ and $n = 3$ satisfy the equation

$$\Delta \, \mathcal{G}_0^{(d)}(\bar{r}) = -\delta^d(\bar{r}), \qquad (2.4.9)$$

where, with $\delta(x)$ denoting the usual one dimensional delta function, then $\delta^d(\bar{r}) = \delta(x_1) \ldots, \delta(x_d)$ is the d–dimensional delta function. Also from the previous section, if $\mathcal{B}$ denotes an open subset of $\mathbb{R}^d$, $d = 2, 3$, then a particular solution of the equation

$$\Delta \, u(\bar{r}) = -f(\bar{r}), \quad \bar{r} \in \mathcal{B}, \qquad (2.4.10)$$

is given by the integral

$$u(\bar{r}) = \int_{\mathcal{B}} \mathcal{G}_0^{(d)}(\bar{r} - \bar{r}') \, f(\bar{r}') \, d\bar{r}'. \qquad (2.4.11)$$

These Green's functions $\mathcal{G}_0^{(d)}$ are given explicitly by

$$\mathcal{G}_0^{(1)}(x_1) \;=\; -\frac{1}{2}\,|x_1|\,,$$

$$\mathcal{G}_0^{(2)}(\bar{\rho}) \;=\; -\frac{1}{2\pi}\,\log(\rho)\,, \quad \rho = \sqrt{x_1^2 + x_2^2} \qquad (2.4.12)$$

$$\mathcal{G}_0^{(3)}(\bar{r}) \;=\; \frac{1}{4\pi r} \quad r = \sqrt{x_1^2 + x_2^2 + x + 3^2}.$$

In order to obtain explicit Laplace transforms of these Green's function, we use the notation $\bar{\Lambda} = (1/\lambda_1,\, \dots,\, 1/\lambda_d)$, and $\Lambda = \sqrt{1/\lambda_1^2 + \dots + 1/\lambda_d^2}$, where the λ_j are complex numbers, with positive real parts. Taking $\mathbb{R}_+ = (0,\infty)$, we define the modified Laplace transform of $\mathcal{G}_0^{(d)}$ (as for the $d = 1$ case in Equation (1.5.54)) by the integral

$$G_0^{(d)}(\bar{\Lambda}) = \int_{\mathbb{R}_+^d} \mathcal{G}_0^{(d)}(\bar{r})\,\exp(-\bar{\Lambda}\cdot\bar{r})\,d\bar{r}. \qquad (2.4.13)$$

Theorem 2.4.2 *For the case of $d = 1$, we have*

$$G_0^{(1)}(\Lambda) = \lambda_1. \qquad (2.4.14)$$

We omit the proof which involves the evaluation of a simple integral.

Theorem 2.4.3 *For the case of $d = 2$, we have*

$$G_0^{(2)}(\bar{\Lambda})$$

$$= \left(\frac{1}{\lambda_1^2} + \frac{1}{\lambda_2^2}\right)^{-1}.$$

$$\cdot\left(-\frac{1}{4} + \frac{1}{2\pi}\left(\frac{\lambda_2}{\lambda_1}\left(\gamma - \log(\lambda_2)\right) + \frac{\lambda_1}{\lambda_2}\left(\gamma - \log(\lambda_1)\right)\right)\right),$$
$$(2.4.15)$$

where $\gamma = -\Gamma'(1) = 0.577\dots$ is Euler's constant [Na].

Proof. Consider the equation (2.4.9) for the case of $d = 2$. Performing integration by parts, and in this proof, simply denoting $\mathcal{G}_0^{(2)}$ by $\mathcal{G}$, we get

$$\int_0^\infty \exp\left(-\frac{x_1}{\lambda_1}\right) \mathcal{G}_{x_1 x_1}(x_1, x_2) \, dx_1$$

$$= \mathcal{G}_{x_1}(x_1, x_2) \exp\left(-\frac{x_1}{\lambda_1}\right)\bigg|_{x_1=0}^{\infty}$$
$$+ \frac{1}{\lambda_1} \int_0^\infty \exp\left(-\frac{x_1}{\lambda_1}\right) \mathcal{G}_{x_1}(x_1, x_2) \, dx_1 \qquad (2.4.16)$$

$$= 0 + \frac{1}{\lambda_1} \exp\left(-\frac{x_1}{\lambda_1}\right) \mathcal{G}(x_1, x_2)\bigg|_{x_1=0}^{\infty}$$
$$+ \frac{1}{\lambda_1^2} \int_0^\infty \exp\left(-\frac{x_1}{\lambda_1}\right) \mathcal{G}(x_1, x_2) \, dx_1,$$

since $\mathcal{G}_{x_1}(0^+, x_2) = 0$ if $x_2 > 0$. Next, multiplying both sides of this equation by $\exp(-x_2/\lambda_2)$, and integrating with respect to x_2 over $\mathbb{R}_+$, we get

$$-\frac{1}{\lambda_1} \int_{\mathbb{R}_+} \mathcal{G}(0, x_2) \exp\left(-\frac{x_2}{\lambda_2}\right) dx_2$$

$$= \frac{1}{2\pi\lambda_1} \int_{\mathbb{R}_+} \log(x_2) \exp\left(-\frac{x_2}{\lambda_2}\right) dx_2$$

$$\qquad (2.4.17)$$

$$= \frac{\lambda_2}{2\pi\lambda_1} \left(\Gamma'(1) + \log(\lambda_2)\right)$$

$$= \frac{\lambda_2}{2\pi\lambda_1} \left(-\gamma + \log(\lambda_2)\right).$$

Repeating the above steps, starting with $\mathcal{G}_{x_2 x_2}$ instead of $\mathcal{G}_{x_1 x_1}$, and noting that the Laplace transform of $\delta(x_1)\,\delta(x_2)$ taken over $\mathbb{R}_+^2$ is $1/4$ (recall the possible fractional value of $\beta(\bar{\rho})$ in the above derivation of (2.2.13)), we arrive at the statement of the theorem. ∎

Theorem 2.4.4 *If $d = 3$, then*

$$G_0^{(3)}(\bar{\Lambda}) =$$

$$-\frac{1}{\Lambda^2} \left(\frac{1}{8} - H_0(\lambda_1, \lambda_2, \lambda_3) - H_0(\lambda_2, \lambda_3, \lambda_1) - H_0(\lambda_3, \lambda_1, \lambda_2)\right),$$

$$\qquad (2.4.18)$$

where, with

$$\mu_1 = \sqrt{\frac{1}{\lambda_2^2} + \frac{1}{\lambda_3^2}},$$

$$\zeta_1 = \sqrt{\frac{\lambda_3 + i\lambda_2}{\lambda_3 - i\lambda_2}},$$

(2.4.19)

and where, with $Q(a)$ defined as in Theorem 2.4.1,

$$H_0(\lambda_1, \lambda_2, \lambda_3) = -\frac{1}{4\pi\lambda_1\mu_1}\{Q(i\zeta_1) - Q(-i\zeta_1)\}. \qquad (2.4.20)$$

Proof. For simplicity of notation in this proof, we shall write $\mathcal{G}$ in place of $\mathcal{G}_0^{(3)}$.

We start with the one dimensional integral

$$J(\lambda_1, x_2, x_3) = \int_{\mathbb{R}_+} \mathcal{G}_{x_1 x_1}(\bar{r}) \exp(-x_1/\lambda_1) \, dx_1, \qquad (2.4.21)$$

in which we carry out integration by parts; since $\mathcal{G}_{x_1}(0, x_2, x_3) = 0$ for all $(x_2, x_3) \in \mathbb{R}_+^2$, we get

$$J(\lambda_1, x_2, x_3)$$

$$= \mathcal{G}_{x_1}(\bar{r}) \exp\left(-\frac{x_1}{\lambda_1}\right)\Big|_{x_1=0}^{\infty}$$

$$+ \frac{1}{\lambda_1} \int_{\mathbb{R}_+} \mathcal{G}_{x_1}(\bar{r}) \exp\left(-\frac{x_1}{\lambda_1}\right) dx_1$$

$$= \frac{1}{\lambda_1} \mathcal{G}(\bar{r}) \exp\left(-\frac{x_1}{\lambda_1}\right)\Big|_{x_1=0}^{\infty} \qquad (2.4.22)$$

$$+ \frac{1}{\lambda_1^2} \int_{\mathbb{R}_+} \mathcal{G}(\bar{r}) \exp\left(-\frac{x_1}{\lambda_1}\right) dx_1$$

$$= -\frac{1}{4\pi\lambda_1\sqrt{x_2^2 + x_3^2}} + \frac{1}{\lambda_1^2} \int_{\mathbb{R}_+} \mathcal{G}(\bar{r}) \exp\left(-\frac{x_1}{\lambda_1}\right) dx_1.$$

Upon multiplying the last line of this equation by $\exp(-x_2/\lambda_2 - x_3/\lambda_3)$ integrating over $\mathbb{R}_+^2$, we get

$$\int_{\mathbb{R}_+^3} \mathcal{G}_{x_1 x_1}(\bar{r}) \, \exp(-\bar{\Lambda} \cdot \bar{r}) \, d\bar{r} = \frac{1}{\lambda_1^2} \, G_0^{(3)}(\bar{\Lambda}) + H_0(\lambda_1, \lambda_2, \lambda_3),$$

$$(2.4.23)$$

where

$$H_0(\lambda_1, \lambda_2, \lambda_3) = \frac{1}{4\pi\lambda_1} K \qquad (2.4.24)$$

and where, upon setting

$$\bar{\mu} = \left(\frac{1}{\lambda_2}, \frac{1}{\lambda_3}\right), \quad \mu = \sqrt{\frac{1}{\lambda_2^2} + \frac{1}{\lambda_3^2}},$$

$$\zeta = \sqrt{\frac{\lambda_3 + i\lambda_2}{\lambda_3 - i\lambda_2}}, \qquad (2.4.25)$$

$$x_2 + i x_3 = \rho Z, \quad \rho = \sqrt{x_2^2 + x_3^2}, \quad and \;\; Z = e^{i\theta},$$

we have

$$\begin{aligned}
K &= \int_{\mathbb{R}_+^2} \exp\left(-\frac{x_2}{\lambda_2} - \frac{x_3}{\lambda_3}\right) \frac{d\,x_2\, d\,x_3}{\sqrt{x_2^2 + x_3^2}}, \\
&= \int_{\mathcal{C}} \int_0^\infty \exp\left(-\frac{\rho\,\mu}{2}\left(\frac{Z}{\zeta} + \frac{\zeta}{Z}\right)\right) d\rho\, d Z \qquad (2.4.26) \\
&= \frac{2}{i} \int_{\mathcal{C}} \frac{dZ}{Z\,\mu\left(\frac{Z}{\zeta} + \frac{\zeta}{Z}\right)},
\end{aligned}$$

with $\mathcal{C}$ is defined as in (2.4.2) above. Using partial fraction decomposition of the integrand in the last line of (2.4.26) and recalling the definition of $Q(a)$ in (2.4.1), we get

$$K = \frac{-1}{\mu} \{Q(i\zeta) - Q(-i\zeta)\}. \qquad (2.4.27)$$

The result of Theorem 2.4.4 now follows upon inserting this value of K in (2.4.24). ∎

The above proof also establishes the following

Corollary 2.4.5 *Let $\mathcal{G}_0^{(3)}$ be defined as in (2.4.12). Then, in the notation of Theorem 2.4.4,* ´

$$\int_{\mathbb{R}_+^2} \mathcal{G}_0^{(3)}(0, x_2, x_3) \exp\left(-\frac{x_2}{\lambda_2} - \frac{x_3}{\lambda_3}\right) dx_2 \, dx_3$$

$$= \int_{\mathbb{R}_+^2} \exp\left(-\frac{x_2}{\lambda_2} - \frac{x_3}{\lambda_3}\right) \frac{dx_2 \, dx_3}{4\pi\sqrt{x_2^2 + x_3^2}} \qquad (2.4.28)$$

$$= \frac{-1}{4\pi\mu_1}\left(Q(i\zeta_1) - Q(-i\zeta_1)\right),$$

where ζ_1 and μ_1 are defined as in (2.4.19) above.

2.4.2 TRANSFORMS FOR HELMHOLTZ EQUATIONS

We determine here, for k a constant, with $\Re k > 0$, the Laplace transforms of the three Green's functions,

$$\mathcal{G}^{(1)}(x_1, k) = \frac{1}{2k}\exp(-k|x_1|), \quad x_1 \in \mathbb{R},$$

$$\mathcal{G}^{(2)}(\bar{\rho}, k) = \frac{1}{2\pi}K_0(k\rho), \quad \bar{\rho} = (x_1, x_2) \in \mathbb{R}^2, \quad \rho = |\bar{\rho}|,$$

$$\mathcal{G}^{(3)}(\bar{r}, k) = \frac{\exp(-kr)}{4\pi r}, \quad \bar{r} = (x_1, x_2, x_3) \in \mathbb{R}^3, \quad r = |\bar{r}|.$$

$$(2.4.29)$$

Here K_0 denotes the modified Bessel function [Na].

Writing $\bar{r} = (x_1, \dots, x_d)$, these Green's functions satisfy the differential equation

$$\Delta\mathcal{G}^{(d)}(\bar{r}, k) - k^2\mathcal{G}^{(d)}(\bar{r}, k) = -\delta^d(\bar{r}), \quad \bar{r} \in \mathbb{R}^d. \qquad (2.4.30)$$

Assuming that λ_1, λ_2, and λ_3 are constants with positive real parts, we again take $\bar{\Lambda} = (1/\lambda_1, \dots, 1/\lambda_n)$. We then determine explicitly the integrals

$$G^{(d)}(\bar{\Lambda}, k) = \int_{\mathbb{R}_+^n} \exp\left(-\bar{\Lambda} \cdot \bar{r}\right) \mathcal{G}^{(d)}(\bar{r}, k) \, d\bar{r}, \qquad (2.4.31)$$

for $d = 1, 2, 3$.

Theorem 2.4.6 *Let $G^{(1)}$ be defined as in (2.4.31) with $d = 1$. Then*

$$G^{(1)}(\bar{\Lambda}, k) = \frac{1}{2k} \frac{\lambda_1}{1 + k \lambda_1}. \qquad (2.4.32)$$

Proof. We substitute The first equation in (2.4.29) into (2.4.31). The evaluation of the resulting integral is straight forward, and we omit the details. ∎

Theorem 2.4.7 *Let $G^{(2)}$ be defined as in (2.4.31), with $n = 2$. Then*

$$G^{(2)}(\bar{\Lambda}, k)$$

$$= \left(k^2 - \frac{1}{\lambda_1^2} - \frac{1}{\lambda_2^2} \right)^{-1} \left(\frac{1}{4} - H(\lambda_1, \lambda_2, k) - H(\lambda_2, \lambda_1, k) \right), \qquad (2.4.33)$$

where

$$H(\lambda_1, \lambda_2, k) = \frac{1}{2\lambda_1} \left(k^2 - \frac{1}{\lambda_2^2} \right)^{-1/2}. \qquad (2.4.34)$$

Proof. It follows, upon examination of the series expansion of $K_0(z)$ in a neighborhood of $z = 0$, that $\mathcal{G}^{(2)}_{x_1}(0, x_2, k) = 0$ if $x_2 \neq 0$. Hence, taking $\bar{\rho} = (x_1, x_2)$, and performing integration by parts, we get

$$\int_{\mathbb{R}_+} \exp\left(-\frac{x_1}{\lambda_1} \right) \mathcal{G}^{(2)}(\bar{\rho}, k) \, d x_1$$

$$= -\frac{1}{\lambda_1} \mathcal{G}^{(2)}(0, x_2, k) + \frac{1}{\lambda_1^2} \int_{\mathbb{R}_+} \exp\left(-\frac{x_1}{\lambda_1} \right) \mathcal{G}^{(2)}(x_1, x_2, k) \, d x_1,$$

$$= -\frac{1}{2\pi\lambda_1} K_0(k \, |x_2|) + \frac{1}{\lambda_1^2} \int_{\mathbb{R}_+} \exp\left(-\frac{x_1}{\lambda_1} \right) \mathcal{G}^{(2)}(x_1, x_2, k) \, d x_1. \qquad (2.4.35)$$

Now, multiply the last equation in (2.5.35) by $\exp(-x_2/\lambda_2)$, integrate over $\mathbb{R}_+$ with respect to x_2, and then recall (2.4.31), to get

$$\int_{\mathbb{R}^2_+} \exp\left(-\frac{x_1}{\lambda_1} - \frac{x_2}{\lambda_2}\right) \mathcal{G}^{(2)}_{x_1 x_1}(\bar{\rho}, k)\, d\bar{\rho}$$

$$= \frac{1}{\lambda_1^2} G^{(2)}(\lambda_1, \lambda_2, k) - \frac{1}{2\pi\lambda_1} \int_{\mathbb{R}_+} K_0(k\,x_2) \exp\left(-\frac{x_2}{\lambda_2}\right) d x_2.$$

$$(2.4.36)$$

At this point we replace K_0 in (2.4.36) by the following expression taken from [Na, Eq. 4.6.23]:

$$K_0(k\,|y|) = \int_1^\infty \exp(-k\,|y|\,\xi)\left(\xi^2 - 1\right)^{-1/2} d\xi. \qquad (2.4.37)$$

Interchanging the order of integration in the resulting integral in (2.4.36), then evaluating the inner integral, we are able to get

$$\int_{\mathbb{R}^2_+} \exp\left(-\frac{x_1}{\lambda_1} - \frac{x_2}{\lambda_2}\right) \mathcal{G}^{(2)}_{x_1 x_1}\bar{\rho}, k)\, d\bar{\rho}$$
$$= \frac{1}{\lambda_1^2} G^{(2)}(\lambda_1, \lambda_2, k) - \frac{1}{2\pi\lambda_1} \int_1^\infty \left(\xi^2 - 1\right)^{-1/2} \frac{d\xi}{k\xi + 1/\lambda_2}$$

$$= \frac{1}{\lambda_1^2} G^{(2)}(\lambda_1, \lambda_2, k) - \frac{1}{2}\left(k^2 - \frac{1}{\lambda_2^2}\right)^{-1/2}.$$

$$(2.4.38)$$

Now, starting with $\int_{\mathbb{R}_+} \exp(-x_2/\lambda_2)\,\mathcal{G}^{(2)}(\bar{\rho}, k)\, d\bar{\rho}$, proceeding similarly as above, and noting that

$$\int\int_{\mathbb{R}^2_+} \left(\Delta\,\mathcal{G}(\bar{\rho}, k) - k^2\,\mathcal{G}(\bar{\rho}, k)\right) \exp\left(-\frac{x_1}{\lambda_1} - \frac{x_2}{\lambda_2}\right) d\bar{\rho}$$

$$(2.4.39)$$

$$= -\int_{\mathbb{R}^2_+} \delta(x_1)\,\delta(x_2)\, d\bar{\rho} = -\frac{1}{4},$$

we get (2.4.33)–(2.4.34). ∎

Theorem 2.4.8 *Let $G^{(3)}(\bar{\Lambda}, k)$ be defined as in (2.4.31). Then*

$$G^{(3)}(\bar{\Lambda}, k) = \left(k^2 - \frac{1}{\lambda_1^2} - \frac{1}{\lambda_2^2} - \frac{1}{\lambda_3^2} \right)^{-1} \cdot$$

$$\cdot \left\{ \tfrac{1}{8} - H(\lambda_1, \lambda_2, \lambda_3, k) - H(\lambda_2, \lambda_3, \lambda_1, k) - H(\lambda_3, \lambda_1, \lambda_2, k) \right\},$$
$$(2.4.40)$$

where, with Q as in Theorem 2.4.1 and with $z_\pm$ defined by

$$z_\pm = \frac{1}{\frac{1}{\lambda_2} + \frac{i}{\lambda_3}} \left(-k \pm \sqrt{k^2 - \frac{1}{\lambda_2^2} - \frac{1}{\lambda_3^2}} \right), \qquad (2.4.41)$$

and H defined by

$$H(\lambda_1, \lambda_2, \lambda_3, k) = \frac{1}{4\pi i \lambda_1 \sqrt{k^2 - \frac{1}{\lambda_2^2} - \frac{1}{\lambda_3^2}}} \left(Q(z_+) - Q(z_-) \right).$$
$$(2.4.42)$$

Proof. We proceed similarly as we did above, for the case of $\mathcal{G}^{(2)}$, to now evaluate the Laplace transform of $\mathcal{G}^{(3)}_{x_1 x_1}$ with respect to x_1. Since $\mathcal{G}^{(3)}_{x_1}(0, x_2, x_3, k) = 0$ for $(x_2, x_3) \in \mathbb{R}^2_+$, we have, upon integrating by parts,

$$\int_0^\infty e^{-x_1/u_1} \mathcal{G}^{(3)}_{x_1 x_1}(\bar{r}, k)\, dx_1$$

$$= -\frac{1}{\lambda_1} \mathcal{G}^{(3)}(0, x_2, x_3, k) + \frac{1}{\lambda_1^2} \int_0^\infty e^{-x_1/\lambda_1} \mathcal{G}^{(3)}(\bar{r}, k)\, dx_1.$$
$$(2.4.43)$$

Hence, setting

$$H(\lambda_1, \lambda_2, \lambda_3, k)$$

$$= \frac{1}{\lambda_1} \int_{\mathbb{R}^2_+} \exp\left(-\frac{x_2}{\lambda_2} - \frac{x_3}{\lambda_3} \right) \mathcal{G}^{(3)}(0, x_2, x_3, k)\, dx_2\, dx_3,$$
$$(2.4.44)$$

we find, after converting to complex coordinates, $(x_2 + i\, x_3) = \rho\, e^{i\theta} = \rho\, Z$, with $Z = e^{i\theta}$, and

$$\mu = \left(\frac{1}{\lambda_2^2} + \frac{1}{\lambda_3^2} \right)^{1/2}$$

$$\zeta = \left(\frac{\lambda_3 + i \lambda_2}{\lambda_3 - i \lambda_2} \right)^{1/2}$$

(2.4.45)

that $x_2/\lambda_2 + x_3/\lambda_3 = (1/2)\rho\,\mu\,(Z/\zeta + \zeta/Z)$. Then, integrating first with respect to ρ, we get

$$H(\lambda_1, \lambda_2, \lambda_3, k) = \frac{\zeta}{2\pi i \lambda_1 \mu} \int_C \frac{dZ}{Z^2 + \frac{2\zeta k Z}{\lambda} + \zeta^2}. \qquad (2.4.46)$$

where once again,

$$C = \left\{ Z \in \mathbb{C} : Z = e^{i\theta},\ 0 < \theta < \pi/2 \right\}. \qquad (2.4.47)$$

Upon denoting the roots of the quadratic in the denominator of the integrand by z_+ and z_-, we find that

$$z_\pm = \frac{\zeta}{\mu} \left(-k \pm \sqrt{k^2 - \mu^2} \right), \qquad (2.4.48)$$

which is the same as (2.4.41). Using this notation to factor the denominator of the integrand in (2.4.46), performing partial fraction decomposition, and using Theorem 2.4.1, we arrive at (2.4.40)–(2.4.42). ∎

The above proof also establishes the following result, which will be useful later in this section.

Corollary 2.4.9 *In the notation of Theorem 2.4.8,*

$$\int_{\mathbb{R}_+^2} \exp\left(-\frac{x_2}{\lambda_2} - \frac{x_3}{\lambda_3} \right) \mathcal{G}^{(3)}(0, x_2, x_3, k)\, d x_2\, d x_3$$

$$= \int_{\mathbb{R}_+^2} \exp\left(-\frac{x_2}{\lambda_2} - \frac{x_3}{\lambda_3} \right) \frac{\exp\left(-k\sqrt{x_2^2 + x_3^2} \right)}{4\pi \sqrt{x_2^2 + x_3^2}}\, d x_2\, d x_3 \qquad (2.4.49)$$

$$= \frac{1}{4\pi i \sqrt{k^2 - \frac{1}{\lambda_2^2} - \frac{1}{\lambda_3^2}}} \left(Q(z_+) - Q(z_-) \right).$$

2.4.3 TRANSFORMS FOR HYPERBOLIC PROBLEMS

We consider here the transforms of Green's functions for problems in d spacial and one time dimension, for $d = 1, 2, 3$. These Green's functions enable us to obtain a particular convolution integral solution to the wave equation

$$\frac{1}{c^2} \frac{\partial^2 u}{\partial t^2} - \Delta u = f. \tag{2.4.50}$$

The Laplace transforms of each of these Green's function is readily available from our Helmholtz equation results of the previous section. As previously defined, let $\bar{\Lambda} = (1/\lambda_1, \ 1/\lambda_2, \ \ldots, \ 1/\lambda_d)$, $\Lambda = \sqrt{1/\lambda_1^2}$, and we denote the Green's function by $\mathcal{G}^{(d)}(\bar{r}, t)$, and its Laplace transform, by

$$\hat{G}^{(d)}(\bar{\Lambda}, \tau) = \int_{\mathbb{R}_+} \int_{\mathbb{R}^d} \exp\left(-\bar{\Lambda} \cdot \bar{r} - \frac{t}{\tau}\right) \mathcal{G}^{(d)}(\bar{r}, t) \, d\bar{r} \, dt. \tag{2.4.51}$$

Let us recall from the previous section, that the Green's function $\mathcal{G}^{(d)}(\bar{r}, t)$ satisfies

$$\frac{1}{c^2} \frac{\partial^2 \mathcal{G}^{(d)}(\bar{r}, t)}{\partial t^2} - \Delta \mathcal{G}^{(d)}(\bar{r}, t) = \delta(t) \, \delta^d(\bar{r}), \quad (\bar{r}, t) \in \mathbb{R}^d \times [0, \infty)$$

$$\mathcal{G}^{(d)}(\bar{r}, 0^+) = \frac{\partial \mathcal{G}^{(d)}}{\partial t}(\bar{r}, 0^+) = 0.$$
$$\tag{2.4.52}$$

Hence, by taking the Laplace transform of this equation with respect to t, and denoting the resulting Laplace transform of $\mathcal{G}^{(d)}(\bar{r}, \tau)$ by $\tilde{G}^{(d)}(\bar{r}, \tau)$, we get

$$\Delta \tilde{G}^{(d)}(\bar{r}, \tau) - \frac{1}{c^2 \tau^2} \tilde{G}^{(d)}(\bar{r}, \tau) = -\delta^{(d)}(\bar{r}). \tag{2.4.53}$$

Comparing this equation with (2.4.30), we immediately arrive at

Theorem 2.4.10 *Let $\hat{G}^{(d)}(\bar{\Lambda}, \tau)$ be defined as in (2.4.51) above. Then:*

(a) $\hat{G}^{(1)}(\bar{\Lambda}, \tau) = G^{(1)}(\bar{\Lambda}, 1/(c\tau))$, where $G^{(1)}(\bar{\Lambda}, k)$ is defined as in Theorem 2.4.2;

(b) $\hat{G}^{(2)}(\bar{\Lambda}\,,\tau) = G^{(2)}(\bar{\Lambda}\,,1/(c\,\tau))\,$, *where $G^{(2)}(\bar{\Lambda}\,,k)$ is defined as in Theorem 2.4.3; and*

(c) $\hat{G}^{(3)}(\bar{\Lambda}\,,\tau) = G^{(3)}(\bar{\Lambda}\,,1/(c\,\tau))\,$, *where $G^{(3)}(\bar{\Lambda}\,,k)$ is defined as in Theorem 2.4.4.*

2.4.4   WAVE EQUATION IN $\mathbb{R}^3 \times (0\,,T)$

In this section we derive some integral equation formulations of the wave equation (2.3.5), where $c = c(\bar{r})$ can now depend on $\bar{r}$. We first take the Laplace transform of this equation, which reduces it to a Helmholtz equation that depends on the Laplace transform parameter s. This Helmholtz equation leads readily to its *Lippmann–Schwinger* Fredholm integral equation formulation over a region in $\mathbb{R}^3$. We next reduce the 3–d Lippmann–Schwinger Fredholm integral equation to 1–d Fredholm integral equation by taking a two–d Fourier transform. Finally we reduce this 1–d Fredholm integral equation to a Volterra integral equation, which lends itself to solution via Sinc methods.

(i.) *Laplace Transform of Equation (2.3.5).* Taking the Laplace transform of Equation (2.3.5) with respect to t, we get the Helmholtz equation,

$$\nabla^2 U - \frac{1}{c^2(\bar{r})\,s^2}\,U = -q(\bar{r}\,,s)\,, \qquad (2.4.54)$$

where the Laplace transform $U(\bar{r}, s) = \int_0^\infty u(\bar{r}, t)\,e^{-t/s}\,dt\,$, and where $q(\bar{r}\,,s) = F(\bar{r}\,,s) + (g(\bar{r})/s + h(\bar{r}))/c^2(\bar{r})$. Here $F(r\,,s)$ denotes the Laplace transform of $f(r\,,t)$ with respect to t with $f(r\,,t)\,$, $g(\bar{r})$ and $h(\bar{r})$ are as in Equation (2.3.5).

(ii.) *Fredholm Integral Equation Formulation.* Instead of (2.4.54), we now consider the similar, but more popularly studied equation of *ultrasonic tomography*,

$$\nabla^2 f - \kappa^2\,(1+\gamma)f = 0\,, \quad \kappa = (c_0\,s)^{-1}\,, \qquad (2.4.55)$$
$$\bar{r} = (x\,,y\,,z) \in V$$

with V a region in $\mathbb{R}^3\,$, $\Re\kappa > 0\,$, and where we assume that $\gamma = 0$ on $\mathbb{R}^3 \setminus V$, and that Γ is bounded on V. A *source f^{in}* is any function that satisfies in V the equation

$$\nabla^2 f^{in} - \kappa^2 f^{in} = 0. \qquad (2.4.56)$$

In (2.4.55), f is the total field. One can set

$$f = f^{in} + f^{sc} \qquad (2.4.57)$$

where f^{sc} is the scattered field.

We shall solve the problem (1.1) subject to a *plane-wave like source* f^{in} of the form

$$f^{in}(\bar{r}) = e^{-\kappa x}. \qquad (2.4.58)$$

This source f^i does in fact satisfy (2.4.56) in V as well as in $\mathbb{R}^3$.

Under the influence of the source (2.4.58), the solution to (2.4.55) takes the form

$$f(\bar{r}) + \kappa^2 \int\int\int_V G((\bar{r} - \bar{r}')\,\gamma(\bar{r}')\,f(\bar{r}')\,d\bar{r}' = f^{in}(\bar{r}), \quad (2.4.59)$$

where

$$G(\bar{r}) = \frac{e^{-\kappa r}}{4\pi r}. \qquad (2.4.60)$$

Substituting $f = f^{sc} + f^{in}$ into (2.4.5.9), one gets an equivalent equation for the scattered field,

$$f^{sc}(\bar{r}) + \kappa^2 \int\int\int_V G(\bar{r} - \bar{r}')\,\gamma(\bar{r}')\left(f^{sc}(\bar{r}') + f^{in}(\bar{r}')\right)\,d\bar{r}' = 0.$$
$$(2.4.61)$$

(iii.) $2 - d$ *Fourier Transform of (2.4.61)*. We will now take the $2 - d$ Fourier transform with respect to the variables (y, z) in the (2.4.61). We first set

$$\bar{\rho} = (y,z), \quad \bar{\Lambda} = (\lambda_2,\lambda_3)$$

$$\rho = |\bar{\rho}| = \sqrt{y^2 + z^2}, \quad \Lambda = |\bar{\lambda}| = \sqrt{\lambda_2^2 + \lambda_3^2}, \qquad (2.4.62)$$

$$\mu = \sqrt{\Lambda^2 + \kappa^2}.$$

Then, denoting the $2-d$ Fourier transform of a function $u \in \mathbf{L}^2(\mathbb{R}^2)$, by $\tilde{u}$, using the result

$$\int\int_{\mathbb{R}^2} e^{i\,\bar{\rho}\cdot\bar{\lambda}}\, G(x,\bar{\rho})\, d\bar{\rho} = \frac{e^{-\mu\,|x|}}{2\,\mu}, \qquad (2.4.63)$$

and using the notations

$$\begin{aligned}
U(x) &= \widetilde{f^{sc}}(x,\bar{\Lambda}) \\
W(x) &= \widetilde{f^{sc}}\gamma(x,\bar{\Lambda}) + f^{in}(x)\tilde{\gamma}(x,\bar{\Lambda}),
\end{aligned} \qquad (2.4.64)$$

we find that

$$\begin{aligned}
U''(x) &- \mu^2\, U(x) = -\kappa^2\, W(x), \\
U(x) &= -\kappa^2 \int_0^a \frac{e^{-\mu|x-\xi|}}{2\,\mu}\, W(\xi)\, d\xi.
\end{aligned} \qquad (2.4.65)$$

(iv.) *Volterra Integral Equation Formulations.* Next, by setting $x = 0$ and $x = a$ in the second of (2.4.65), and then differentiating (2.4.65), and again setting $x = 0$ and $x = a$ in the result, we can deduce that

$$U'(a) = -\mu\, U(a) \quad \text{and} \quad U'(0) = \mu\, U(0). \qquad (2.4.66)$$

On the other hand, the solution to the ODE in (2.4.65) can also be expressed in terms of either of the two Volterra equations,

$$
\begin{aligned}
U(x) &= -\kappa^2 \int_0^x \frac{\sinh(\mu(x-\xi))}{\mu} W(\xi)\, d\xi \\
&\quad + A \cosh(\mu x) + B \frac{\sinh(\mu x)}{\mu},
\end{aligned}
$$

$$
\begin{aligned}
U(x) &= -\kappa^2 \int_x^a \frac{\sinh(\mu(\xi-x))}{\mu} W(\xi)\, d\xi \\
&\quad + C \cosh(\mu(a-x)) + D \frac{\sinh(\mu(a-x))}{\mu}.
\end{aligned}
$$

$$
(2.4.67)
$$

We can then deduce, using the results of (2.4.66), that $A = U(0)$, $B = U'(0)$, $C = U(a)$, and $D = -U'(a)$. These results enable us to simplify the equations in (2.4.67) to the following:

$$
U(x) = -\kappa^2 \int_0^x \frac{\sinh(\mu(x-\xi))}{\mu} W(\xi)\, d\xi + U(0)\, e^{\mu x}
$$

$$
U(x) = -\kappa^2 \int_x^a \frac{\sinh(\mu(\xi-x))}{\mu} W(\xi)\, d\xi + U(a)\, e^{\mu(a-x)}.
$$

$$
(2.4.68)
$$

The solution to the second equation in (2.4.64) is discussed in §4.3.1.

2.4.5 TRANSFORMS FOR PARABOLIC PROBLEMS

We obtain here, the $n + 1$–dimensional Laplace transforms of the free space Green's functions on $\mathbb{R}^d \times [0, \infty)$. As above, we use the notations $\bar{r} = (x_1,\, x_2,\, \ldots,\, x_n)$, $r = |\bar{r}| = \sqrt{x_1^2 + x_2^2 + \ldots + x_n^2}$, $\bar{\Lambda} = (1/\lambda_1,\, 1/\lambda_2,\, \ldots,\, 1/\lambda_n)$, and $\Lambda = \sqrt{1/\lambda_1^2,\, 1/\lambda_2^2,\, \ldots,\, 1/\lambda_n^2}$. The Green's function for parabolic problems is the function which with ε a positive constant solves the equation

$$
\frac{\partial \mathcal{G}^{(d)}(\bar{r}, t)}{\partial t} - \varepsilon\, \Delta_{\bar{r}} \mathcal{G}^{(d)}(\bar{r}, t) = \delta(t)\, \delta^d(\bar{r})
$$

$$
(2.4.69)
$$

$$
\mathcal{G}^{(d)}(\bar{r}, 0) = 0.
$$

and which remains bounded as $r \to \infty$. This function is explicitly expressed by

$$\mathcal{G}^{(d)}(\bar{r}, t) = \frac{1}{(4\pi\varepsilon t)^{d/2}} \exp\left(-\frac{r^2}{4\varepsilon t}\right). \tag{2.4.70}$$

Furthermore, taking the Laplace transform of (2.4.70) with respect to t,

$$\tilde{G}^{(d)}(\bar{r}, \tau) = \int_{\mathbb{R}_+} \exp\left(-\frac{t}{\tau}\right) \mathcal{G}^{(d)}(\bar{r}, t)\, dt \tag{2.4.71}$$

we arrive at the Helmholtz differential equation

$$\Delta_{\bar{r}}\, \tilde{G}^{(d)}(\bar{r}, \tau) - \frac{1}{\varepsilon\tau}\tilde{G}^{(d)}(\bar{r}, \tau) = -\frac{1}{\varepsilon}\delta^d(\bar{r}). \tag{2.4.72}$$

Then, setting

$$\hat{G}^{(d)}(\bar{\Lambda}, \tau) = \int_{\mathbb{R}_+^d} \exp\left(-\bar{\Lambda}\cdot\bar{r}\right) \tilde{G}^{(d)}(\bar{r}, \tau)\, d\bar{r}, \tag{2.4.73}$$

and comparing (2.4.72) and (2.4.30), we arrive at

Theorem 2.4.11 *Let $\hat{G}^{(d)}(\bar{\Lambda}, \tau)$ be defined as in (2.4.73). Then:*

(a) $\hat{G}^{(1)}(\bar{\Lambda}, \tau) = (1/\varepsilon)\, G^{(1)}(\bar{\Lambda}, \sqrt{1/(\varepsilon\tau)})$, where $G^{(1)}(\bar{\Lambda}, k)$ is defined as in Theorem 2.4.6;

(b) $\hat{G}^{(2)}(\bar{\Lambda}, \tau) = (1/\varepsilon)\, G^{(2)}(\bar{\Lambda}, \sqrt{1/(\varepsilon\tau)})$, where $G^{(2)}(\bar{\Lambda}, k)$ is defined as in Theorem 2.4.7; and

(c) $\hat{G}^{(3)}(\bar{\Lambda}, \tau) = (1/\varepsilon)\, G^{(3)}(\bar{\Lambda}, \sqrt{1/(\varepsilon\tau)})$, where $G^{(3)}(\bar{\Lambda}, k)$ is defined as in Theorem 2.4.8.

2.4.6 NAVIER–STOKES EQUATIONS

We discuss here, Navier–Stokes equations on $\{(\bar{r}, t) \in \mathbb{R}^3 \times \mathbb{R}_+\}$, where $\mathbb{R} = (-\infty, \infty)$ and $\mathbb{R}_+ = (0, \infty)$. As above, we set $\bar{r} = (x_1, x_2, x_3)$, $\bar{r}' = (x_1', x_2', x_3')$, and $r = |\bar{r}|$. We shall write g_{x_i} in place of $\frac{\partial g}{\partial x_i}$, and we shall denote by $\mathcal{G} = \mathcal{G}(\bar{r}, t)$, the Green's function of the heat equation discussed above on $\mathbb{R}^3 \times \mathbb{R}_+$, i.e.,

$$\mathcal{G}(\bar{r}, t) = \frac{1}{(4\pi\varepsilon t)^{3/2}} \exp\left(-\frac{r^2}{4\varepsilon t}\right). \tag{2.4.74}$$

Our discussion involves the following PDE as presented in [Ff], i.e.,

$$\frac{\partial u^i}{\partial t} - \varepsilon \Delta u^i = -\sum_{j=1}^{3} u^j \frac{\partial u^i}{\partial x_j} - \frac{\partial p}{\partial x_i} + f^i(\bar{r}, t), \quad (\mathbf{x}, t) \in \mathbb{R}^3 \times \mathbb{R}_+,$$

$$(2.4.75)$$

which is to be solved subject to the condition

$$\operatorname{div} \mathbf{u} = \sum_{i=1}^{3} \frac{\partial u^i}{\partial x_i} = 0, \quad (\bar{r}, t) \in \mathbb{R}^3 \times \mathbb{R}_+, \tag{2.4.76}$$

and subject to initial conditions

$$\mathbf{u}(\bar{r}, 0) = \mathbf{u}^0(\bar{r}) \quad \mathbf{r} \in \mathbb{R}^3. \tag{2.4.77}$$

Here, $\mathbf{u} = (u^1, u^2, u^3)$ denotes a velocity vector of flow, p denotes the pressure, $\mathbf{u}^0 = (u^{0,1}, u^{0,2}, u^{0,3})$ is a given divergence–free velocity field on $\mathbb{R}^3$, $f^i(\bar{r}, t)$ are the components of a given, externally applied force, such as gravity, ε is a positive coefficient (the viscosity) and $\Delta = \sum_{i=1}^{3} \frac{\partial^2}{\partial x_i^2}$ is the Laplacian in the space variables.

Mathematicians are interested in proving the existence of solutions $\mathbf{u}(\bar{r}, t)$ and $p(\bar{r}, t)$ to (2.4.75) – (2.4.77), (see [Ff]) under the conditions, e.g., that $\mathbf{u}$, p, and $\mathbf{u}^0(\bar{r})$ satisfy for all real K, $\alpha > 0$, and some constants $C_{\alpha K}$ and $C_{\alpha m K}$ the inequalities

$$|\partial_{\bar{r}}^\alpha \mathbf{u}^0(\bar{r})| \ \leq \ C_{\alpha K} (1 + |\bar{r}|)^{-K}, \ \text{on} \ \mathbb{R}^3,$$

$$(2.4.78)$$

$$|\partial_{\bar{r}}^\alpha \partial_t^m p(\bar{r}, t)| \ \leq \ C_{\alpha m K} (1 + r + t)^{-K} \ \text{on} \ \mathbb{R}^3 \times \mathbb{R}_+,$$

and such that

$$p, \ \mathbf{u} \in \mathbf{C}^\infty \left(\mathbb{R}^3 \times \mathbb{R}_+ \right). \tag{2.4.79}$$

A detailed analysis of a Sinc method of solving (2.4.75)–(2.4.77) is given in [ST]. We shall here neither go into the lengthy mathematical details covered in [ST], nor shall we present the Sinc method of solution covered there. Instead, we mention a sufficient number of results obtained in [ST] to enable us to describe another method of solution, also based on Sinc.

Proceeding in the spirit of this manual, we use the Green's function $\mathcal{G}$ of (2.4.74) to transform the system (2.4.75) of three PDE, with $f \equiv 0$, into the following system of three integral equations, for $i = 1, 2, 3$,

$$u^i(\bar{r}, t) = \int_{\mathbb{R}^3} \mathcal{G}(\bar{r} - \bar{r}', t)\, u^{0,i}(\bar{r})\, d\bar{r}'$$

$$- \int_0^t \int_{\mathbb{R}^3} \mathcal{G}(\bar{r} - \bar{r}', t - t') \left(\sum_{j=1}^3 u^j \frac{\partial u^i}{\partial x'_j} + \frac{\partial p}{\partial x'_i} \right) d\bar{r}'\, dt'.$$

$$\text{(2.4.80)}$$

Let us observe that if g is any function that satisfies the inequalities of (2.4.78), and if γ is defined by either of the integrals

$$\gamma(\bar{r}, t) = \int_{\mathbb{R}^3} \mathcal{G}(\bar{r} - \bar{r}', t)\, g(\bar{r}', t)\, d\bar{r}',$$

$$\text{(2.4.81)}$$

$$\gamma(\bar{r}, t) = \int_0^t \int_{\mathbb{R}^3} \mathcal{G}(\bar{r} - \bar{r}', t - t')\, g(\bar{r}', t')\, d\bar{r}'\, dt',$$

then γ also satisfies (2.4.78), and we can readily verify the following identities applied to γ as defined in the first of (2.4.81) :

$$\gamma_{x_i}(\bar{r}, t) = \int_{\mathbb{R}^3} \mathcal{G}_{x_i}(\bar{r} - \bar{r}', t)\, g(\bar{r}', t)\, d\bar{r}'$$

$$= - \int_{\mathbb{R}^3} \mathcal{G}_{x'_i}(\bar{r} - \bar{r}', t)\, g(\bar{r}', t)\, d\bar{r}' \qquad \text{(2.4.82)}$$

$$= \int_{\mathbb{R}^3} \mathcal{G}(\bar{r} - \bar{r}', t)\, g_{x'_i}(\bar{r}', t)\, d\bar{r}'.$$

These identities enable us to replace (2.4.80) by the easier to solve equivalent system of equations

$$\mathbf{u}(\bar{r}, t) = \int_{\mathbb{R}^3} \mathcal{G}(\bar{r} - \bar{r}', t)\, \mathbf{u}^0(\bar{r}')\, d\bar{r}'$$

$$+ \int_0^t \int_{\mathbb{R}^3} \{ ((\nabla' \mathcal{G}) \cdot \mathbf{u}(\bar{r}', t'))\, \mathbf{u}(\bar{r}', t') + \qquad \text{(2.4.83)}$$

$$+ p(\bar{r}', t')\, (\nabla' \mathcal{G}) \}\, d\bar{r}'\, dt'.$$

In the second integral in (2.4.83) we have written $\mathcal{G}$ for $\mathcal{G}(\bar{r}-\bar{r}', t-t')$.

Eliminating the Pressure

By differentiating (2.4.80) with respect to x_i, summing over i, and in that way, "forcing" (2.4.76), we get

$$\sum_{i=1}^{3} u_{x_i}^i = 0 = \sum_{i=1}^{3} \left(A^i - B^i - C^i \right), \qquad (2.4.84)$$

where

$$
\begin{aligned}
A^i &= \int_{\mathbb{R}^3} \mathcal{G}_{x_i}(\bar{r} - \bar{r}', t)\, u^{0,i}(\bar{r}')\, d\bar{r}' \\
B^i &= \int_0^t \int_{\mathbb{R}^3} \mathcal{G}_{x_i}(\bar{r} - \bar{r}', t - t') \sum_{j=1}^{3} u^j(\bar{r}', t')\, u_{x_j}^i(\bar{r}', t')\, d\bar{r}'\, dt' \\
C^i &= \int_0^t \int_{\mathbb{R}^3} \mathcal{G}_{x_i}(\bar{r} - \bar{r}', t - t')\, p_{x_i'}\, d\bar{r}'\, dt'.
\end{aligned}
$$
$$(2.4.85)$$

By combining (2.4.84) and (2.4.82) we arrive at the equation

$$0 = \int_0^t \int_{\mathbb{R}^3} \mathcal{G}(\bar{r} - \bar{r}', t - t') \left\{ g(\bar{r}', t') + \Delta_{(\bar{r})'}\, p(\bar{r}', t') \right\} d\bar{r}'\, dt'. \qquad (2.4.86)$$

in which

$$g(\bar{r}, t) = \sum_{i=1}^{3} \sum_{j=1}^{3} u_{x_j}^i(\bar{r}, t)\, u_{x_i}^j(\bar{r}, t). \qquad (2.4.87)$$

That is, the solution $\mathbf{u}$ of (2.4.80) will also satisfy (2.4.76), if we determine $p(\bar{r}, t)$ such that

$$\Delta_{\bar{r}}\, p(\bar{r}, t) = -\, g(\bar{r}, t), \qquad (\bar{r}, t) \in \mathbb{R}^3 \times R_+, \qquad (2.4.88)$$

with g as in (2.4.87). There is only one solution to this PDE problem which satisfies the second inequality in (2.4.78) and (2.4.79) (See [ST]); this solution is given by the integral

$$p(\bar{r}, t) = \int_{\mathbb{R}^3} \mathcal{G}_0(\bar{r} - \bar{r}')\, g(\bar{r}, t)\, d\bar{r}', \qquad (2.4.89)$$

where (see §2.4.1))

$$\mathcal{G}_0(\bar r) = \frac{1}{4 \pi r}, \qquad (2.4.90)$$

and where g is given in (2.4.87).

We may note, that upon substituting the expression (2.4.87) into (2.4.83), we are faced with the evaluation of a very high dimensional integral involving the product of $\mathcal{G}_0$ and $\mathcal{G}$. However, the following identity enables us to circumvent this:

$$
\begin{aligned}
\mathcal{K}(\bar r, t) &= \int_{\mathbb{R}^3} \mathcal{G}(\bar r - \bar r', t)\, \mathcal{G}_0(\bar r')\, d\bar r', \\[2mm]
&= \frac{1}{4\, \pi^{3/2} (\varepsilon\, t)^{1/2}\, r} \int_0^r \exp\left(-\frac{\rho^2}{4\,\varepsilon\, t}\right) d\rho,
\end{aligned}
\qquad (2.4.91)
$$

Indeed, we now have the integral equation

$$
\begin{aligned}
\mathbf{u}(\bar r, t) &= \int_{\mathbb{R}^3} \mathcal{G}(\bar r - \bar r', t)\, \mathbf{u}^0(\bar r')\, d\bar r' \\[2mm]
&\quad + \int_0^t \int_{\mathbb{R}^3} \left((\nabla'\, \mathcal{G}) \cdot \mathbf{u}(\bar r', t')\right) \mathbf{u}(\bar r', t')\, d\bar r'\, dt' \\[2mm]
&\quad + \int_0^t \int_{\mathbb{R}^3} \left(\nabla'\, \mathcal{K}(\bar r - \bar r', t - t')\right) g(\bar r', t')\, d\bar r'\, dt'.
\end{aligned}
\qquad (2.4.92)
$$

Let us now discuss the solution to (2.4.81).

We write

$$\mathbf{u} = \mathbf{T}\,\mathbf{u}, \qquad (2.4.93)$$

where

$$\mathbf{T}\,\mathbf{u} = \mathbf{Q}\,\mathbf{u}^0 + \mathbf{R}\,\mathbf{w} + \mathbf{S}\,g, \qquad (2.4.94)$$

and where,

$$(\mathbf{Q}\,\mathbf{u}^0)\,(\bar{r},t) \;=\; \int_{\mathbb{R}^3} \mathcal{G}(\bar{r}-\bar{r}',t-t')\,\mathbf{u}^0(\bar{r}')\,d\bar{r}'\,dt'$$

$$(\mathbf{R}\,\mathbf{w})(\bar{r},t) \;=\; \int_0^t \int_{\mathbb{R}^3} \left((\nabla'\,\mathcal{G})\cdot\mathbf{u}(\bar{r}',t')\right)\,\mathbf{u}(\bar{r}',t')\,d\bar{r}'\,dt'$$

$$(\mathbf{S}\,g)(\bar{r},t) \;=\; \int_0^t \int_{\mathbb{R}^3} \left(\nabla'\,\mathcal{K}(\bar{r}-\bar{r}',t-t')\right)\,g(\bar{r}',t')\,d\bar{r}'\,dt'\,,$$

$$\tag{2.4.95}$$

with g again defined as in (2.4.83). We intend to solve (2.4.93) by successive approximation. In this regard, we have left out some of the mathematics (which will appear in [ST]). Briefly, a Sinc function space of the type used in §2.6 is defined; this is shown to be dense in the $\mathbf{C}_0^\infty$ family of functions of (2.4.78). It is then shown that if each of the components of $\mathbf{u}^0$ and $\mathbf{u}$ belongs to this space, and if $\mathbf{u}^0$ is also divergence–free, then the operator $\mathbf{T}$ maps this space back into itself, and moreover, it is a contraction whenever the time interval $(0,T)$ is sufficiently small. We have not included any details or proofs some of which are lengthy. Also, the Sinc method of solution used in [ST] is more efficient, in part, because it uses the DFT procedure of §1.4.9 (see Problem 2.4.6).

Let us now comment on each of the operators $\mathbf{Q}$, $\mathbf{R}$, and $\mathbf{S}$.

$\mathbf{Q}$: The three dimensional Laplace transform of $\mathcal{G}(\bar{r},t)$ taken with respect to the spacial variables cannot be explicitly expressed, while the four dimensional one, $\hat{G}(\bar{\Lambda},\tau) = \varepsilon^{-1}\,G^{(3)}\bar{\Lambda},1/\sqrt{\varepsilon\,\tau})$ as defined in Eq. (2.4.73) with $n=3$, is given in Theorem 2.4.11 part (c). We thus start with this four dimensional Laplace transform $G^{(3)}(\bar{\Lambda},1/\sqrt{c\tau})$. Examination of the Laplace transform inversion procedure of §1.5.10, shows that we can use the Laplace transform $(\varepsilon\,\tau)^{-1}\,G^{(3)}(\bar{\Lambda},1/\sqrt{c\tau})$ to evaluate $(\mathbf{Q}\,\mathbf{u}^0)\,(\bar{r},t)$ using a four dimensional convolution, by treating the components of $\mathbf{u}^0$ as functions defined on $\mathbb{R}^3 \times (0,T)$, even though these components take the same value for all $t \in (0,T)$.

$\mathbf{R}$: Here we use the Laplace Transform $-\widehat{G_{x_j}}(\bar{\Lambda},\tau)$ of $-\mathcal{G}_{x_j}(\bar{r},t)$ over $\mathbb{R}_+^4$. To get the Laplace transform of $\mathcal{G}_{x_j}(\bar{r},t)$, we first take a Laplace transform with respect to t, thus replacing $\mathcal{G}(\bar{r},t)$ with $\bar{G}(\bar{r},\tau)$. Here we consider only the case of $j=1$. Performing integration by parts, we get

$$- \int_0^\infty \bar{G}_{x_1}(\bar{r}, \tau) \exp(-x_1/\lambda_1) \, dx_1$$

$$= - \bar{G}(\bar{r}, \tau) \exp(-x_1/\lambda_1) \big|_{x_1=0}^{x_1=\infty}$$
$$- \frac{1}{\lambda_1} \int_0^\infty \hat{G}(\bar{r}, \tau) \exp(-x_1/\lambda_1) \, dx_1, \qquad (2.4.96)$$

$$= \bar{G}(\bar{r}, \tau) \big|_{x_1=0} - \frac{1}{\lambda_1} \int_0^\infty \bar{G}(\bar{r}, \tau) \exp(-x_1/\lambda_1) \, dx_1.$$

We now multiply each equation by $\exp(-x_2/\lambda_2 - x_3/\lambda_3)$, and integrate over $\mathbb{R}_+^2$ with respect to $\bar{\rho}$, where $\bar{\rho} = (x_2, x_3)$ to deduce, via Corollary 2.4.5, that

$$- \widehat{G_{x_1}}(\bar{\Lambda}, \tau) = - \int_{\mathbb{R}_+^3} \bar{G}_{x_1}(\bar{r}, \tau) \exp(-\bar{r} \cdot \bar{\Lambda}) \, d\bar{r}$$

$$= - \frac{1}{\lambda_1 \, \varepsilon} G^{(3)}(\bar{\Lambda}, 1/\sqrt{\varepsilon \tau}) + \lambda_1/\varepsilon \, H(\lambda_1, \lambda_2, \lambda_3, 1/\sqrt{\varepsilon \tau}),$$
$$(2.4.97)$$

where $\hat{G}(\bar{\Lambda}, \tau)$ and $H(\lambda_1, \lambda_2, \lambda_3, k)$ are defined in Theorem 2.4.11 and Theorem 2.4.8 respectively.

S: The Laplace transform of $\mathcal{K}$ is already given above, in the paragraph following (2.4.81). The Laplace transform of $-\mathcal{K}_{x_j}(\bar{\Lambda}, \tau)$ is obtainable by proceeding similarly as for the case of **R** above, and we get

$$- \int_{\mathbb{R}_+^4} \exp(-\bar{\Lambda} \cdot \bar{r} - t/\tau) \, \mathcal{K}_{x_1}(\bar{r}, t) \, d\bar{r} \, dt$$

$$= - \frac{\tau}{\lambda_1} \left(G_0^{(3)}(\bar{\Lambda}) - G^{(3)}(\bar{\Lambda}, 1/\sqrt{\varepsilon \tau}) \right) \qquad (2.4.98)$$

$$+ \tau \lambda_1 \left(H_0(\lambda_1, \lambda_2, \lambda_3) - H(\lambda_1, \lambda_2, \lambda_3, 1/\sqrt{\varepsilon \tau}) \right),$$

where $G_0^{(3)}$ and $H_0(\lambda_1, \lambda_2, \lambda_3)$ are defined as in Theorem 2.4.4, $\mathcal{G}$ is defined as in (2.4.70), and where $G^{(3)}(\bar{\Lambda}, k)$ and $H(\lambda_1, \lambda_2, \lambda_3)$ are defined as in Theorem 2.4.6.

Remarks. (1) Assuming that $\mathbf{u}^0$ is appropriately selected so that the conditions (2.4.82), (2.4.84) and (2.4.85) are satisfied with respect to the spacial variables, it follows, that a solution $\mathbf{u}$ and p obtained via the above scheme can be expressed in terms of integrals involving $\mathcal{G}$, and $\mathcal{G}_0$. By (2.4.87) and (2.4.88), these functions then also satisfy (2.4.72), (2.4.84), and (2.4.85).

(2) If an initial value $\mathbf{u}^0$ is selected, and we determine the Sinc parameters in each dimension (there is only one parameter, N, in the program given in §4.3.2) to achieve a uniformly accurate approximation of $\mathbf{T}\mathbf{u}^0$ on $\mathbb{R}^3 \times (0,T)$, for arbitrary $T > 0$, then it is easy to see, since the norm of the indefinite integration matrix with respect to the map $\varphi(z) = \log(z/(T-z))$, which transforms $(0,T)$ to $\mathbb{R}$, is proportional to T, that the operator $\mathbf{T}$ is a contraction for all T sufficiently small.

(3) The vector $\mathbf{u}^0 = \left(u^{0,1}, u^{0,2}, u^{0,3}\right)$ may be selected in several ways. One way is to select two arbitrary functions v and w belonging to $\mathbf{C}_0^\infty$, and which also satisfy the first of (2.4.78), and then take $u^{0,1} = v_{z_3}$, $u^{0,2} = w_{z_3}$, and $u^{0,3} = -(v_{z_1} + w_{z_2})$. It follows div $\mathbf{u}^0 = 0$.

2.4.7 TRANSFORMS FOR BIHARMONIC GREEN'S FUNCTIONS

Here we derive transforms for obtaining particular solutions to the time independent problem

$$\nabla^4 U(\bar{\rho}) = f(\bar{\rho}), \qquad \bar{\rho} = (x,y) \in \mathcal{B}, \qquad (2.4.99)$$

as well as to the wave problem

$$\frac{1}{c^2}\frac{\partial^2 U(\bar{\rho},t)}{\partial t^2} + \nabla^4 U(\bar{\rho},t) = f(\bar{\rho},t), \qquad (\bar{\rho},t) \in \mathcal{B} \times (0,T). \quad (2.4.100)$$

The Time Independent Problem. A particular solution to the problem (2.4.87) is given by

$$U(\bar{\rho}) = \int\int_{\mathcal{B}} \mathcal{G}(\bar{\rho} - \bar{\rho}' - \eta)\, f(\bar{\rho}')\, d\bar{\rho}', \qquad (2.4.101)$$

where the Green's function, $\mathcal{G}$, is given by [Hc]

$$\mathcal{G}(\bar{\rho}) = \frac{1}{8\pi}\rho^2 \log(\rho). \qquad (2.4.102)$$

Setting $\mathcal{R} = x^2 + y^2$, it follows directly by differentiating $\mathcal{G}$, that

$$
\begin{aligned}
\mathcal{G}_x(\bar{\rho}) &= \frac{x}{8\,\pi}\,(\log \mathcal{R} + 1) \\
\mathcal{G}_{xx}(\bar{\rho}) &= \frac{1}{8\,\pi}\left(\log \mathcal{R} + 1 + \frac{2\,x^2}{\mathcal{R}}\right) \\
\mathcal{G}_{xxx}(x\bar{\rho}) &= \frac{1}{8\,\pi}\left(\frac{6\,x}{\mathcal{R}} - \frac{4\,x^3}{\mathcal{R}^2}\right) \\
\mathcal{G}_{xxy}(\bar{\rho}) &= -\frac{y}{4\,\pi\,\mathcal{R}}\left(1 - \frac{2\,x^2}{\mathcal{R}}\right),
\end{aligned}
\tag{2.4.103}
$$

and similarly for differentiation with respect to y. Starting with the identity

$$
\nabla^4\,\mathcal{G}(\bar{\rho}) = \delta^2(\bar{\rho}),
\tag{2.4.104}
$$

and proceeding as in the proof of Theorem 2.4.2, we can use the identities (2.4.91) to arrive at

Theorem 2.4.12 *The Laplace transform of the function $\mathcal{G}$ defined in (2.4.90) above is*

$$
\begin{aligned}
G(s,\sigma) &= \int_0^\infty \int_0^\infty \mathcal{G}(x,y)\,\exp(-x/s - y/\sigma)\,dx\,dy \\
&= \left(\frac{1}{s^2} + \frac{1}{\sigma^2}\right)^{-2}\left\{\frac{1}{4} + \frac{1}{8\,\pi}\,(R(s,\sigma) + R(\sigma,s))\right\},
\end{aligned}
\tag{2.4.105}
$$

where

$$
R(s,\sigma) = 2\,\log(s)\left(\frac{3\,s}{\sigma} + \frac{s^3}{\sigma^3}\right) + (7 - 6\,\gamma)\,\frac{s}{\sigma} + (3 - 2\,\gamma)\,\frac{s^3}{\sigma^3}.
\tag{2.4.106}
$$

Moreover,

$$
\begin{aligned}
G_x(s,\sigma) &= \frac{1}{s}\,G(s,\sigma) - \frac{\sigma^3}{8\,\pi}\,(2\,\log(\sigma) + 3 - 2\,\gamma), \\[2mm]
G_{xx}(s,\sigma) &= \frac{1}{s}\,G_x(s,\sigma) \\[2mm]
&= \frac{1}{s^2}\,G(s,\sigma) - \frac{\sigma^3}{8\,\pi\,s}\,(2\,\log(\sigma) + 3 - 2\,\gamma).
\end{aligned}
\tag{2.4.107}
$$

Note here that $\mathcal{G}_x$ is an odd function of x, (and similarly, $\mathcal{G}_y$ is an odd function of y) and this requires special care in the programming for the evaluation of the convolution of $\mathcal{G}_x$ (and similarly, $\mathcal{G}_y$) with another function. See the "biharm" programs listed in Chapter 5.

The Time Dependent Problem. A particular solution to the problem (2.4.88) is given by (see [Bk, p.152])

$$U(\bar{\rho}, t) = \int_B \int_0^t S(\bar{\rho} - \bar{\rho}', \, t - \tau) \, f(\bar{\rho}', \, \tau) \, d\tau \, d\bar{\rho}', \qquad (2.4.108)$$

where

$$S(x, y, t) \;=\; \frac{1}{4\pi ct} \sin\left(\frac{\rho^2}{4ct}\right)$$

$$\;=\; \Im\mathcal{E}(\bar{\rho}, t), \qquad (2.4.109)$$

and where

$$\mathcal{E}(\bar{\rho}, t) = \frac{1}{4\pi ct} \exp\left(i\,\frac{\rho^2}{4ct}\right). \qquad (2.4.110)$$

The following Theorem states conditions which yield the correct Laplace transform of S under the assumption that the principal value of the square root is used.

Theorem 2.4.13 *Let $\mathcal{G}(x, y, t)$ be defined as in Eq. (2.4.198), let $\Re\tau > 0$, $\Re s > 0$, $\Re\sigma > 0$, and let $|1/(c\tau)| > \sqrt{|1/s|^2 + |1/\sigma|^2}$. Then*

$$\int_0^\infty \int_0^\infty \int_0^\infty \exp\left(-\frac{x}{s} - \frac{y}{\sigma} - \frac{t}{\tau}\right) \mathcal{E}(\bar{\rho}, t) \, d\bar{\rho} \, dt = E(s, \sigma, \tau, i),$$

$$(2.4.111)$$

where, with $a = \pm i$,

$$E(s, \sigma, \tau, a)$$

$$= \frac{1}{c}\left(\frac{a}{c\tau} - \frac{1}{s^2} - \frac{1}{\sigma^2}\right)\left(\frac{1}{4} - H(s, \sigma, \tau, a) - H(\sigma, s, \tau, a)\right),$$

$$(2.4.112)$$

and where

$$H(s, \sigma, \tau, a) = \frac{1}{2 \, s \, \sqrt{\frac{a}{c\,\tau} - \frac{1}{\sigma^2}}}. \tag{2.4.113}$$

Proof. The result of this lemma is based, in essence on two integral formulas: the well known formula for the modified Bessel function, (see Problem 2.4.3)

$$\int_0^\infty \exp\left(-z\,t - \frac{\zeta}{t}\right) \frac{dt}{t} = 2\,K_0\left(2\,(z\,\zeta)^{1/2}\right), \tag{2.4.114}$$

which is valid for $\Re z > 0$, and $\Re\zeta \geq 0$ (or $\Re z \geq 0$, and $\Re\zeta > 0$), as well as the result of Theorem 2.4.7. We omit the details. ∎

Corollary 2.4.14 *Let $\mathcal{G}(x, y, t)$ be defined as in 2.4.105, and let G denote the Laplace transform of $\mathcal{G}$, i.e.,*

$$S(u, v, \tau) = \int_{\mathbb{R}_+^3} \exp\left(-\frac{x}{s} - \frac{y}{\sigma} - \frac{t}{\tau}\right) \mathcal{S}(\bar\rho, t) \, d\bar\rho \, dt, \tag{2.4.115}$$

let E be defined as in Theorem 2.4.13, and let $\mathcal{S}$ be defined as in (2.4.99)–(2.4.100). Then the Laplace transform of $\mathcal{S}$ is

$$S(s, \sigma, \tau)$$

$$= \frac{E(s, \sigma, \tau, i) - E(s, \sigma, \tau, -i)}{2\,i}. \tag{2.4.116}$$

Remark By comparing Equations (2.4.98) and the equation

$$V(\bar\rho, t) = \int \int_{\mathcal{B}} \int_0^t \mathcal{E}(\bar\rho - \bar\rho', t - \tau) \, f(\bar\rho', \tau) \, d\bar\rho' \, d\tau, \tag{2.4.117}$$

we can deduce that if f in these expressions is a real valued function, then this function U defined in (2.4.98) is also given by $U = \Im V$. Furthermore, this function V satisfies the equation

$$\frac{\partial V(\bar\rho, t)}{\partial t} - ic\,\nabla^2\,V(\bar\rho, t) = f(\bar\rho, t), \quad (\bar\rho, t) \in \mathcal{B} \times (0, T). \tag{2.4.118}$$

A word of warning: If $\mathcal{B}$ is a bounded domain and if f is a real valued function, then the function U defined by (2.4.98) is continuous on the closure of $\mathcal{B}$ even if

$$f(\bar{\rho}, t) = \mathcal{O}\left(|\bar{\rho} - \bar{c}|^{\alpha-4}\right)$$

as $\bar{r} \to c$, for any $\alpha > 0$, at a boundary point $\bar{c}$ of $\mathcal{B}$. On the other hand, this continuity property holds only for the case of the imaginary part of V; the resulting real part is continuous on the closure of $\mathcal{B}$ only if

$$f(\bar{r}, t) = \mathcal{O}\left(|\bar{r} - \bar{c}|^{\alpha-2}\right)$$

as $\bar{\rho} \to \bar{c}$, for any $\alpha > 0$, at a boundary point $\bar{c}$ of $\mathcal{B}$. On the other hand, this continuity property holds only for the case of the imaginary part of V; the resulting real part is continuous on the closure of $\mathcal{B}$ only if

$$f(\bar{\rho}, t) = \mathcal{O}\left(|\bar{\rho} - \bar{c}|^{\alpha-2}\right), \quad \bar{c} \in \partial\mathcal{B}$$

as $|bar\rho \to \bar{c}$, for any $\alpha > 0$.

That is, the procedure of using (2.4.118) in order to compute U by first computing V via (2.4.108) and then getting U using the expression $U = \Im V$ might yield inaccurate results if the real-valued function f is highly singular.

PROBLEMS FOR SECTION 2.4

2.4.1 *Laplace Transformed Heat Equation* Let $\varepsilon = \varepsilon(\bar{r})$ in (2.3.3), $\varepsilon = \varepsilon_0$ in (2.3.1), and assume that $\varepsilon(\bar{r}) = \varepsilon_0$ on $\mathbb{R}^d \setminus \mathcal{B}$, where $\mathcal{B}$ is a subregion of $\mathbb{R}^d$. Take the Laplace transform with respect to t of equations (2.3.1) and (2.3.3), find an explicit expression for the Laplace transform $\hat{G}$ of $\mathcal{G}$, and obtain an integral equation over $\mathcal{B}$ for the Laplace transform $\hat{u}$ of u.

2.4.2 (a) Devise a procedure for approximating the nine integrals in (2.4.86) via use of Midordinate integration (see (1.4.31)).

(b) Given a vector function $\mathbf{b}(\mathbf{x})$, devise an algorithm for approximating an integral over B of the form

$$\int_B \mathbf{G}(\mathbf{x} - \mathbf{y})\, \mathbf{b}(\mathbf{y})\, dy$$

via use of Sinc convolution, using your result for the (a)–Part of this problem, using the Laplace transform of $1/|\mathbf{x}|$, and combined with the methods of §1.5.12.

2.4.3 Let $\Re a > 0$, $\Re b \geq 0$, let $\Re c > 0$, and set

$$I(a, b, c) = \int_0^\infty t^{-c} \exp\left(-a\,t - \frac{b}{t}\right) dt.$$

Prove that

$$I(a, b, 1/2) = \sqrt{\tfrac{\pi}{a}}\,\exp\left(-2\sqrt{ab}\right);$$

$$I(a, b, 3/2) = I(b, a, 1/2)$$
$$I(a, b, 1) = 2\,K_0(2\sqrt{a\,b}),$$

where K_0 denotes the modified Bessel function or order 0 [Na].

2.4.4 Let $\bar{x} = |\bar{x}|\,\hat{\bar{x}}$ and $\bar{y} = |\bar{y}|\,\hat{\bar{y}}$ denote vectors in $\mathbb{R}^d$, with $\hat{\bar{x}}$ and $\hat{\bar{y}}$ denoting unit vectors, and let S denote the surface of the unit ball in $\mathbb{R}^d$. Prove that

$$\int_S \exp(i\,\bar{x}\cdot\bar{y})\,dA(\hat{\bar{y}}) = \begin{cases} 2\,\pi\,J_0(|\bar{x}|\,|\bar{y}|) & \text{if } n = 2, \\[2mm] 4\,\pi\,\dfrac{\sin(|\bar{x}|\,|\bar{y}|)}{|x|\,|y|} & \text{if } n = 3. \end{cases}$$

where J_0 denotes the Bessel function or order 0 [Na].

2.4.5 (a) Let $\mathcal{G}$ be defined as in (2.4.59), and prove that, with $\bar{y} = (y_1, y_2, y_3)$, the three dimensional Fourier transform of $\mathcal{G}$ is given by the identity

$$\exp\left(-\varepsilon\,t\,|\bar{y}|^2\right) = \int_{\mathbb{R}^3} \exp(i\,\bar{y}\cdot\bar{r})\,\mathcal{G}(\bar{r}, t)\,d\bar{r}.$$

(b) Determine the Fourier transform $\hat{K}(\bar{y}, t)$ of $\mathcal{K}(\bar{r}, t)$ with respect to $\bar{r}$. [Ans.: $|\bar{y}|^{-2}\,\exp\left(-\varepsilon\,t\,|\bar{y}|^2\right)$.]

(c) Use the answer in the (b)–Part of this problem to determine an explicit expression for the Green's function $K(\bar{r}, t)$.

2.4.6 Let α and d denote positive constants, let $\mathbf{L}_{\alpha,d}(\mathrm{id})$ be defined as in Definition 1.3.3, and assume that $f = f(x,y,z)$ is a function defined on $\mathbb{R}^3$, such that:

  – $f(\cdot,y,z) \in \mathbf{L}_{\alpha,d}(\mathrm{id})$ uniformly, for $(y,z) \in \mathbb{R}^2$;
  – $f(x,\cdot,z) \in \mathbf{L}_{\alpha,d}(\mathrm{id})$ uniformly, for $(x,z) \in \mathbb{R}^2$; and
  – $f(x,y,\cdot) \in \mathbf{L}_{\alpha,d}(\mathrm{id})$ uniformly, for $(x,y) \in \mathbb{R}^2$.

Let $\hat{f}$ denote the three dimensional Fourier transform of f. Define U by

$$U(\bar{r}) = \int_{\mathbb{R}^3} \mathcal{G}(\bar{r} - \bar{r}') \, f(\bar{r}') \, d\bar{r}'.$$

Set $\bar{k} = (k_1, k_2, k_3)$, and prove that if h is selected corresponding to any positive integer N, by $h = (\pi\, d/(\alpha\, N))^{1/2}$, then there exists a positive constant C that is independent of N, such that

$$\left| U(\bar{r}) - \frac{h^3}{8\pi^3} \sum_{|k_i| \le N} \exp\left(-\varepsilon\, t\, h\, |\bar{k}|^2\right) \exp(-i\bar{r} \cdot \bar{k})\, \hat{f}(h\,\bar{k}) \right|$$

$$\le C \exp\left(-(\pi\, d\, \alpha\, N)^{1/2}\right),$$

uniformly, for $\bar{r} \in \mathbb{R}^3$.

2.4.7 (i) Let a, b, and c be arbitrary complex constants. Prove that the solution to the ODE initial value problem

$$\frac{d^2 y}{(d x)^2} + 2\frac{d y}{d x} + c y = 0,$$

$$y(0) = a, \quad y'(0) = b,$$

is given by

$$y(x) = e^{-x}\left(a\,\cosh\left(x\sqrt{1-c}\right) + (b-a)\,\frac{\sinh\left(x\sqrt{1-c}\right)}{\sqrt{1-c}}\right).$$

(ii) Discuss what happens as c changes continually along a straight line in $\mathbb{C}$ through the point $c = 1$.

2.4.8 (i) In the notation of Definition 1.5.2, let let us write $z << \zeta$ whenever z and ζ are points on Γ which one meets in the order a, z, ζ, b as one traverses Γ from a to b. Let $\mathcal{G}$ be defined by

$$\mathcal{G}(z, \zeta) = \begin{cases} \dfrac{1}{2}(\zeta - z) & \text{if } \zeta << z, \\ \dfrac{1}{2}(z - \zeta) & \text{if } z << \zeta. \end{cases}$$

Prove that the function

$$u(z) = \int_\Gamma \mathcal{G}(z, \zeta) f(\zeta) \, d\zeta$$

satisfies on Γ the equation

$$\frac{d^2 u}{d z^2} = -f.$$

2.4.9 Let $\mathcal{G}_0^{(3)}(\bar{\Lambda})$ and $\mathcal{G}^{(3)}(\bar{\Lambda}, k)$ be defined as in Theorems 2.4.4 and 2.4.8 respectively. Prove that $\lim_{k \to 0} \mathcal{G}^{(3)}(\bar{\Lambda}, k) = \mathcal{G}_0^{(3)}(\bar{\Lambda})$.

2.5 Multi–Dimensional Convolution Based on Sinc

In this section we extend the one dimensional convolution method to the approximation of convolution integrals in more than one dimension. To this end, one dimensional convolution extends directly to more than one dimension for the case of rectangular regions. However, we shall also derive such extensions for approximating multidimensional convolution integrals over curvilinear regions.

2.5.1 RECTANGULAR REGION IN $2 - d$

Let us illustrate the approximation of a two dimensional convolution integral of the form

$$p(x, y) = \int_{a_2}^{y} \int_{x}^{b1} f(x - \xi, \eta - y) \, g(\xi, \eta) \, d\xi d\eta, \tag{2.5.1}$$

where we seek an approximation over the region $\mathcal{B} = (a_1, b_1) \times (a_2, b_2)$, and with $(a_\ell, b_\ell) \subseteq \mathbb{R}$, $\ell = 1, 2$. We assume that the mappings $\varphi_\ell : \Gamma_\ell = (a_\ell, b_\ell) \to \mathbb{R}$ have been determined. We furthermore assume that positive integers M_ℓ, N_ℓ as well as positive numbers h_ℓ have been selected, and we set $m_\ell = M_\ell + N_\ell + 1$. We then define the Sinc points by $z_j^{(\ell)} = \left(\varphi^{(\ell)}\right)^{-1}(j\, h_\ell)$, for $j = -M_\ell, \ldots, N_\ell$. Next, we determine matrices A_ℓ, X_ℓ, S_ℓ and X_ℓ^{-1}, such that

$$
\begin{aligned}
A_1 &= h_1 \left(I_{m_1}^{(-1)}\right)^T D(1/\phi_1') &= X_1 S_1 X_1^{-1}, \\
A_2 &= h_2 I_{m_2}^{(-1)} D(1/\phi_2') &= X_2 S_2 X_2^{-1}.
\end{aligned}
\tag{2.5.2}
$$

In (2.5.2), the matrix D is defined as in §1.5.4, while the matrices I_{m_ℓ}, A_ℓ are defined as in §1.5.8, and the matrices S_ℓ are diagonal matrices,

$$
S_\ell = \operatorname{diag}\left[s_{-M_\ell}^{(\ell)}, \ldots, s_{N_\ell}^{(\ell)}\right].
\tag{2.5.3}
$$

We require the two dimensional Laplace transform

$$
F(s^{(1)}, s^{(2)}) = \int_{E_1} \int_{E_2} f(x, y)\, e^{-x/s^{(1)} - y/s^{(2)}}\, dx\, dy,
\tag{2.5.4}
$$

which we assume to exist for all $s^{(\ell)} \in \Omega_+$, with Ω_+ denoting the open right half plane, and with E_ℓ any subset of $\mathbb{R}$ such that $(0, b_\ell - a_\ell) \subseteq E_\ell \subseteq \mathbb{R}_+ = (0, \infty)$. We remark here that it is frequently possible to just take $E_\ell = \mathbb{R}_+$, in which case we might be able to look up the requisite Laplace transform in a table, and then get the Laplace transform by simply replacing s in the tabulated Laplace transform by $1/s$. At times this will not be possible, when one or both of the intervals (a_ℓ, b_ℓ) is necessarily finite, e.g., when $f(x, y) = \exp\left(x^2 + 2\, y^3\right)$.

It is then possible to show (see [S1, Sec. 4.6]) that the values $p_{i,j}$ which approximate $p(z_i^{(1)}, z_j^{(2)})$ can be computed via the following succinct algorithm. In this algorithm the we use the notation, e.g., $\mathbf{h}_{i,\cdot} = (h_{i,-M_2}, \ldots, h_{i,N_2})^T$, and similarly for $\mathbf{h}_{\cdot,j}$. We may note in passing, the obvious ease of adaptation of this algorithm to parallel computation.

Algorithm 2.5.1 Convolution over $2-d$ Rectangle

1. Form the arrays $z_j^{(\ell)}$, and $\frac{d}{dx}\varphi^{(\ell)}(x)$ at $x = z_j^{(\ell)}$ for $\ell = 1, 2$, and $j = -M_\ell, \ldots, N_\ell$, and then form the rectangular matrix block of numbers $[g_{i,j}] = [g(z_i^{(1)}, z_j^{(2)})]$.

2. Determine A_ℓ, S_ℓ, X_ℓ, and X_ℓ^{-1} for $\ell = 1, 2$, as defined in (2.5.2).

3. Form $h_{\cdot,j} = X_1^{-1} g_{\cdot,j}$, $j = -M_2, \ldots, N_2$;

4. Form $k_{i,\cdot} = X_2^{-1} h_{i,\cdot}$, $i = -M_1, \ldots, N_1$;

5. Form $r_{i,j} = F(s_i^{(1)}, s_j^{(2)}) k_{i,j}$,
   $i = -M_1, \ldots, N_1$, $j = -M_2, \ldots, N_2$;

6. Form $q_{i,\cdot} = X_2 r_{i,\cdot}$, $i = -M_1, \ldots, N_1$;

7. Form $p_{\cdot,j} = X_1 q_{\cdot,j}$, $j = -M_2, \ldots, N_2$.

Remark 2.5.1   1. The integral expression (2.5.1) for $p(x, y)$ is a repeated integral, where we first integrate with respect to ξ and then with respect to η. Algorithm 2.5.1 is obtained by applying the one dimensional convolution procedure to the inner integral, with y and η fixed, and then again applying the one dimensional procedure to the remaining integral. A different algorithm results by reversing the order of integration, although numerically the results are virtually the same (see Problem 2.5.2).

2. This algorithm follows directly from the one–dimensional procedure of §1.5.9 (see [S1, §4.6] or [S7]).

3. If we conveniently set $[p_{i,j}] = P$, $X_1 = X$, $X^{-1} = Xi$, $X_2 = Y$, $Y^{-1} = Yi$, $[g_{i,j}] = G$, and $\left[F\left(s_i^{(1)}, s_j^{(2)}\right)\right] = F$ then the steps 4. to 8. in this algorithm can be written as a one–line MATLAB® statement, as follows:

$$P = X * (F. * (Xi * G * Yi.')) * Y.'; \qquad (2.5.5)$$

4. Once the entries $p_{i,j}$ of $P = [p_{ij}]$ have been computed, we can use these numbers to approximate p on the region $\mathcal{B}$ via Sinc basis approximation. To this end, we set $\rho^{(\ell)} = e^{\phi^{(\ell)}}$, which enables us to define the functions (see Definitions 1.5.2 and 1.5.12)

$$
\begin{aligned}
\gamma_i^{(\ell)} &= \operatorname{sinc}\{[\phi^{(\ell)} - ih]/h\}, \quad \ell = 1, 2; \quad i = -N_\ell, \dots, N_\ell, \\
w_i^{(\ell)} &= \gamma_i^{(\ell)}, \quad \ell = 1, 2; \quad i = -N_\ell + 1, \dots, N_\ell - 1, \\
w_{-M_\ell}^{(\ell)} &= \frac{1}{1 + \rho^{(\ell)}} - \sum_{j=-M_\ell+1}^{N_\ell} \frac{1}{1 + e^{jh_\ell}} \gamma_j^{(\ell)}, \\
w_{N_\ell}^{(\ell)} &= \frac{\rho^{(\ell)}}{1 + \rho^{(\ell)}} - \sum_{j=-M_\ell}^{N_\ell-1} \frac{e^{jh_\ell}}{1 + e^{jh_\ell}} \gamma_j^{(\ell)}.
\end{aligned}
$$

$$(2.5.6)$$

Then

$$
p(x, y) \approx \sum_{i=-M_1}^{N_1} \sum_{j=-M_2}^{N_2} p_{i,j}\, w_i^{(1)}(x) w_j^{(2)}(y). \qquad (2.5.7)
$$

This last equation can also be written in a symbolic notation via use of our previous definitions of §1.5.4. Upon setting $w^{(j)} = \left(w_{-M_j}^{(j)}, \dots, w_{N_j}^{(j)} \right)$, for $j = 1, 2$, and letting P defined as above, we can write

$$
p(x, y) = w^{(1)}(x)\, P \left(w^{(2)}(y) \right)^T. \qquad (2.5.8)
$$

Let us briefly discuss the complexity of the above procedure. We can easily deduce that if the above two dimensional Laplace transform F is either known explicitly, or if the evaluation of this transform can be reduced to the evaluation of a one–dimensional integral, and if:

$p(\cdot, y)$ is analytic and of class $\mathbf{Lip}_\alpha$ on (a_1, b_1) for all fixed $y \in (a_2, b_2)$, and

$p(x, \cdot)$ is analytic and of class $\mathbf{Lip}_\alpha$ on (a_2, b_2) for all fixed $x \in (a_1, b_1)$, then the complexity, i.e., the total amount of work required to achieve an error ε when carrying out the computations of the above algorithm (to approximate $p(x,y)$ at $(2N+1)^2$ points) on a sequential machine, is $\mathcal{O}([\log(\varepsilon)]^6)$.

It is easily seen that the above algorithm extends readily to d dimensions, in which case, under similar conditions on the functions, the complexity for evaluating a d–dimensional convolution integral over a rectangular region (at $(2N+1)^d$ points) to within an error of ε is of the order of $[\log(\varepsilon)]^{2d+2}$.

2.5.2 RECTANGULAR REGION IN $3-d$

We illustrate here an explicit procedure for approximating a three dimensional convolution integral of the form

$$p(x_1, x_2, x_3) =$$

$$\int_{a_3}^{x_3} \int_{x_2}^{b_2} \int_{a_1}^{x_1} f(x_1 - \xi_1, \xi_2 - x_2, x_3 - \xi_3)\, g(\xi_1, \xi_2, \xi_3)\, d\xi_1 d\xi_2 d\xi_3 ,$$

$$(2.5.9)$$

where the approximation is sought over the region $\mathbf{B} = \Pi_{\ell=1}^3 \times (a_\ell, b_\ell)$, and with $(a_\ell, b_\ell) \subseteq \mathbb{R}$. We select M_ℓ, N_ℓ, and h_ℓ, $\ell = 1, 2, 3$, set $m_\ell = M_\ell + N_\ell + 1$, specify the Sinc points by $z_j^{(\ell)} = \left(\phi^{(\ell)}\right)^{-1}(jh_\ell)$, for $j = -M_\ell, \cdots, N_\ell$. Next, we determine matrices A_ℓ, X_ℓ and S_ℓ,

$$
\begin{aligned}
A_\ell &= h_\ell I_{m_\ell}^{(-1)} D(1/\phi'_\ell) &= X_\ell S_\ell X_\ell^{-1}, \quad \ell = 1, 3 \\
A_2 &= h_2 \left(I_{m_2}^{(-1)}\right)^T D(1/\phi'_2) &= X_2 S_2 X_2^{-1}.
\end{aligned}
\tag{2.5.10}
$$

Here, $I_{m_\ell}^{(-1)}$ is defined as in the previous subsection, and the S_ℓ are diagonal matrices,

$$S_\ell = \operatorname{diag}[s_{-M_\ell}^{(\ell)}, \cdots, s_{N_\ell}^{(\ell)}]. \tag{2.5.11}$$

Arbitrarily taking $c_\ell \in [(b_\ell - a_\ell), \infty]$, we define the Laplace transform,

$$H(s^{(1)}, s^{(2)}, s^{(3)})$$

$$= \int_0^{c_1} \int_0^{c_2} \int_0^{c_3} f(x, y, z) \, \exp(-x/s^{(1)} - y/s^{(2)} - z/s^{(3)}) \, dz \, dy \, dx.$$

$$(2.5.12)$$

Algorithm 2.5.2 Convolution over $3 - d$ Rectangle

1. Form the array $z_i^{(\ell)}$, and $\frac{d}{dx}\phi^{(\ell)}(x)$ at $x = z_i^{(\ell)}$ for $\ell = 1, 2, 3$, and $i = -M_\ell, \cdots, N_\ell$, and then form the array

$$[g_{i,j,k}] = [g(z_i^{(1)}, z_j^{(2)}, z_k^{(3)})].$$

2. Determine A_ℓ, S_ℓ, and X_ℓ, for $\ell = 1, 2, 3$, as defined in (2.5.10).

3. Form $\mathbf{h}_{\cdot,j,k} = X_1^{-1} \mathbf{g}_{\cdot,j,k}$;

4. Form $\mathbf{k}_{i,\cdot,k} = X_2^{-1} \mathbf{h}_{i,\cdot,k}$;

5. Form $\mathbf{m}_{i,j,\cdot} = X_3^{-1} \mathbf{k}_{i,j,\cdot}$;

6. Form $t_{i,j,k} = H(s_i^{(1)}, s_j^{(2)}, s_k^{(3)}) \, m_{i,j,k}$;

7. Form $\mathbf{r}_{i,j,\cdot} = X_3 \mathbf{t}_{i,j,\cdot}$;

8. Form $\mathbf{q}_{i,\cdot,k} = X_2 \mathbf{r}_{i,\cdot,k}$;

9. Form $\mathbf{p}_{\cdot,j,k} = X_1 \mathbf{q}_{\cdot,j,k}$.

2.5.3   CURVILINEAR REGION IN $2 - d$

A big asset of finite element methods is that they readily enable the solution of PDE over curvilinear regions. Here we derive a procedure for approximating a convolution integral over a curvilinear region which will enable us to solve PDE over curvilinear regions via Sinc methods. Following this derivation, we also give an explicit version of an analogous three dimensional algorithm.

As witnessed in §2.5.1 and §2.5.2, multidimensional Laplace transforms of Green's functions enable very efficient procedures for evaluating convolutions of Green's function integrals over rectangular regions. And although the transformation (2.5.17) below (which is a transform the square $(0,1)^2$) to the region defined in (2.5.14)) destroys a convolution, we were fortunate that the more general one dimensional convolutions of §1.5.11 can be applied. That is, we can approximate the transformed integral by *using the Laplace transform of the original free space Green's function.*

a. The General Case. We discuss here approximation of the integral

$$p(x,y) = \int \int_{\mathcal{B}} f(x - x', y - y', x', y') \, dx' \, dy'. \qquad (2.5.13)$$

Here $\mathcal{B}$ is the region

$$\mathcal{B} = \{(x, y) : a_1 < x < b_1, \; a_2(x) < y < b_2(x)\} \qquad (2.5.14)$$

where for purposes of this illustration a_1 and b_1 are finite constants and a_2 and b_2 are analytic arcs defined on (a_1, b_1) (See §2.6 below.).

One procedure for solving a PDE over $\mathcal{B}$ is to first transform and then solve, the PDE into another one over $(0,1) \times (0,1)$. This indeed has been done in [Pa], via Sinc collocation of the derivatives of the PDE. While such a procedure is accurate, it nevertheless leads to large matrices. However, Sinc convolution enables us to avoid large matrices. We mention, when the PDE over $\mathcal{B}$ is solved on $\mathcal{B}$ via a convolution integral expression involving Green's functions, we must exercise skill in order to be able to use the same Laplace transforms of the Green's functions after making the transformation of the region (2.4.14) into a rectangular one.

We first split the integral representing p into four integrals:

$$p = r_1 + r_2 + r_3 + r_4 \qquad (2.5.15)$$

with

$$r_1 = \int_{a_1}^{x} \int_{a_2(x')}^{y} f(x - x', \, x', \, y - y', \, y') \, dy' \, dx',$$

$$r_2 = \int_{x}^{b_1} \int_{a_2(x')}^{y} f(x - x', \, x', \, y - y', \, y') \, dy' \, dx',$$

$$(2.5.16)$$

$$r_3 = \int_{a_1}^{x} \int_{y}^{b_2(x')} f(x - x', \, x', \, y - y', \, y') \, dy' \, dx',$$

$$r_4 = \int_{x}^{b_1} \int_{y}^{b_2(x')} f(x - x', \, x', \, y - y', \, y') \, dy' \, dx'.$$

Each of these can be treated in exactly the same manner, and we shall therefore consider only the first of these, namely, $r_1(x, y)$.

The integral representing r_1 over $\mathcal{B}$ can be readily be transformed into an integral over the unit square by means of the transformation

$$\begin{aligned} x' &= a_1 + (b_1 - a_1)\,\xi' \\ y' &= a_2(x') + (b_2(x') - a_2(x'))\,\eta'. \end{aligned}$$

$$(2.5.17)$$

In view of this transformation, we now set

$$\begin{aligned} \beta &= b_1 - a_1, \\ \gamma &= \gamma(\xi') = b_2(x') - a_2(x'). \end{aligned}$$

$$(2.5.18)$$

This transformation enables us to replace the above integral r_1 by

$$P(\xi, \eta) =$$

$$\int_{0}^{\xi} \int_{0}^{\eta} f(\beta\,(\xi - \xi'), \, a_1 + \beta\,\xi', \, \gamma\,(\eta - \eta'), \, a_2(x') + \gamma\,\eta')\,\beta\,\gamma\,d\eta'\,d\xi'.$$

$$(2.5.19)$$

It is convenient, for purposes of explaining the details of the procedure, to use the notations

$$\tilde{f}(x, x', \sigma, y') = \int_{0}^{c_2} f(x, x', y, y')\,e^{-y/\sigma}\,dy$$

$$(2.5.20)$$

$$\hat{f}(s, x', \sigma, y') = \int_{0}^{c_1} \tilde{f}(x, x', \sigma, y')\,e^{-x/s}\,dx,$$

where $c_1 \geq \beta$ and where $c_2 \geq \max_{\xi \in [0,1]} |\gamma(\xi)|$.

We now approximate the repeated integral P by first performing the integration with respect to η (for which we show only the dependence on η, the remaining variables being fixed). Thus, we set

$$p(\eta) \equiv \int_0^\eta f(\beta(\xi - \xi'), \ \beta\xi', \ \gamma(\eta - \eta'), \ \gamma\eta') \beta\gamma \, d\eta'. \quad (2.5.21)$$

This integral can be approximated in exactly the same manner as the integral p of (1.5.61). To this end, we first require the following Laplace transform,

$$\int_0^{c_2'} f(\beta\xi, a_1 + \beta\xi', \gamma\eta, a_2(x') + \gamma\eta') \beta\gamma \, e^{-\eta/\sigma} \, d\eta$$
$$= \int_0^{c_2} f(\beta\xi, \beta\xi', \ y, a_2(x') + \gamma\eta') \gamma \, e^{-y/(\gamma\sigma)} \, dy \quad (2.5.22)$$

$$\equiv \tilde{f}(\beta\xi, \ a_1 + \beta\xi', \ \gamma\sigma, \ a_2(x') + \gamma\eta') \beta.$$

where in the above, $c_2' \geq c_2 / \max_{\xi \in [0,1]} |\gamma(\xi)|$.

Then, denoting $B = A_2$ (as in (2.5.10) above) to be the Sinc indefinite integration matrix over $(0,1)$ with respect to η, diagonalizing $B = X_2 \, S_2 \, X_2^{-1} \equiv Y \, T \, Y^{-1}$, where $T = S_2$ is the diagonal matrix,

$$T = \text{diag}\left[\sigma_{-M_2}, \ \ldots \ldots \ \sigma_{N_2}\right],$$

and then with $Y = [y_{ij}]$, $Y^{-1} = [y^{ij}]$, and using the last equation in (1.5.64), we get

$$p(\eta_n) = \sum_{k=-M_2}^{N_2} y_{n,k} \cdot$$

$$\cdot \sum_{\ell=-M_2}^{N_2} y^{k,\ell} \, \tilde{f}(\beta(\xi - \xi'), \ a_1 + \beta\xi', \ a_2(x') + \gamma\sigma_k, \ \gamma\eta_\ell) \beta.$$
$$(2.5.23)$$

Next, integrating p with respect to ξ' from 0 to ξ, letting $A = A_1$ denote the Sinc indefinite integration matrix of order $m_1 = M_1 + N_1 + 1$ over $(0,1)$ with respect to ξ, and diagonalizing $A = X_1 \, S_1 \, X_1^{-1} \equiv X \, S \, X^{-1}$, with $X = [x_{ij}]$, $X^{-1} = [x^{ij}]$, and

$S = \text{diag}(s_{-M_1} \ \cdots , \ s_{N_1})$, and again applying the last equation in (1.5.64), we get

$$P(\xi_m , \ \eta_n) \approx q_{mn} \equiv \sum_{i=-M_1}^{N_1} x_{mi} \sum_{j=M_1}^{N_1} x^{ij} \sum_{k=-M_2}^{N_2} y_{n,k} \sum_{\ell=-M_2}^{N_2} y^{k\ell} \cdot$$

$$\cdot \hat{f}(\beta \, s_i , \ \beta \, \xi_j , \ \gamma(\xi_j) \, \sigma_k , \ \gamma(\xi_j) \, \eta_\ell) \, \beta \, \gamma(\xi_j).$$

(2.5.24)

Algorithm 2.5.3 Nonlinear $2 - d - P$ Convolution

1. Determine parameters M_ℓ, N_ℓ, h_ℓ, set $m_\ell = M_\ell + N_\ell + 1$, set $J_\ell = \{-M_\ell \ \cdots , \ N_\ell\}$, $\ell = 1, 2$, determine Sinc points $\{\xi_j\}_{j=-M_1}^{N_1}$, and $\{\eta_j\}_{j=-M_2}^{N_2}$, and function values $\beta_j = \beta(\xi_j)$, with $j = -M_1 , \ \ldots , \ N_1$.

2. Form the integration matrices

$$
\begin{aligned}
A &= h_1 \, I^{(-1)} \, D\left(e^w / (1 + e^w)^2\right); \quad w = k \, h_1 , k \in J_1 , \\
&= X \, S \, Xi; \quad X = [x_{ij}], \quad Xi = X^{-1} = [x^{ij}] \\
S &= \text{diag}\left(s_{-M_1} , \ \ldots , \ s_{N_1}\right), \\
B &= h_2 \, I^{(-1)} \, D\left(\frac{e^w}{(1 + e^w)^2}\right); \quad w = k \, h_2 , k \in J_2 , \\
&= Y \, T \, Yi, \quad Y = [y_{ij}], \quad Yi = Y^{-1} = [y^{ij}], \\
T &= \text{diag}\left(\sigma_{-M_2} , \ \ldots , \ \sigma_{N_2}\right).
\end{aligned}
$$

(2.5.25)

3. Assuming an explicit expression for $\hat{f}$ defined in (2.5.20) above, form the array $\{K(j, i, k, \ell)\}$, with

$K(j, i, k, \ell)$

$= \hat{f}(\beta \, s_i , a_1 + \beta \, \xi_j , \ \gamma(\xi_j) \, \sigma_k , a_2(a_1 + \beta \, \xi_j , \ \gamma(\xi_j) \, \eta_\ell) \, \beta \, \gamma(\xi_j).$

where $(i, j) \in J_1 \times J_1$ and $(k, \ell) \in J_2 \times J_2$.

Remark. Setting $\nu = \max(m_1, m_2)$ this step requires $\mathcal{O}(\nu^4)$ operations.

4. Set $P(j, i, k) = K(j, i, k, :) \, (Y i(k, :))^T$;

5. Set $L(i, k) = X(i, :) \, P(:, i, k)$;

6. Set $Q = [q_{m,n}] = X \, L \, Y^T$.

Remark. At this point the entries $q(m, n)$ approximate the values $P(\xi_m, \eta_n)$, i.e.,

$$P(\xi_m, \eta_n) \approx q(m, n). \qquad (2.5.26)$$

Note, also, that under assumptions of analyticity of function f in each variable, with the remaining variables held fixed and real, the complexity of evaluating the above "general" two dimensional convolution integral (2.5.13) for p over the curvilinear region $\mathcal{B}$ of (2.5.14) at $(2N+1)^2$ points to within an error of ε is of the order of $(\log(\varepsilon))^8$.

b. A Specific Case. Let us now also consider the special, practical, more important case of

$$r(x, y) = \int \int_{\mathcal{B}} \mathcal{G}(x - x', \, y - y') \, g(x', \, y') \, dy' \, dx', \qquad (2.5.27)$$

where $\mathcal{B}$ is again defined as in (2.5.14). The Sinc approximation of this practically important case requires less work than that of $p(x, y)$ above.

In this case, again splitting the integral as in (2.5.15)–(2.5.16), we consider only the transformations (2.5.17)–(2.5.18) applied to the integral r_1 from (2.5.16), and arrive at the integral

$$R(\xi, \eta) =$$
$$\qquad (2.5.28)$$
$$\int_0^\xi \int_0^\eta \mathcal{G}(\beta \, (\xi - \xi'), \gamma \, (\eta - \eta')) \, g(x', y') \, \beta \, \gamma \, d\eta' \, d\xi',$$

where, by (2.5.18), $x' = a_1 + \beta \xi'$ and $y' = a_2(x') + \gamma \eta'$. Instead of considering each of the equations (2.5.20) separately, we can consolidate into a single step, and write

$$\hat{G}(s, \sigma) = \int_0^\infty \int_0^\infty \mathcal{G}(x, y) \, e^{-x/s - y/\sigma} \, dy \, dx, \qquad (2.5.29)$$

where for convenience of notation, we have chosen the upper limits in the Laplace transform to be infinite, although finite limits are required for some functions $\mathcal{G}$.

The notation of (2.5.29) then yields

$$\int_0^\infty \int_0^\infty \mathcal{G}(\beta\,\xi\,,\ \gamma\,\eta)\, e^{-\xi/s - \eta/\sigma}\beta\,\gamma\,d\xi\,d\eta = \hat{G}(\beta\,s,\ \gamma\,\sigma). \quad (2.5.30)$$

Proceeding like in (2.5.13) above, we get

$$R(\xi_m\,,\eta_n)$$

$$\approx r_{mn} \equiv \sum_{i=-M_1}^{N_1} x_{mi} \sum_{j=M_1}^{N_1} x^{ij} \sum_{k=-M_2}^{N_2} y_{n,k} \sum_{\ell=-M_2}^{N_2} y^{k\ell}\,\cdot \quad (2.5.31)$$

$$\cdot\,\hat{G}(\beta\,s_i\,,\ \gamma(\xi_j)\,\sigma_k)\,g(\beta\,\xi_j\,,\ \gamma(\xi_j)\,\eta_\ell).$$

Algorithm 2.5.4 Nonlinear $2-d-R$ Convolution

1. Determine parameters M_ℓ, N_ℓ, h_ℓ, set $m_\ell = M_\ell + N_\ell + 1$, $\ell = 1, 2$, determine Sinc points $\{\xi_j\}_{j=-M_1}^{N_1}$, and $\{\eta_k\}_{k=-M_2}^{N_2}$, and values $\gamma_j = \gamma(\xi_j)$, $j = -M_1\,,\ \ldots\,,\ N_1$.

2. Form the integration matrices

$$
\begin{aligned}
A &= I^{(-1)}\,D\left(\frac{h_1\,e^w}{(1+e^w)^2}\right); \ w = k\,h_1\,, k \in [-M_1\,, N_1]\,,\\
&= X\,S\,Xi; \quad X = [x_{ij}], \quad Xi = X^{-1} = [x^{ij}]\\
S &= \mathrm{diag}\,(s_{-M_1}\,,\ \ldots\,,\ s_{N_1})\,,\\
B &= I^{(-1)}\,D\left(\frac{h_2\,e^w}{(1+e^w)^2}\right); \ w = k\,h_2\,, k \in [-M_2\,, N_2]\,,\\
&= Y\,T\,Y^{-1}, \quad Y = [y_{ij}], \quad Yi = Y^{-1} = [y^{ij}]\,,\\
T &= \mathrm{diag}\,(\sigma_{-M_2}\,,\ \ldots\,,\ \sigma_{N_2})\,.
\end{aligned}
$$

$$(2.5.32)$$

3. Assuming that $\hat{G}$ in (2.5.29) is explicitly defined, form the arrays $\left\{G^{[2]}(i, j, k)\right\}$, and $g^{[2]}(j, \ell)$, where

$$G^{[2]}(i,j,k) = \hat{G}(\beta\, s_i,\; \gamma_j\, \sigma_k)$$
$$g^{[2]}(j,\ell) = g\,(a_1 + \beta\,\xi_j,\; a_2(a_1 + \beta\xi_j) + \gamma_j\,\eta_\ell)\,.$$

Note: Setting up the array $G^{[2]}(i,j,k)$ is an $\mathcal{O}(\nu^3)$ operation, for the case of $\nu = \max(m_1,\; m_2,\; m_3)$.

4. Eliminate the index "ℓ", via the tensor product taken with respect to ℓ,

$$h^{[2]}(j,k)) = g^{(2)}(j,\ell)\,(Y\,i(k,\ell))\,;$$

5. Set $H^{[2]}(j,i,k) = G[2](i,j,k)\,h^{[2]}(j,k)$ for all (i,j,k). (This is a "Hadamard product" calculation which requires $\mathcal{O}(\nu^3)$ operations.)

6. Eliminate the index "j" by performing the following tensor product with respect to j, i.e., setting
$K(i,k) = X\,i(i,j)\,H^{[2]}(j,i,k)$.

7. Set $\mathbf{r} = X\,K\,Y^T$, with $\mathbf{r} = [r_{mn}]$.

Remark. The entries r_{mn} computed via this algorithm approximate the values $R(\xi_m,\eta_n)$ with $R(\xi,\eta)$ defined as in (2.5.28).

2.5.4   CURVILINEAR REGION IN $3-d$

We now describe an analogous procedure for approximating three dimensional indefinite convolution integrals of the form

$$p(x,y,z) = \int\int\int_{\mathcal{B}} f(x-x',\; x',\; y-y',\; y',\; z-z',\; z')\,dx'\,dy'\;dz'.$$
$$(2.5.33)$$

and also, the more special but practically more important integral,

$$q(x,y,z) = \int\int\int_{\mathcal{B}} f(x-x',y-y',z-z')\,g(x',y',z')\,dx'\,dy'\,dz'.$$
$$(2.5.34)$$

We shall assume that the region $\mathcal{B}$ takes the form

$$\mathcal{B} = \{(x, y, z) :$$

$$a_1 < x < b_1, \; a_2(x) < y < b_2(x), \; a_3(x, y) < z < b_3(x, y)\} \,.$$
(2.5.35)

a. *Approximation of* $p(x, y, z)$.

The definite convolution integral $p(x, y, z)$ can be split up into 8 indefinite convolution integrals, the approximation of each of which is, carried out in the same procedure, except for use of different indefinite integration matrices. We now give an explicit algorithm for approximating one of these, namely, that of the integral

$$p_1(x, y, z) =$$

$$\int_{a_1}^{x} \int_{a_2(x')}^{y} \int_{a_3(x', y')}^{z} f(x - x', x', y - y', y', z - z', z') \, dz' \, dy' \, dx'.$$
(2.5.36)

The region $\mathcal{B}$ can be readily be transformed to the unit cube by means of the transformation

$$
\begin{aligned}
x' &= a_1 + [b_1 - a_1]\,\xi' \\
y' &= a_2(x') + [b_2(x') - a_2(x')]\,\eta', \\
z' &= a_3(x', y') + [b_3(x', y') - a_3(x', y')]\,\zeta'.
\end{aligned}
$$
(2.5.37)

Once again, we use the more compact notation

$$
\begin{aligned}
\beta &= b_1 - a_1 \\
\gamma &= \gamma(\xi') = b_2(x') - a_2(x') \\
\delta &= \delta(\xi', \eta') = b_3(x', y') - a_3(x', y').
\end{aligned}
$$
(2.5.38)

by which the integral (2.5.36) becomes

$$p_1(x, y, z) \equiv P(\xi, \eta, \zeta)$$

$$P(\xi, \eta, \zeta) = \int_{0}^{\xi} \int_{0}^{\eta} \int_{0}^{\zeta} f(\beta(\xi - \xi'), \; a_1 + \beta\,\xi', \; \gamma(\eta - \eta'), \; \cdot$$
(2.5.39)

$$\cdot \; a_2 + \gamma\,\eta', \; \delta(\zeta - \zeta'), \; a_3 + \delta\,\zeta')\,\beta\,\gamma\,\delta \, d\zeta' \, d\eta' \, d\xi'.$$

Let us now set[3]

$$\hat{f}(r, x', s, y', t, z')$$

$$= \int_0^\infty \int_0^\infty \int_0^\infty f(x, x', y, y', z, z') \exp\left(-\frac{x}{r} - \frac{y}{s} - \frac{z}{t}\right) dz\, dy\, dx.$$

$$(2.5.40)$$

Then

$$\int_0^\infty \int_0^\infty \int_0^\infty \exp(-\xi/r - \eta/s - \zeta/t) \cdot$$

$$\cdot f(\beta\,\xi, a_1 + \beta\xi', \gamma\,\eta, a_2 + \gamma\,\eta', \delta\,\zeta, a_3 + \delta\,\zeta')\, \beta\,\gamma\,\delta\, d\zeta\, d\eta\, d\xi$$

$$= \hat{f}(\beta\, r, a_1 + \beta\, x', \gamma\, s,\ a_2 + \gamma\, y', \delta\, t,\ a_3 + \delta\, z').$$

$$(2.5.41)$$

Let us now determine parameters $M_{n'}$, $N_{n'}$, $m_{n'} = M_{n'} + N_{n'} + 1$ and $h_{n'}$, $n' = 1, 2, 3$, let us determine Sinc points $\{\xi_{i'}\}_{-M_1}^{N_1}$, $\{\eta_{j'}\}_{-M_2}^{N_2}$, and $\{\zeta_{k'}\}_{-M_3}^{N_3}$, and coordinate values $x'_{i'} = a_1 + \beta\xi_{i'}$, $y'_{i',j'} = a_2(x'_{i'}) + \gamma_{i'}\eta_{j'}$, and $z'_{i',j',k'} = a_3(x_{i'}, y_{i',j'}) + \delta_{i',j'}\zeta_{k'}$.

Let us also form the matrices

$$
\begin{aligned}
A &= h_1\, I^{(-1)}\, D\left(\frac{e^w}{(1 + e^w)^2}\right); \quad w = j\,h_1,\ \ j \in [-M_1, N_1] \\
&= X\, R\, X^{-1},\quad X = [x_{ij}],\ \ X^{-1} = Xi = [x^{ij}] \\
R &= \operatorname{diag}\left[r_{-M_1}, \ldots, r_{N_1}\right]. \\
B &= h_2\, I^{(-1)}\, D\left(\frac{e^w}{(1 + e^w)^2}\right); \quad w = k\,h_2,\ \ k \in [-M_2, N_2] \\
&= Y\, S\, Y^{-1},\quad Y = [y_{ij}],\ \ Y^{-1} = Yi = [y^{ij}], \\
S &= \operatorname{diag}\left[s_{-M_2}, \ldots, s_{N_1}\right], \\
C &= h_3\, I^{(-1)}\, D\left(\frac{e^w}{(1 + e^w)^2}\right); \quad w = \ell\,h_3,\ \ \ell \in [-M_3, N_3] \\
&= Z\, T\, Z^{-1},\quad Z = [z_{ij}],\ \ Z^{-1} = Zi = [z^{ij}] \\
T &= \operatorname{diag}\left[t_{-M_3}, \ldots, t_{N_3}\right].
\end{aligned}
$$

$$(2.5.42)$$

---

[3]For convenience, we take the upper limits in the Laplace transform of f to be infinite, although this may not always be possible.

In these definitions, the dimensions of the matrices $I^{(1)}$ are $m_n \times m_n$ with $n = 1$, 2, 3 for A, B, and C respectively.

Then, setting

$$G^{[2]}(i, i', j, j', k, k')$$

$$= \hat{f}(\beta\, r_i\,, \beta\, \xi_{i'}\,, \ \gamma_{i'}\, s_j\,, \gamma_{i'}\, \eta_{j'}\,, \ \delta_{i',j'}\, t_k\,, \delta_{i',j'}\, \zeta_{k'}). \tag{2.5.43}$$

we get the approximation $P(\xi_\ell\,, \eta_m\,, \zeta_n) \approx p_{\ell\, m\, n}$, where

$$p_{\ell\,,m\,,n} =$$

$$\sum_i x_{\ell,i} \sum_{i'} x^{i,i'} \sum_j y_{m,j} \sum_{j'} y^{j,j'} \sum_k z_{n,k} \sum_{k'} z^{k,k'} \cdot \tag{2.5.44}$$

$$\cdot G^{[2]}(i, i', j, j', k, k').$$

Algorithm 2.5.5 Nonlinear $3 - d - P$ Convolution

1. Determine parameters $M_{n'}$, $N_{n'}$, $m_{n'} = M_{n'} + N_{n'} + 1$ and $h_{n'}$, set $m_{n'} = M_{n'} + N_{n'} + 1$, determine Sinc points $\xi_{i'}$, $\eta_{j'}$, and $\zeta_{k'}$ as well as the numbers $\gamma_{i'}$ and $\delta_{i',j'}$ as above.

2. Form the matrices A, B, C, and determine their eigenvalue decompositions as in (2.5.42) above.

   In these definitions, the dimensions of the matrices $I^{(1)}$ are $m_{n'} \times m_{n'}$ with $n' = 1$, 2, 3 for A, B, and C respectively.

3. Form the array $\left\{ G^{[2]}(i, i',\ j, j',\ k, k') \right\}$, with

   $$G^{[2]}(i, i', j, j', k, k')$$

   $$= \hat{f}(\beta\, r_i\,, \beta\, \xi_{i'}\,, \ \gamma_{i'}\, s_j\,, \gamma_{i'}\eta_{j'}\,, \ \delta_{i',j'}\, t_k\,, \delta_{i',j'}\, \zeta_{k'}).$$

   where i and i' are integers in $[-M_1\,, N_1]$, j and j' are integers in $[-M_2\,, N_2]$, and k and k' are integers in $[-M_3\,, N_3]$.

4. Eliminate k', by forming the following tensor product with respect to k':

   $$H^{[2]}(i', i, j, j', k) = G^{[2]}(i, i', j, j', k, k')\, Z(k, k').$$

5. Eliminate i' by performing the tensor following product with respect to i':

$$K^{[2]}(i,j,k,j') = X(i,i')\,H^{[2]}(i',i,j,j',k).$$

6. Eliminate j' by forming the followings tensor product with respect to j':

$$L^{[2]}(i,j,k) = K^{[2]}(i,j,k,j')\,Y(j,j');$$

7. Eliminate k by taking the following tensor product with respect to k:

$$M^{[2]}(i,n,j) = L^{[2]}(i,j,k)\,Z(n,k);$$

8. Eliminate j by taking the following tensor product with respect to j:

$$N^{[2]}(i,m,n) = M^{[2]}(i,n,j)\,Y(m,j));$$

9. Eliminate i by taking the following tensor product with respect to i:

$$p_{\ell,m,n} = X(\ell\,i)\,N^{[2]}(i,m,n).$$

Remark. Having computed the array $\{p_{\ell,m,n}\}$, we can now approximate $p(x,y,z)$ at any point of $\mathcal{B}$ by means of the formula

$$p_1(x,y,z) \approx \sum_{\ell=-N_1}^{N_1} \sum_{m=-N_2}^{N_2} \sum_{n=-N_3}^{N_3} p_{\ell,m,n}\, w_\ell^{(1)}(\xi)\, w_m^{(2)}(\eta)\, w_n^{(3)}(\zeta),$$

(2.5.45)

where the $\mathbf{w}^{(d)} = (w_{-M_n}^{(d)},\dots,w_{N_n}^{(d)})$, are defined as in Definition 1.5.12, but with (ξ,η,ζ) defined as in (2.5.37). Of course, to get $p(x,y,z)$, we need to traverse the above algorithm seven more times, in order to similarly compute seven additional integrals, $p_2,\dots,p_8$ of the same type, and then add these to p_1.

The formation of $G^{[2]}$ in Step 3. of the above algorithm requires $\mathcal{O}(\nu^6)$ operations, and consequently, the complexity of approximation of the function p of (2.5.33) via use of the Algorithm 2.5.5 to within an error of ε is $\mathcal{O}(|\log(\varepsilon)|^{12})$.

We next consider the special but more important case of q needed for applications, which can be accomplished with less work.

b. *Approximation of* $q(x, y, z)$.

We again split the integral q as a sum of eight integrals, and first consider

$$q_1(x, y, z) =$$

$$\int_{a_1}^{x} \int_{a_2(x')}^{y} \int_{a_3(x',y')}^{z} f(x - x', y - y', z - z')\, g(x', y', z')\, dz'\, dy'\, dx'.$$
(2.5.46)

Under the transformation (2.5.37), q_1 becomes Q, where

$$Q(\xi, \eta, \zeta)$$
$$= \int_0^{\xi} \int_0^{\eta} \int_0^{\zeta} f(\beta(\xi - \xi'),\ \gamma(\eta - \eta'),\ \delta(\zeta - \zeta'))\ \cdot$$
(2.5.47)

$$\cdot\, g(a_1 + \beta\xi',\ a_2 + \gamma\eta',\ a_3 + \delta\zeta')\, \beta\,\gamma\,\delta\, d\zeta'\, d\eta'\, d\xi',$$

where $\beta = b_1 - a_1$, $\gamma(x') = b_2(x') - a_2(x')$, and $\delta(x', y') = b_3(x', y') - a_3(x', y')$.
Let us set

$$F(r, s, t) = \int_0^{\infty} \int_0^{\infty} \int_0^{\infty} f(x, y, z)\, \exp\left(-\frac{x}{r} - \frac{y}{s} - \frac{z}{t}\right) dz\, dy\, dx.$$
(2.5.48)

Then, we get

$$\int_0^{\infty} \int_0^{\infty} \int_0^{\infty} f(\beta\xi, \gamma\eta, \delta\zeta)\, \beta\,\gamma\,\delta \exp\left(-\frac{\xi}{r} - \frac{\eta}{s} - \frac{\zeta}{t}\right) d\zeta\, d\eta\, d\xi$$

$$= F(\beta r,\ \gamma s,\ \delta t).$$
(2.5.49)

Let us now form the arrays

$$\Phi(i, i', j, j', k) = F(\beta r_i, \gamma_{i'} s_j, \delta_{i',j'} t_k)$$

$$\psi(i', j', k') = g(a_1 + \beta \xi_{i'}, a_2(x'_{i'} + \gamma_{i'} \eta_{j'}, a_3(x'_{i'}, y'_{i',j'}) + \delta_{i',j'} \zeta_{k'}).$$
$$(2.5.50)$$

By proceeding as for $p_1(x, y, z)$ defined in (2.5.39) above, we now get the approximation $Q(\xi_\ell, \eta_m, \zeta_n) \approx \mathbf{q}(\ell, m, n)$ where

$$q_{\ell, m, n} =$$

$$\sum_i x_{\ell,i} \sum_{i'} x^{i,i'} \sum_j y_{m,j} \sum_{j'} y^{j,j'} \sum_k z_{n,k} \sum_{k'} z^{k,k'}. \qquad (2.5.51)$$

$$\cdot \Phi(i, i', j, j', k) \, \psi(i', j', k').$$

We can evaluate $q_{\ell,m,n}$ via the following algorithm.

Algorithm 2.5.6 Nonlinear $3 - d - Q$ Convolution

1. Determine parameters $M_{n'}$, $N_{n'}$, $m_{n'} = M_{n'} + N_{n'} + 1$ and $h_{n'}$, Sinc points $\xi_{i'}$, $\eta_{j'}$, and $\zeta_{k'}$ as well as $\gamma_{i'}$ and $\delta_{i',j'}$ as above.

2. Form the matrices A, B, C, and determine their eigenvalue decompositions as in (2.4.42) above.

   In these definitions, the dimensions of the matrices $I^{(1)}$ are $m_n \times m_n$ with $n = 1, 2, 3$ for A, B, and C respectively.

3. Form the arrays

$$\left\{ F^{[2]}(i, i', j, j', k) \right\}, \quad \left\{ g^{[2]}(i', j', k') \right\},$$

with

$$F^{[2]}(i, i', j, j', k) = F(\beta r_i \ \gamma_{i'} s_j, \ \delta_{i',j'} t_k)$$

and with

$$g^{[2]}(i', j', k')$$

$$= g(a_1 + \beta\,\xi_{i'}, \ a_2(x_{i'} + \gamma_{i'}\,\eta_{j'}), \ a_3(x'_{i'}, y'_{i',j'}) + \delta_{i',j'}\,t_{k'})$$

Here i and i' are integers in $[-M_1, N_1]$, j and j' are integers in $[-M_2, N_2]$, and k and k' are integers in $[-M_3, N_3]$.

Note: The formation of the array $\{F^{[2]}(i, i', j, j', k)\}$ requires $\mathcal{O}(\nu^5)$ operations.

4. Eliminate k', using tensor product multiplication with respect to k',

$$h^{[2]}(i', \ j', \ k) = g^{[2]}(i', j', k')\, Zi(k, k').$$

5. With Hadamard product operations, form the set

$$G^{[2]}(i', i, j, j', k) = F^{[2]}(i, i', j, j', k)\, h^{[2]}(i', j', k).$$

6. Eliminate i', by taking the tensor product with respect to i',

$$H^{[2]}(i, j, k, j') = Xi(i, i')\, G^{[2]}(i', i, j, j', k).$$

7. Eliminate j' using the tensor product with respect to j',

$$K^{[2]}(i, j, k) = H^{[2]}(i, j, k, j')\, (Yi(j, j'))^T.$$

8. Evaluate $L^{[2]}(\ell, j, k)$ using the tensor product with respect to i,

$$L^{[2]}(\ell, j, k) = X(\ell, i)\, K^{[2]}(i, j, k).$$

9. Recall that $m_n = M_n + N_n + 1$. For each $\ell = -M_1, \ \ldots, \ N_1$, form the $m_2 \times m_3$ matrix $U = [u_{j,k}]$ with $u_{j,k} = L^{[2]}(\ell, j, k)$, and then form a matrix $V = [v_{m\,nk}]$ by taking the product

$$V = Y U Z^T.$$

Then form the array

$$q_{\ell m n} = v_{m,n}.$$

Note: At this point we have formed the array $q_{\ell,m,n}$ Under suitable assumptions of analyticity with respect to each variable on the functions f and g in (2.5.34), the complexity of approximation of q to within an error of ε with this algorithm is $\mathcal{O}(|\log(\varepsilon)|^{10})$. Once this array, as well as seven similar ones, corresponding to the approximation of the other seven indefinite convolutions have been formed, we would sum, to get an array $q^*_{\ell,m,n}$. We can then approximate the function $q(x,y,z)$ to within ε for any (x,y,z) in $\mathcal{B}$ via the expression

$$q(x,y,z) \approx \sum_{\ell=-N_1}^{N_1} \sum_{m=-N_2}^{N_2} \sum_{n=-N_3}^{N_3} q^*_{\ell,m,n}\, w_\ell^{(1)}(\xi)\, w_m^{(2)}(\eta)\, w_n^{(3)}(\zeta).$$

$$(2.5.52)$$

2.5.5 BOUNDARY INTEGRAL CONVOLUTIONS

Our approximations of convolution integrals are based on Laplace transforms, for which we have assumed that the original integral is along a straight line segment. We have already illustrated an extension of Sinc convolutions from a line segment to an arc in our approximation of the Hilbert and Cauchy transforms at the end of §1.5.12. We shall here similarly illustrate the procedure for convolutions over surfaces that are smooth boundaries of three dimensional regions.

Specifically, we assume that $\mathcal{B}$ is a simply connected region in $\mathbb{R}^3$ with boundary $\partial\mathcal{B}$, such that the number $\beta(\bar{r})$ defined as in (2.2.20) exists and is positive for all $\bar{r} \in \partial\mathcal{B}$. We will furthermore assume that $\partial\mathcal{B}$ consists of the union of p patches,

$$\partial\mathcal{B} = \bigcup_{J=1}^{p} S_J.$$

$$(2.5.53)$$

with each patch S_J having the representation

$$S_J = \{\bar{r} = (u^j(\bar{\sigma}), v^j(\bar{\sigma}), w^j(\bar{\sigma})) : \bar{\sigma} \in (0,1)^2\}$$

$$(2.5.54)$$

with $\bar{\sigma} = (\xi, \eta)$. Each of the functions u^j, v^j and w^j belonging to the space $\mathbf{X}^{(2)}((0,1)^2)$ of all functions $g(\bar{\sigma})$ such that both functions $g(\cdot, \eta)$ for each fixed $\eta \in [0,1]$, and $g(\xi, \cdot)$ for each fixed $\xi \in [0,1]$ belong to $\mathbf{M}'$. These spaces are discussed below in §2.6.

Let us first consider the approximation of the convolution integral

$$f(\bar{r}) = \int \int_S \frac{g(\bar{r}')}{|\bar{r} - \bar{r}'|} dA(\bar{r}'), \quad \bar{r} \in S, \qquad (2.5.55)$$

where $dA(\bar{r}')$ denotes the element of surface area (see §2.1.3).

Transformation of S_J of Eq. (2.5.54) to $(0,1)^2$ enables us to replace the integral (2.5.55) with a system of p integrals over $(0,1)^2$ of the form

$$F_J(\bar{\sigma}) = \sum_{K=1}^{p} \int \int_{(0,1)^2} \frac{G_K(\bar{\sigma}')}{|\bar{r}(\bar{\sigma}) - \bar{r}(\bar{\sigma}')|} \mathcal{J}_K(\bar{\sigma}') d\bar{\sigma}', \qquad (2.5.56)$$

for $J = 1, \ldots, p$, with F_J the transformed restriction f_J of f to S_J, with G_K the transformed restriction g_K of g to S_K, and with $\mathcal{J}_K$ the Jacobian of the transformation.

It is recommended to use repeated Sinc quadrature to approximate the integrals on the right hand side of (2.4.56) for the case when $K \neq J$.

We can use Sinc convolution for the case when $K = J$. To this end, set

$$\mathcal{F}_J(\bar{\sigma}) = \int \int_{(0,1)^2} \frac{\mathcal{G}_J(\bar{\sigma}, \bar{\sigma}')}{|\bar{\sigma} - \bar{\sigma}|} d\bar{\sigma}, \qquad (2.5.57)$$

with

$$\mathcal{G}_J(\bar{\sigma}, \bar{\sigma}') = G_J(\bar{\sigma}, \bar{\sigma}') \mathcal{J}_J(\bar{\sigma}') \frac{|\bar{\sigma} - \bar{\sigma}'|}{|\bar{r}(\bar{\sigma}) - \bar{r}(\bar{\sigma}')|}. \qquad (2.5.58)$$

Now with note that $\mathcal{G}_J(\xi, \xi', \eta, \eta') = \mathcal{G}_J(\bar{\sigma}, \bar{\sigma}')$ belongs to $\mathbf{X}^{(2)}((0,1)^2)$ as a function of (ξ, ξ'), and similarly, with (ξ, ξ') fixed, it belongs to $\mathbf{X}^{(2)}((0,1)^2)$ as a function of (η, η'). Furthermore, the integral (2.5.57) taken over $(0,1)^2$ can be split into four indefinite convolution integrals, of the form,

$$\int_0^\xi \int_\eta^1 \kappa(\xi - \xi', \xi', \, \eta - \eta' \, \eta') \, d\eta' \, d\xi'. \qquad (2.5.59)$$

Each of the one dimensional integrals in this repeated integral can now be approximated in a manner similar to the way the integral p in (1.5.61) was approximated. The algorithmic details are the same as those of Algorithm 2.5.3.

PROBLEMS FOR SECTION 2.5

2.5.1 Justify the details of Steps #3 to #9 of Algorithm 2.5.1.

2.5.2 (a) Derive an explicit algorithm similar to Algorithm 2.5.1, but based on changing the order of integration in the definition of $p(x, y)$ in (2.5.1).
(b) Prove that the algorithm obtained in (a) yields the same solution $[p_{i,j}]$ as that of Algorithm 2.5.1.

2.5.3 Expand Algorithm 2.5.4 in full detail.

2.5.4 Prove that the function $\mathcal{G}_J(\bar{\sigma}, \bar{\sigma}')$ defined as in the last paragraph of §2.5 does, in fact, belong to the space $\mathbf{X}^{(2)}((0,1)^2)$.

2.6   Theory of Separation of Variables

In this section we discuss the mathematics of solutions of partial differential equations (PDE) via boundary integral equation (BIE) methods. We shall make assumptions on analyticity that enable us to construct exponentially convergent and uniformly accurate solutions using a small number of one dimensional Sinc matrix operations. This is our *separation of variables* method. We achieve this by first transforming the PDE problem set on a curvilinear region into a problem over rectangular region. We then obtain an approximate solution via one dimensional Sinc methods. For purposes of pedagogy, our discussion is limited mainly to standard linear elliptic, as well as to linear and nonlinear parabolic, and hyperbolic PDE.

Partial differential equations are nearly always modeled using ordinary functions and methods of calculus, and we are thus nearly always granted the presence of analyticity properties that fit naturally with the methods of solution enabled by Sinc methods. When solving PDE using Sinc methods, we would, on the one hand, like to achieve the high exponential accuracy of one dimensional Sinc methods of §1.5, and on the other hand, we would like to have a fast method of execution. We shall spell out in this section how we can, in fact achieve both of these goals. In order to simplify our proofs, we shall assume in this section that the region over which the PDE problem is given is both simply connected and bounded. We will then be able to transform all n–dimensional PDE problems to the multidimensional cube $(0, 1)^d$.

§2.6.1 below contains a discussion of the type of piecewise analyticity possessed by coefficients and solutions of a PDE when an engineer or scientist models the PDE. We also briefly mention the function spaces associated with solving PDE via the popular finite difference, finite element and spectral methods. All–in–all, the number of regions can expand out of hand, and the process can become complex and costly.

By contrast, with Sinc methods, a relatively few number of regions will suffice. We introduce simple arcs, curves, contours, surfaces and quasi–convex regions that can be conveniently transformed to the unit cube without destroying the requisite analyticity of the associated function spaces, where separation of variables and exponentially convergent Sinc approximation is possible.

§2.6.2 carries out two fundamental steps:

1. A given PDE with boundary of domain, with coefficients, and with initial and/or boundary data belonging to appropriate spaces **X** of the type discussed in §2.6.1 , has solutions that also belong to **X** , and

2. Solution of the PDE by one dimensional Sinc methods, our separation of variables method.

Our discussions of this type in §2.6.2 thus cover two and three dimensional elliptic PDE, as well as that of parabolic and hyperbolic PDE.

2.6.1   Regions and Function Spaces

Function spaces for studying PDE have been created largely by mathematicians (e.g., Friedman, Hormander, Sobolev, Yosida), who use them to to prove *existence* of solutions. Most practical solvers of PDE bypass creating proper function spaces necessary for efficient and accurate approximation of solution of their own problems. They instead settle for established spaces, such as those of Sobolev, who developed an elaborate notation involving functions and their derivatives and expended considerable effort in the approximation of solutions over Cartesian products of one dimensional Newton–Cotes polynomials that interpolate at equi–spaced points. It is perhaps unfortunate that Sobolev spaces have become popular both for constructing solutions using finite difference methods, and for constructing finite element type solutions. This widespread application of Sobolev's theory generates in general poorly accurate and inefficient methods of solution. Whereas most PDE problems in applications have solutions that are analytic, Sobolev spaces hold a much larger class of functions than is necessary for applied problems. In short, methods based on this approach converge considerably more slowly than Sinc methods.

In short, spaces born from pure mathematics (e.g., Sobolev spaces) do not take advantage of PDE's that arise in the real world, this is because calculus is nearly always at the heart of the modeling process. Therefore, the PDE which emerge will exist in a union of a small number of regions on each of which the functional components of the PDE as well as the boundary of the region, and the initial and/or boundary data, show up as being analytic on arcs which themselves are analytic. Consequently, the solutions to the PDE also have this analyticity property. An engineer, scientist, and even an applied mathematician would be hard put to find an instance for which the functional form of the PDE, the boundary of the region, or the given data is everywhere of class $\mathbf{C}^\infty$ and not analytic, or for which the solution may have only one smooth first derivative, but with second derivative at most measurable, and not piecewise analytic on arcs in open domains.

While it is true should a finite difference or a finite element solution exactly meet the requirement of, for example, a certain Sobolev space, it will converge at an optimal rate, as argued above. It is highly unlikely such solutions will ever appear in practice. The Sinc approach assumes from the outset analyticity as described above and designs algorithms for solutions accordingly.

The payoff with Sinc is huge. In d dimensions, assuming a solution is of class $\mathbf{Lip}^1$ on a region, with finite difference or finite element method using m equal to n^d points, the convergence of approximation is $\mathcal{O}\left(m^{-1/d}\right)$, whereas Sinc is able to achieve an $\exp\left(-c\,m^{1/(2\,d)}\right)$ rate of convergence.

Should a particular Sobolev space be chosen for the purpose of analysis of a finite difference or a finite element algorithm, then one achieves optimal convergence to solutions that exactly meet the requirements of the space. However because finite difference or finite element methods do not take advantage of any existing analyticity, this optimal convergence is nonetheless considerably slower than what can be achieved with Sinc.

For example, consider the space of twice differentiable functions on the interval $(0,1)$ with second derivative non–smooth on all subintervals. Let suppose that an engineer actually encounters such a function f (an unlikely occurrence). Approximating f via the $(n+1)$ *Chapeau spline*

$$(S^{(n)} f)(x) = \sum_{j=0}^{n} f(j/n)\, S_j^{(n)}(x)\,, \qquad (2.6.1)$$

with hat–like triangular basis function S_j^n yields a $\mathcal{O}\left(n^{-2}\right)$ rate of convergence. Using an basis Sinc approximation instead, in (2.6.1), results in the same $\mathcal{O}\left(n^{-2}\right)$ rate of convergence. However altering f to be analytic on $(0,1)$ and with second derivative continuous on $[0,1]$, keeps $\|f - S^{(n)} f\|$ at $\mathcal{O}\left(n^{-2}\right)$, while using a Sinc approximation drives this error down to $\mathcal{O}\left(\exp(-c\,n^{1/2})\right)$. Furthermore, this convergence is preserved when singularities exist at the end–points of $(0,1)$, even when these singularities are not explicitly known, as long as f is analytic on $(0,1)$ and of class $\mathbf{Lip}_\alpha[0,1]$, with α a constant, $0 < \alpha \leq 1$.

In general *spectral methods* can outperform Sinc and achieve an $\mathcal{O}\left(\exp(-c\,n)\right)$ of convergence when functions are analytic in an open domain containing $[0,1]$. Yet, for a function such as x^α, with $\alpha \in (0,1)$ (arising frequently in solutions of PDE), which does not have a first derivative bounded on $[0,1]$, convergence of Sinc methods is still $\mathcal{O}\left(\exp\left(-c\,n^{1/2}\right)\right)$, whereas that of spectral methods as well as that of the above spline approximation drops to $\mathcal{O}\left(n^{-\alpha}\right)$.

We should mention if the nature of a singularity in the solution of the PDE is known (e.g., with reference to the above, the exact value of α) in two dimensions it is possible to choose basis functions such that no loss of convergence results with the use of finite element or spectral methods. In three or higher dimensions the nature of the singularity cannot be known explicitly and convergence degenerates as above. In fact Fichera's theorem [Fr] tells us that an elliptic PDE in 3 dimensions has $\mathbf{Lip}_\alpha$–type singularities at corners and edges, even though the data is smooth, and the exact nature of these singularities while computable, are not explicitly known for three dimensional problems.

One of the successes of Sinc approximations is the rapid convergence these approximations provide when used properly and under the right conditions. We have already identified in §1.5.2 the one dimensional spaces of analytic functions $\mathbf{L}_{\alpha,\beta,d}(\varphi)$ and $\mathbf{M}_{\alpha,\beta,d}(\varphi)$, in which Sinc methods converge exponentially for approximation over Γ, with Γ denoting a finite, a semi–infinite, an infinite interval, indeed, even an analytic arc even in the presence of relatively unknown–type end–point singularities on Γ. (See Definition 1.5.2.) In this section we extend the definition of these spaces in a coherent manner, for approximation over more than one such arc in one dimension, and to regions in more than one dimension. Our main objective is to be able to solve PDE via one dimensional Sinc methods, and thus gain drastic reduction in computation time for solutions to many types of applied problems, especially for solutions of PDE and IE. The success of applying separation of variables for evaluation of multidimensional convolution–type integrals in §2.5 carries over with equal success by reducing the solutions of PDE and IE to a finite sequence of one dimensional Sinc matrix additions and multiplications.

As already stated, problems will be set in bounded regions $\mathcal{B}$. We are thus able to reduce approximation over $\mathcal{B}$ in d dimensions to approximation over the d–cube $(0,1)^d$. This is no restriction for Sinc methods. For, if φ_1 maps bounded arc Γ_1 to $\mathbb{R}$, and φ_2 maps unbounded arc Γ_2 to $\mathbb{R}$, and if $f \in \mathbf{M}_{\alpha,\beta,d}(\varphi_2)$, then $g = f \circ \varphi_2^{-1} \circ \varphi_1 \in \mathbf{M}_{\alpha,\beta,d}(\varphi_1)$, then Sinc approximation over unbounded arc Γ_2 is accomplished by Sinc approximation over Γ_1 (Theorem 1.5.3).

We shall restrict our analysis to bounded regions, reducing all approximations to approximations over $(0,1)$ in one dimension, and to the n–cube, $(0,1)^d$ in n dimensions. This is no restriction for Sinc methods however, since if, e.g., φ_1 is a map defined as in Definition 1.5.12 of a bounded arc Γ_1 to $\mathbb{R}$, if φ_2 is a map defined as in Definition 1.5.2 of an unbounded Γ_2 to $\mathbb{R}$, and if $f \in \mathbf{M}_{\alpha,\beta,d}(\varphi_2)$, then $g = f \circ \varphi_2^{-1} \circ \varphi_1 \in \mathbf{M}_{\alpha,\beta,d}(\varphi_1)$, i.e., the requisite properties for accurate Sinc approximation on Γ_2 are, in effect, the same as those required for accurate Sinc approximation on Γ_1 (see Theorem 1.5.3).

Definition 2.6.1 *Let ϕ denote the mapping $\phi(z) = \log(z/(1-z))$, defined in Example 1.5.5, and for some constants $\alpha \in (0,1]$ and $\delta \in (0,\pi)$, let $\mathbf{M} = \mathbf{M}_{\alpha,\alpha,\delta}(\phi)$ be defined as in Definition 1.5.2. Let $\mathbf{M}'$ denote the family of all functions in $\mathbf{M}$, which are analytic and uniformly bounded in the region*

$$\mathcal{E}_\delta = \left\{ z \in \mathbb{C} : \left| \arg\left(\frac{z}{1-z} \right) \right| < \delta \right\}. \tag{2.6.2}$$

Note here, our letting ϕ, with $\phi(z) = \log(z/(1-z))$ denote the conformal map of $\mathcal{E}_d$ onto the strip $\mathcal{D}_d = \{z \in C : |\Im z| < \delta\}$. We wish to distinguish this function ϕ from the more general function φ which maps an arbitrary region $\mathcal{D}$ onto $\mathcal{D}_d$.

Analytic Arcs, Curves, Contours and Regions.

Let us begin with definitions of *analytic arcs*, *curves*, and *contours*.

Definition 2.6.2 *An analytic arc Γ in $\mathbb{R}^d$ is defined by*

$$\bar{\rho} : [a,,b] \to \Gamma = \{\mathbf{x}(t) = (x^1(t), \ldots, x^d(t)) : a \le t \le b\} \tag{2.6.3}$$

with $[a,b]$ a closed and bounded interval, such that for some positive constants $\alpha \in (0,1]$ and $\delta \in (0,\pi)$, $x^j \in \mathbf{M} = \mathbf{M}_{\alpha,\delta}$, for $j = 1,\ 2,\ \ldots\ d$, such that $0 < \sum_{j=1}^d (\dot{x}^j(t))^2$, for all $t \in (a,b)$, and such that the length of the arc is finite. Here the dot indicates differentiation with respect to t.

We shall simply write $\Gamma \in \mathbf{M}$, in place of $\bar{\rho} \in \mathbf{M}^d$.

It is easy to show (see Problem 2.6.1) that if $g \in \mathbf{M}_{\alpha,\beta,\delta}(\mathcal{D})$, for some $\alpha \in (0,1]$, $\beta \in (0,1]$ and $\delta \in (0,\pi)$, and with $\mathcal{D} = \phi^{-1}(\mathcal{D}_\delta)$ (see Eq.

(1.4.36) and Definition 1.5.2), and if $\rho \in \mathbf{M}$ denotes an analytic arc, $\rho : (0,1) \to \Gamma = \phi^{-1}(\mathbb{R})$, then $f = g \circ \rho \in \mathbf{M}$.

One must take care for computational purposes, to ensure that the resulting numbers δ and α are not unduly small. Consider, for example, the case of $\Gamma = [-1,1]$, and $f(x) = (10^{-6} + x^2)^{-1/2}$. After transforming the interval $[-1,1]$ to the interval $(0,1)$, we would find that f is analytic on $(0,1)$ and of class $\mathbf{Lip}_1$ on $[0,1]$, and also, that $f \in \mathbf{M}_{1,\delta}$, with $\delta = 10^{-3}$. Such a small δ results in poor Sinc approximation. However, if we were now to split the arc $[-1,1]$ into the union of the two arcs, $[-1,0]$ and $[0,1]$, then f would again be analytic and of class $\mathbf{Lip}_1$ and moreover, $f \in \mathbf{M}_{\alpha,\delta}$ with $\delta = \pi/2$ on each of the arcs. We can similarly ensure that α is not unduly small in the definition of the arcs. That is, if α is very small, we can nearly always increase α by splitting the arc into a relatively small union of smaller arcs.

Definition 2.6.3 *A curve, $\mathcal{B}^{(1)}$, is a union of ν_1 analytic arcs Γ_ℓ, with the property that if $\bar{\rho}^\ell : [0,1] \to \Gamma^\ell$, then $\bar{\rho}^\ell(1) = \bar{\rho}^{\ell+1}(0)$, for $\ell = 1, 2, \ldots, \nu_1 - 1$.*

We note that the curve defined in this manner is both oriented and connected. It readily follows that a problem on a curve can be transformed into a system of ν_1 problems on $(0,1)$.

Definition 2.6.4 *A contour, $\mathcal{B}^{(1)}$ is just a* curve, *for which $\rho^{\nu_1}(1) = \rho^1(0)$.*

Contours $\mathcal{B}^{(1)}$ can be boundaries of two dimensional bounded regions, or boundaries of three dimensional surfaces, $\mathcal{B}^{(2)}$. In such cases we shall require that the interior angles at the junctions of two arcs of Γ are defined and positive. We do not otherwise restrict the size of the interior angles. For example, when the arcs are in a plane, we would have a *Riemann surface* emanating from a junction of two arcs with interior angle greater than 2π. Indeed, we have already successfully obtained Sinc constructions of conformal maps in such circumstances [SSc].

1. *The Space* $\mathbf{X}^{(1)} \left(\mathcal{B}^{(1)} \right)$ *in One Dimension.*

Let us next define a class of functions on $\mathcal{B}^{(1)}$, which is suitable for Sinc approximation.

Definition 2.6.5 *Let $\mathcal{B}^{(1)}$ be defined as above, in terms of ν_1 analytic arcs ρ_ℓ, $\ell = 1, , 2, \ldots, \nu_1$. The space $\mathbf{X}^{(1)}\left(\mathcal{B}^{(1)}\right)$ denotes the family of all functions F defined on $\mathcal{B}^{(1)}$, such that if $\bar{\rho}_\ell : (0,1) \to \Gamma_\ell$, and if F_ℓ denotes the restriction of F to Γ_ℓ, then $F_\ell \circ \bar{\rho}_\ell \in \mathbf{M}'$, $\ell = 1, 2, \ldots, \nu_1$.*

Remark. While the spaces $\mathbf{M}'$ and $\mathbf{X}^{(1)}\left(\mathcal{B}^{(1)}\right)$ suffice for proofs in this section, we may wish to describe more specific spaces for solving particular problems, although we shall always require piecewise analyticity. For example:

- For purposes of quadrature or indefinite integration of g over $\mathcal{B}^{(1)}$, we could require $\gamma_\ell / \phi' \in \mathbf{L}_{\alpha,\delta}(\phi)$, where $\gamma_\ell = g_\ell \circ \bar{\rho}_\ell$, and where $\phi(t) = \log(t/(1-t))$, so that g can be discontinuous or even unbounded at the junctions of the Γ_j;

- For solving a Dirichlet problem in a region with boundary that is a *contour*, we could require only that $\gamma_j \in \mathbf{M}'$ in order to get a uniformly accurate approximation over the region so that g can be discontinuous at the junctions of two arcs, Γ_j and Γ_{j+1}; and

- When approximating the Hilbert transform of a function g, and requiring that this Hilbert transform taken over a contour $\mathcal{B}^{(1)}$ should map a space $\mathbf{X}\left(\mathcal{B}^{(1)}\right)$ back into $\mathbf{X}\left(\mathcal{B}^{(1)}\right)$, we would let $\mathbf{X}\left(\mathcal{B}^{(1)}\right)$ denote the family of all functions g defined on $\mathcal{B}^{(1)}$, such that g_j, the restriction of g to Γ_j belongs to $\mathbf{M}'$.

Our definitions enable our using proofs only in spaces $\mathbf{M}'$. These spaces $\mathbf{M}'$ are dense in all of the "larger" spaces of this paragraph.

2. *The Space $\mathbf{X}^{(2)}\left(\mathcal{B}^{(2)}\right)$ in Two Dimensions.*

Let us briefly motivate the type of analyticity we shall require in more than one dimension. For the two dimensional case, let $F = F(x, y)$ be a given function defined on the rectangular region $(0,1)^2$, let

T^x and T^y be one–dimensional operations of calculus in the variables $x \in (0,1)$ and $y \in (0,1)$ respectively, and let $\mathcal{T}^x$ and $\mathcal{T}^y$ be their corresponding Sinc approximations. These operators typically commute with one another, and we can thus bound the error, $T^x T^y F - \mathcal{T}^x \mathcal{T}^y F$ in the following usual "telescoping" and "triangular inequality" manner:

$$\|T^x T^y F - \mathcal{T}^x \mathcal{T}^y F\|$$

$$\leq \|T^y (T^x F - \mathcal{T}^x F)\| + \|\mathcal{T}^x (T^y F - \mathcal{T}^y F)\| , \qquad (2.6.4)$$

$$\leq \|T^y\| \, \|T^x F - \mathcal{T}^x F\| + \|\mathcal{T}^x\| \, \|T^y F - \mathcal{T}^y F\| .$$

The last line of this inequality shows that we can still get the exponential convergence of one dimensional Sinc approximation, provided that there exist positive constants c_1 and c_2, such that:

1. For each fixed $y \in [0,1]$, the function $F(\cdot, y)$ of one variable belongs to the appropriate space $\mathbf{M}'$, and

2. the norm $\|T^y\|$ is bounded;

and likewise, provided that

1. For each fixed $x \in [0,1]$, the function $F(x, \cdot)$ of one variable belongs to the appropriate space $\mathbf{M}'$, and

2. the norm $\|T^x\|$ is bounded.

According to §1.5, we would then be able to bound the terms on the extreme right hand sides of (2.6.4) by

$$\mathcal{O}\left(\exp(-c_1 \, n_1^{1/2})\right) + \mathcal{O}\left(\exp(-c_2 \, n_2^{1/2})\right),$$

where n_1 and n_2 are the orders of the Sinc approximation operators $\mathcal{T}^x$ and $\mathcal{T}^y$ respectively.

Although the norm bounds of these operators $\mathcal{T}_x$ and $\mathcal{T}^y$ do not always remain bounded as a function of their order n_ℓ, they increase at a rather slow, polynomial function rate, as compared with a rate such as $\mathcal{O}\left(\exp(-c\,n_\ell^{1/2})\right)$. Thus they do not appreciably change the exponential rate of convergence[4]. (See, e.g., Theorem 6.5.2 and Problem 3.1.5 of [S1]).

Let us now consider the case when the region $\mathcal{B}^{(2)} \subset \mathbb{R}^2$ is not necessarily rectangular, although we shall assume for sake of simplicity of proofs in this section, that $\mathcal{B}^{(2)}$ is simply connected and bounded. Furthermore, we shall assume that the boundary of the region $\mathcal{B}^{(2)}$ is a curve $\mathcal{B}^{(1)}$ of the type described above, and moreover, that the region $\mathcal{B}^{(2)}$ can be represented as the union of rotations of a finite number ν_2 of regions of the form

$$\mathcal{B}_J^{(2,*)} = \{(x',y') : a_J^1 < x' < b_J^1, \ a_J^2(x') < y' < b_J^2(x')\}. \qquad (2.6.5)$$

for $J = 1, \ 2, \ \ldots, \ \nu_2$. That is, the original region $\mathcal{B}_J^{(2)}$ can be expressed in terms of (x,y) based on the rotation $(x,y) = \mathcal{R}_J^2(x',y')$, where $\mathcal{R}_J^2$ is defined by the matrix transformation

$$\mathcal{R}_J^2\,(x,y) = (x',y')\begin{pmatrix} \cos(\theta_J) & \sin(\theta_J) \\ -\sin(\theta_J) & \cos(\theta_J) \end{pmatrix}, \qquad (2.6.6)$$

and where θ_J is the angle of rotation. We then have

$$\mathcal{B}^{(2)} = \bigcup_{J=1}^{\nu_2} \mathcal{B}_J^{(2)}. \qquad (2.6.7)$$

We assume furthermore that any two regions $\mathcal{B}_J^{(2)}$ and $\mathcal{B}_K^{(2)}$ with $K \neq J$ share at most a common analytic arc. In the above definition of $\mathcal{B}_J^{(2,*)}$, a_J^1 and b_J^1 are constants, while with $\beta = b_J^2 - a_J^2$, $a_J^2(a_J^1 + \beta\,t) :$ $(0,1) \to \mathbb{R}^2$ and $b_J^2(a_J^1 + \beta\,t) : (0,1) \to \mathbb{R}^2$ represent analytic arcs of the type defined above. Such regions $\mathcal{B}_J^{(2,*)}$ can easily be represented as transformations of a square $(0,1)^2$ via the transformation

$$(x',y') \ = \ T_J^{2,*}(\xi,\eta)$$

$$\begin{aligned} x' &= a1_J + (b_J^1 - a_J^1)\,\xi \\ y' &= a_J^2(x') + (b_J^2(x') - a_J^2(x'))\,\eta. \end{aligned} \qquad (2.6.8)$$

---

[4]We must, of course take special care for operations such as taking Hilbert transforms; these cases can be readily dealt with.

It thus follows in terms of the above transformations $\mathcal{R}_J^2$ and $\mathcal{T}_J^{(2,*)}$ that any function F defined on $\mathcal{B}_J^{(2)}$ can be represented as a function G on $(0,1)^2$ via the representation $G = F \circ \mathcal{R}_J^2 \circ \mathcal{T}_J^{(2,*)}$.

Note that a problem over $\mathcal{B}^{(2)}$ can be transformed into a system of ν_2 problems over $(0,1)^2$. Note also, that we have not excluded the possibilities that $a_J^2(a^1) = b_J^2(a^1)$ or that $a_J^2(b^1) = b_J^2(b^1)$.

Definition 2.6.6 (a) A function G defined on $(0,1)^2$ will be said to belong to $\mathbf{X}^{(2)}\left((0,1)^2\right)$ if $G(x,\cdot) \in \mathbf{M}'$ for all $x \in [0,1]$, and if $G(\cdot,y) \in \mathbf{M}'$ for all $y \in [0,1]$.

(b) Let $\mathcal{B}^{(2)}$, $\mathcal{R}_J^2$, $\mathcal{B}_J^{(2,*)}$ and $\mathcal{T}_J^{2,*}$ be defined as above, with $J = 1, 2, \ldots, \nu_2$. Let a function F be defined on $\mathcal{B}^{(2)}$, and let F_J denote the restriction of F to $\mathcal{B}_J^{(2)}$. This function F will be said to belong to $\mathbf{X}^{(2)}\left(\mathcal{B}^{(2)}\right)$ if and only if $G_J \in \mathbf{X}^{(2)}\left((0,1)^2\right)$ for $J = 1, 2, \ldots, \nu_2$, where $G_J = F_J \circ \mathcal{R}_J^2 \circ \mathcal{T}_J^{2,*}$.

Suppose, now that $F \in \mathbf{X}^{(2)}\left(\mathcal{B}^{(2)}\right)$, with $\mathcal{B}^{(2)}$ represented as a union of the ν_2 regions $\mathcal{B}_J^{(2)}$ as above, and that F_J is the function F restricted to $\mathcal{B}_J^{(2)}$. Let $\gamma \in \overline{\mathcal{B}_J^{(2)}}$ be an analytic arc of finite length, such that $\bar{\rho} : (0,1) \to \gamma$, with $\bar{\rho} \in \mathbf{M}'$. It then follows that $F_J \circ \bar{\rho} \in \mathbf{M}'$ (see Problem 2.6.1).

We can then solve the problem over $(0,1)^2$, via a Sinc approximation to get a "mapped" approximate solution which for each J can be represented on $(0,1)^2$ by

$$U^J(\xi,\eta) = \sum_{i=-M}^{M} \sum_{j=-N}^{N} \omega_i^{(1)}(\xi)\, U_{i,j}^J\, \omega_j^{(2)}(\eta). \qquad (2.6.9)$$

and where the functions $\omega_j^{(k)}$ are defined as in §1.5.4. We will furthermore retain the desired exponential convergence of Sinc approximation, provided that each of the functions $U^J(\cdot,\eta)$ for all fixed $\eta \in [0,1]$ and $U^J(\xi,\cdot)$ for all fixed $\xi \in [0,1]$ belong to $\mathbf{M}'$.

To get back to approximation on $\mathcal{B}_J^{(2)}$ we simply solve for ξ and η in (2.6.8), i.e.,

$$\xi = \frac{x' - a_J^1}{b_J^1 - a_J^1}$$

$$\eta = \frac{y' - a^2(x')}{b^2(x') - a^2(x')}.$$

(2.6.10)

to get the results in terms of (x', y'), from which we can then get the solutions in terms of (x, y) via (2.6.6).

3. The Space $\mathbf{X}_{\alpha,\delta}^{(2)}(S)$ on a Surface S.

We also need to be concerned with analyticity on surface patches.

Definition 2.6.7 *Let $\bar{\rho}^{(2)} : (0,1)^2 \to S$ denote a transformation defined by $\bar{\rho}^{(2)}(\bar{\sigma}) = (\xi(\bar{\sigma}), \eta(\bar{\sigma}), \zeta(\bar{\sigma}))$, where $\bar{\sigma} = (t, \tau)$, and where each of the functions $\xi(\cdot, \tau)$, $\eta(\cdot, \tau)$, and $\zeta(\cdot, \tau)$, belong to $\mathbf{M}'$ for all $\tau \in [0,1]$, and also, such that each of the functions $\xi(t, \cdot)$, $\eta(t, \cdot)$, and $\zeta(t, \cdot)$, belong to $\mathbf{M}'$ for all $t \in [0,1]$. Let $\mathbf{X}^{(2)}(S)$ denote the family of all functions $F = F(x, y, z)$ such that $F \circ \bar{\rho}^{(2)} \in \mathbf{X}^{(2)}((0,1)^2)$.*

4. The Space $\mathbf{X}^{(3)}\left(\mathcal{B}^{(3)}\right)$ in Three Dimensions.

Similarly, as for the two dimensional case, we shall assume that regions in three dimensions, can be represented as a finite union of rotations of quasi-convex regions of the form

$$\mathcal{B}_J^{(3,*)} = \{(x', y', z') : a_J^1 < x' < b_J^1,$$

$$a_J^2(x') < y' < b_J^2(x'), \ a_J^3(x', y') < z' < b_J^3(x', y')\}.$$

(2.6.11)

That is,

$$\mathcal{B}^{(3)} = \bigcup_{J=1}^{\nu_3} \mathcal{B}_J^{(3)},$$

(2.6.12)

where $\mathcal{B}_J^{(3)}$ and $\mathcal{B}_J^{(3,*)}$ are related by a three dimensional rotational transformation $\mathcal{R}_J$

$$\mathcal{B}_J^{(3)} = \mathcal{R}_J \mathcal{B}_J^{(3,*)},$$

(2.6.13)

where, with $c = \cos(\theta_J)$, $s = \sin(\theta_J)$, $C = \cos(\phi_J)$, and $S = \sin(\phi_J)$,

$$(x', y', z') = \mathcal{R}_J(\theta_J, \phi_J) \iff$$

$$\begin{pmatrix} x' \\ y' \\ z' \end{pmatrix} = \begin{pmatrix} -s & -Cc & Sc \\ c & -Cs & Ss \\ 0 & S & C \end{pmatrix} \begin{pmatrix} x \\ y \\ z \end{pmatrix}. \tag{2.6.14}$$

The region $\mathcal{B}_J^{(3,*)}$ can be represented as a one-to-one transformation of the cube $(0,1)^3$ via the transformation

$$(x', y', z') = \mathcal{T}_{3,J}(\xi, \eta, \zeta)$$

$$\begin{aligned} x' &= a_J^1 + (b_J^1 - a_J^1)\xi \\ y' &= a_J^2(x') + (b_J^2(x') - a_J^2(x'))\eta \\ z' &= a_J^3(x', y') + (b_J^3(x', y') - a_J^3(x', y'))\zeta. \end{aligned} \tag{2.6.15}$$

In the transformation (2.6.15) we have not excluded the possibility that $a_J^2(a^1) = b_J^2(a^1)$ or that $a_J^2(b^1) = b_J^2(b^1)$. Also, we have tacitly assumed here that the functions a_J^2 and b_J^2 are analytic arcs, and that the functions a^3 and b^3 are "surface patch" functions belonging to the space $\mathbf{X}^{(2)}\left((0,1)^2\right)$.

Definition 2.6.8 *Let* $\mathcal{B}^{(3)}$, $\mathcal{B}_J^{(3)}$, $\mathcal{R}_J$, *and* $\mathcal{T}_J^{(3,*)}$ *be defined as above, for* $j = 1, \ldots, \nu_3$. *Let* F_J *denote the function* F *restricted to* $\mathcal{B}_J^{(3)}$, *and set* $G_J = F_J \circ \mathcal{T}_J \circ \mathcal{R}_J$. *The function* F *will be said to belong to* $\mathbf{X}^{(3)}\left(\mathcal{B}^{(3)}\right)$ *if the following three conditions are met for* $J = 1, 2, \ldots, \nu_3$:

$G_J(\cdot, \eta, \zeta) \in \mathbf{M}'$ *for all* $(\eta, \zeta) \in (0,1)^2$;

$G_J(\xi, \cdot, \zeta) \in \mathbf{M}'$ *for all* $(\xi, \zeta) \in (0,1)^2$; *and*

$G_J(\xi, \eta, \cdot) \in \mathbf{M}'$ *for all* $(\xi, \eta) \in (0,1)^2$.

We can then solve the problem over $(0,1)^3$, via a Sinc approximation to get a "mapped" solution which can be represented on $(0,1)^3$ via the Sinc interpolant

$$U^J(\xi,\eta,\zeta) = \sum_{i=-L}^{L} \sum_{j=-M}^{M} \sum_{k=-N}^{N} U_{i,j,k}^J \, \omega_i^{(1)}(\xi) \, \omega_j^{(2)}(\eta) \, \omega_k^{(3)}(\zeta),$$

(2.6.16)

for each $J = 1, \ldots, \nu_3$, where the functions $\omega_j^{(k)}$ are defined as in §1.5.4. Then we can get back to the region $\mathcal{B}_J^{(3,*)}$ by setting

$$
\begin{aligned}
\xi &= \frac{x' - a_J^1}{b_J^1 - a_J^1} \\
\eta &= \frac{y' - a_J^2(x)}{b_J^2(x') - a_J^2(x')} \\
\zeta &= \frac{z' - a_J^3(x',y')}{b_J^3(x',y') - a_J^3(x',y')}.
\end{aligned}
$$

(2.6.17)

Finally, we can then get back to the regions $\mathcal{B}_J^{(3)}$ via use of (2.6.13), and hence also to the region $\mathcal{B}^{(3)}$.

By proceeding in the way we did from two to three dimensions, we can similarly define spaces $\mathbf{X}^{(d)}\left(\mathcal{B}^{(d)}\right)$ in d dimensions, with $d > 3$. Chapter 4 of this handbook does, in fact, contain examples of solutions to problems in more than 3 dimensions.

2.6.2   ANALYTICITY AND SEPARATION OF VARIABLES

We show here, for linear elliptic, linear and nonlinear parabolic and hyperbolic PDE in two and three dimensions, that if the coefficients of a PDE, along with the boundary of a region on which a PDE is defined, and the given initial and boundary data are functions that are analytic and of class $\mathbf{Lip}_\alpha$ in each variable, assuming that these functions belong to spaces of the type defined in the previous subsection, then the solution of the PDE also has these properties.

Let us begin with the classic Cauchy–Kowalevsky theorem:

Theorem 2.6.9 *Let the differential equation*

$$\sum a^\beta D^\beta u = f$$

(2.6.18)

of order m be given, with the coefficients a^β and f analytic in a neighborhood of the origin, and with the coefficient of $D_{x_1}^m$ is not

$= 0$ when $x(x_1, x_2, \dots, x_d) = 0$. Then, for every function w that is analytic in each variable in a neighborhood of the origin, there exists a unique solution u of (2.6.18) analytic in each variable in a neighborhood of the origin and satisfying the boundary conditions

$$D_1^j(u - w) = 0 \quad when \quad x_1 = 0 \quad and \quad j < m. \qquad (2.6.19)$$

A nonlinear version involving time is the following:

Theorem 2.6.10 *(Cauchy-Kowalevsky II) If the functions F, f_0, f_1, $\dots$, f_{k-1} are analytic near the origin, then there is a neighborhood of the origin where the following Cauchy problem (initial value problem) has a unique analytic solution $u = u(x, y, z, t)$:*

$$\begin{aligned}
\frac{\partial^k}{\partial t^k} u(x, y, z, t) &= F(x, y, z, t, u, u_x, u_y, u_z), \\
&\qquad \text{a } k^{th} \text{ order PDE} \qquad\qquad (2.6.20) \\
\frac{\partial^j}{\partial t^j} u(x, y, z, 0) &= f_j(x, y, z) \quad \text{for all } 0 \le j < k
\end{aligned}$$

We next consider separately, each of the following cases:

A. *Two and three dimensional Dirichlet and Neumann problems of Poisson's equation;*

B. *Two and three dimensional linear and nonlinear heat problems;* and

C. *Two and three dimensional linear and nonlinear wave problems.*

The mathematical area of differential equations is *open ended*, and it is therefore impossible to give proofs of separation of variables, indeed, even to give a proof of the requisite analyticity of solutions as discussed above, for all types of PDE. In this section we thus restrict our considerations to those PDE whose solutions can be written either as an integral, or the solution of a linear or nonlinear integral equation (IE) with kernel that is a product of an explicitly known function and one of the Green's functions discussed in §2.3. All proofs of analyticity are similar for such equations, these being variations of the method first proposed in Theorem 6.5.6 of [S1, §6.5]. Below, we sketch some of these proofs.

Laplace–Dirichlet Problems in Two Dimensions.

We discuss here in detail, analyticity and separation of variables for two dimensional *Dirichlet* problems. These take the form

$$\nabla^2 u(\bar r) = 0, \quad \bar r \in \mathcal{B}^{(2)},$$

$$u(\bar r) = g(\bar r), \bar r \in \mathcal{B}^{(1)} = \partial \mathcal{B}^{(2)},$$

$$(2.6.21)$$

where $\mathcal{B}^{(2)}$ and $\mathcal{B}^{(1)} = \partial \mathcal{B}^{(2)}$ are defined in §2.6.1 above, and where g is a given real–valued function defined on $\mathcal{B}^{(1)}$.

The solution of (2.6.21) is, of course, a harmonic function, whenever g is such that a solution exists. But for simplicity of proof, we shall admit all functions g that are continuous on $\mathcal{B}^{(1)}$, and such that if we denote by g_J the function g restricted to Γ_J, where $\bar\rho_J : [0,1] \to \Gamma_J$, then $g_j \circ \bar\rho_J \in \mathbf{M}'$. That is, $g \in \mathbf{X}^{(1)}\left(\mathcal{B}^{(1)}\right)$.

Let $\zeta \in \mathcal{B}^{(1)} = \partial \mathcal{B}^{(2)}$, and for given $F \in \mathbf{X}^{(1)}\left(\mathcal{B}^{(1)}\right)$, let us denote the Hilbert transform of F taken over $\mathcal{B}^{(1)}$ by

$$(\mathcal{S}F)(\zeta) = \frac{P.V.}{\pi i} \int_{\mathcal{B}^{(1)}} \frac{F(t)}{t - \zeta} \, dt. \qquad (2.6.22)$$

Let v denote a conjugate harmonic function of the solution u of (2.6.21), and let $f = u + iv$ denote a function that is analytic in $D = \left\{ z = x + iy \in \mathbb{C} : (x,y) \in \mathcal{B}^{(2)} \right\}$. Let $z \in D$, and let us set

$$f(z) = \frac{1}{\pi i} \int_{\mathcal{B}^{(1)}} \frac{\mu(\tau)}{\tau - z} \, d\tau \qquad (2.6.23)$$

where μ is a real valued function on $\mathcal{B}^{(1)}$, which is to be determined. Upon letting $z \to \zeta \in \mathcal{B}^{(1)}$, with ζ not a corner point of Γ, and taking real parts, we get the equation [Hn]

$$\mu(\zeta) + (K\,\mu)(\zeta) = q(\zeta), \qquad (2.6.24)$$

with $K\mu = \Re\mathcal{S}\,\mu$, and with $q(\zeta) = q(x + iy) = g(x,y)$.

The integral equation operator K defined by $K\,u = \Re\mathcal{S}\,u$ arises for nearly every integral equation that is used for constructing the conformal maps. It has been shown in [Aga, Ah. I, §3] that this operator K has a simple eigenvalue 1, for which the corresponding eigenfunction is also 1. Furthermore, the other eigenvalues λ such that the equation $Kv = \lambda v$ has non–trivial solutions v are all less than 1 in absolute value.

Writing $\kappa = (K-1)/2$ we can rewrite (2.6.24) as

$$\mu(\zeta) + (\kappa\mu)(\zeta) = q(\zeta)/2. \tag{2.6.25}$$

It now follows that the norm of κ is less than one in magnitude, and so the series

$$\sum_{p=0}^{\infty} (-1)^p \, \kappa^p \, q/2 \tag{2.6.26}$$

converges to the unique solution of (2.6.25). Since $q \in \mathbf{Lip}_\alpha(\Gamma)$, we know that ([AK]) $\mathcal{S}q \in \mathbf{Lip}_\alpha(\Gamma)$, so that $\kappa q \in \mathbf{Lip}_\alpha(\Gamma)$. It thus follows by Theorem 1.5.3, that $\kappa : \mathbf{X}^{(1)}\left(\mathcal{B}^{(1)}\right) \to \mathbf{X}^{(1)}\left(\mathcal{B}^{(1)}\right)$. The sum (2.6.26) thus converges to a function $\mu \in \mathbf{X}^{(1)}\left(\mathcal{B}^{(1)}\right)$.

That is, since $\bar\rho_J : [0,1] \to \Gamma_J$, it follows, by Theorem 1.5.3, that the function f defined in (2.6.23) above is not only analytic in the region D but in the larger region

$$D_\mathcal{E} \equiv \bigcup_{J=1}^{n} \zeta_J^{-1}(\mathcal{E}) \bigcup D \tag{2.6.27}$$

This means of course, that if ζ is any bounded analytic arc in the closure of the region $\mathcal{B}^{(2)}$, then $u \circ \zeta(t)$ is an analytic function of t in $\mathcal{E}$, and moreover, it belongs to $\mathbf{M}'$. That is, $u \in \mathbf{X}^{(2)}\left(\mathcal{B}^{(2)}\right)$.

We mention that the one dimensional methods of §1.5.12 and §1.5.13 can be used to solve these two dimensional Dirichlet problems, i.e., we have can obtain a solution via separation of variables.

Neumann Problems in Two Dimensions.

We consider here the solution to the boundary value problem

$$\nabla^2 u(\bar r) = 0, \quad \bar r \in \mathcal{B}^{(2)},$$

$$\frac{\partial u}{\partial n} = g \quad \text{on} \quad \mathcal{B}^{(1)} = \partial\mathcal{B}^{(2)}, \tag{2.6.28}$$

where $\mathbf{n}$ denotes the unit outward normal at points of smoothness of $\mathcal{B}^{(1)}$, and where we assume that $g \in \mathbf{X}^{(1)}\left(\mathcal{B}^{(1)}\right)$. Let g_J denote the restriction of g to $\mathcal{B}_J^{(1)}$ and let $\gamma_J(t) = g_J(\bar{\rho}_J(t))$, where $\bar{\rho}_J :$ $(0,1) \to \mathcal{B}_J^{(1)}$ is an analytic arc. Then we have $\gamma_J \in \mathbf{M}'$. If u denotes the solution to (2.6.28), then a function v that is conjugate to u, is given on $\mathcal{B}^{(1)}$ by $v = \mathcal{S}\,u$, and we also have $u = \mathcal{S}\,v$, where $\mathcal{S}$ is defined as in (2.6.22). Furthermore, the Cauchy–Riemann equations imply that

$$\frac{\partial u}{\partial n} = \frac{\partial v}{\partial t} = g\,, \tag{2.6.29}$$

where $\mathbf{t}$ denotes the unit tangent at points of smoothness of $\mathcal{B}^{(1)}$. Given $g(x,y) = q(x + i\,y)$ on $\mathcal{B}^{(1)}$, we can thus determine $p(z) =$ $c + \int_a^z q(t)\,dt$, where c is an arbitrary constant, where the integrations are taken along $\mathcal{B}^{(1)}$, and where we can accurately carry out such indefinite integrations via use of Sinc indefinite integration approximation along each segment $\mathcal{B}_J^{(1)}$ of $\mathcal{B}^{(1)}$. We can thus solve a Dirichlet problem to determine a function ν, as we determined μ above, and having determined ν, we can get $\mu = \mathcal{S}\,\nu$ through a Sinc method to approximate the Hilbert transform on each arc Γ_J. We can then determine u in the interior of $\mathcal{B}$ through use of a Sinc method to approximate a Cauchy transform on each $\Gamma_J = \mathcal{B}_J^{(1)}$.

These outlined steps thus reduce the solution of two dimensional Neumann problems to the solution of two dimensional Dirichlet problems using one dimensional indefinite integration and Hilbert transforms. These are one dimensional procedures that can be carried out efficiently with Sinc–Pack, The solution does of course depend on the arbitrary constant c, and is thus not unique, a result predicted by the *Fredholm Alternative*.

Solution of a Poisson Problem on $\mathcal{B}^{(2)}$

A particular solution to Poisson's equation

$$\nabla^2 w(\bar{r}) = -f(\bar{r})\,, \quad \bar{r} \in \mathcal{B}^{(2)}\,, \tag{2.6.30}$$

is given by

$$w(\bar{r}) = \int_{\mathcal{B}^{(2)}} \mathcal{G}(x - \xi\,,\; y - \eta)\,f(\xi\,,\;\eta)\,d\xi\,d\eta\,, \tag{2.6.31}$$

where $\bar{r} = (x\,,\;y)$, and where $\mathcal{G}$ denotes the Green's function

$$G(x, y) = \frac{1}{2\pi} \log \left\{ \frac{1}{\sqrt{x^2 + y^2}} \right\}. \qquad (2.6.32)$$

It will now be assumed that $f \in \mathbf{X}^{(2)} \left(\mathcal{B}^{(2)} \right)$. A solution to (2.6.30) subject to, e.g., given Dirichlet boundary conditions, can be obtained by first obtaining the above particular solution w, and then adding a solution of (2.6.21). Similar remarks apply for the case of a solution to (2.6.30) subject to Neumann conditions.

Analyticity of w. Let us assume that $\mathcal{B}^{(2)}$ has been subdivided into a union of a finite number of regions $\mathcal{B}_J^{(2)}$ as in (2.6.7) above, and that $f \in \mathbf{X}^{(2)} \left(\mathcal{B}^{(2)} \right)$. The above integral expressing $w(\bar{r})$ can then be written in the form

$$w(x, y) = \sum_{K=1}^{\nu_2} \int_{\mathcal{B}_K^{(2)}} G(x - x', \, y - y') f_K(x', \, y') \, dx' \, dy'. \qquad (2.6.33)$$

where f_K denotes the restriction of f to $\mathcal{B}_K^{(2)}$.

Theorem 2.6.11 *If $f \in \mathbf{X}^{(2)} \left(\mathcal{B}^{(2)} \right)$, then $w \in \mathbf{X}^{(2)} \left(\mathcal{B}^{(2)} \right)$.*

Proof. Consider first the case when $(x, y) \in \mathcal{B}_J^{(2)}$, and we are integrating over $\mathcal{B}_K^{(2)}$, with $J \neq K$. If $\mathcal{B}_J^{(2)}$ and $\mathcal{B}_K^{(2)}$ do not have a common boundary, then, after transformation of both $(x, y) \to (\xi, \eta) \in (0, 1)^2$ and $(x', y') \to (\xi', \eta') \in (0, 1)^2$ via (2.6.8), we arrive at an integral of the form

$$\omega_{JK}(\xi, \eta) = \int_{(0,1)^2} \mathcal{G}_{JK}(\xi, \xi', \eta, \eta') F_K(\xi', \eta') \, \mathcal{J}_K(\xi', \eta') d\xi' \, d\eta'$$
$$(2.6.34)$$

where $\mathcal{G}_{JK}$ is the transformed Green's function, F_K is the transformed f_K, and $\mathcal{J}_K$ is the Jacobian of the transformation (2.6.8). We then find that $\mathcal{G}(\cdot, \xi', \eta, \eta')$ belongs to $\mathbf{M}'$ for each fixed $(\xi', \eta, \eta') \in [0, 1] \times [0, 1] \times [0, 1]$, and $\mathcal{G}(\xi, \cdot, \xi', \eta')$ belongs to $\mathbf{M}'$ for each fixed $(\xi, \xi', \eta') \in [0, 1] \times [0, 1] \times [0, 1]$. It thus follows, that the transformed Green's function belongs to $\mathbf{X}^{(2)} \left(\mathcal{B}^{(2)} \right)$ as a function of

(ξ, η), for each fixed (ξ', η') on $[0, 1] \times [0, 1]$. We thus conclude that $\omega_{JK} \in \mathbf{X}^{(2)}\left(\mathcal{B}_J^{(2)}\right)$.

If $J \neq K$ but $\mathcal{B}_J^{(2)}$ and $\mathcal{B}_K^{(2)}$ do have a common boundary, then $\mathcal{G}(x - \xi, \; y - \eta)$ is not uniformly bounded. In this case we again transform over $(0, 1)^2$ as above. However, instead of integrating over $(0, 1)^2$, we integrate (in theory) over

$$(0, 1)_\ell^2 = \{(\xi', \eta') : 1/4^\ell \leq \xi' \leq 1 - 1/4^\ell, \; 1/4^\ell \leq \eta'' \leq 1 - 1/4^\ell\}$$

$$\ell = 1, \, 2, \, \ldots.$$

$$(2.6.35)$$

We get a sequence of functions $\left\{\omega_{JK}^\ell\right\}_{\ell=1}^\infty$. Upon fixing an arbitrary $\eta \in [0, 1]$ and setting $u^\ell = \omega_{JK}^\ell(\cdot, \eta)$, it is clear that the sequence $\{u^\ell\}_1^\infty$ is an infinite sequence of analytic functions that is uniformly bounded on $\mathcal{E}$ for all fixed $y \in [0, 1]$. By Montel's theorem, this sequence has a convergent subsequence. It is clear, also, that the sequence $\{u^\ell(\xi)\}_1^\infty$ converges for every fixed $\xi \in [1/4, 3/4]$, so that in particular, it converges at an infinite sequence of discrete points on $[1/4, 3/4]$ that has a limit point on $[1/4, 3/4]$. It thus follows by Vitali's theorem (see [S1, Corollary 1.1.17]) that the sequence $\{u^\ell\}_1^\infty$ converges to a function u that is analytic and bounded on $\mathcal{E}$ for all fixed $\eta \in [0, 1]$. It can furthermore be readily verified by proceeding as in the proof of Theorem 6.5.6 of [S1], that $u \in \mathbf{Lip}_\alpha[0, 1]$ whenever $f \in \mathbf{X}^{(2)}\left(\mathcal{B}^{(2)}\right)$.

Similar statements can be made for the sequence of functions $\{v^\ell\}_1^\infty$, where $v^\ell = \omega_{JK}^\ell(x, \cdot)$.

It follows, therefore, that the sequence of functions $\left\{\omega_{JK}^\ell\right\}_{\ell=1}^\infty$ converges to a function $\omega \in \mathbf{X}^{(2)}\left(\mathcal{B}_J^{(2)}\right)$.

We skip the proof for the case when $J = K$, leaving it to Problem 3.2.1, since it is nearly exactly the same as the proof analyticity discussed above. ∎

Separation of Variables Evaluation of the Convolution. Let G denote the two dimensional Laplace transform of $\mathcal{G}$,

$$G(s,\,\sigma) = \int_0^\infty \int_0^\infty \mathcal{G}(x,y) \exp\left(-\frac{x}{s} - \frac{y}{\sigma}\right) dx\,dy. \qquad (2.6.36)$$

This function G is explicitly evaluated in the proof of Lemma 2.4.2.

We consider two cases, that for a rectangular, and that for a curvilinear region.

1. *The Case of a Rectangular Region $\mathcal{B}$.* We consider, for simplicity, the case of $\mathcal{B} = (0,1)^2$.

   For purposes of indefinite integration on $(0,1)$, let $\phi(z) = \log(z/(1-z))$, $\mathbf{w} = (\omega_{-N},\,\ldots,\,\omega_N)$ denote a Sinc basis for $(0,1)$, and in the notation of (1.5.49), select $N_x = N$, $N_y = N$, $h_x = h_y = h$, take $m = 2N+1$, and use the same Sinc points $z_{-N},\,\ldots,\,z_N$ for approximation in both variables x and y. We use the notation of Definition 1.5.12,

$$
\begin{aligned}
(\mathcal{J}^+ f)(x) &= \int_0^x f(t)\,dt, \\
(\mathcal{J}^- f)(x) &= \int_x^1 f(t)\,dt, \\
(\mathcal{J}_m^+ f)(x) &= \mathbf{w}(x)\,A\,V\,f, \qquad\qquad (2.6.37) \\
A^+ &= h\,I^{(-1)}D(1/\phi'), \\
(\mathcal{J}_m^- f)(x) &= \mathbf{w}(x)\,B\,V\,f, \\
A^- &= h\,(I^{(-1)})^T D(1/\phi'),
\end{aligned}
$$

and similarly for integration with respect to y, in which case we replace A by B. Let us diagonalize these matrices, in the form

$$A^+ = X\,S\,X^{-1}, \qquad A^- = Y\,S\,Y^{-1} \qquad (2.6.38)$$

Here S is a diagonal matrix,

$$S = \mathrm{diag}\,(s_{-N},\,\ldots,\,s_N).$$

For convenience, we set $Xi = X^{-1}$ and $Yi = Y^{-1}$.

The above integral expression for w can be written as a sum of four indefinite product integrals in the form

$$w = \left(\int_0^y \int_0^x + \int_0^y \int_x^1 + \int_y^1 \int_0^x + \int_y^1 \int_x^1 \right) \mathcal{G} \, f d\xi \, d\eta.$$

$$(2.6.39)$$

Each of these four product integrals can be evaluated using Algorithm 2.5.1. Indeed, the second integral in (2.6.39) – call it v – is of exactly the same form as the integral (2.5.1). With G denoting the two dimensional Laplace transform of $\mathcal{G}$, with F the matrix having $(i,j)^{th}$ entry $f(z_i, z_j)$, and $V = [V_{ij}]$, then by (2.5.5) the seven–step Algorithm 2.5.1 can be reduced to the following single line matrix (MATLAB®) expression

$$V = Y * (G. * (Yi * F * Xi.')) * X.'$$ (2.6.40)

where X and Y are square matrices of eigenvectors of A and B defined as above, and where we have denoted by G the $m \times m$ matrix with entries $G(s_i, s_j)$. This yields an accurate approximation $v(z_i, z_j) \approx V_{i,j}$.

That is, we can evaluate an $m \times m$ matrix $W = [W_{i,j}]$ (with $m = 2N + 1$) via the four line *Matlab* program

```
W =      X*(G.*(Xi*F*Xi.'))*X.';
W = W + Y*(G.*(Yi*F*Xi.'))*X.';
W = W + X*(G.*(Xi*F*Yi.'))*Y.';
W = W + Y*(G.*(Yi*F*Yi.'))*Y.';
```

to get the accurate approximation

$$w(z_i, z_j) \approx W_{i,j}.$$ (2.6.41)

We are thus able to approximate the above integral w applying one dimensional Sinc methods and thus obtaining separation of variables.

2. *The Case of a Curvilinear $\mathcal{B}^{(2)}$.* We consider here the case when $\mathcal{B}^{(2)} = \bigcup_{J=1}^{\nu_2} \mathcal{B}_J^{(2)}$ is a curvilinear region, with each $\mathcal{B}_J^{(2)}$ a rotation (2.6.6) of the form of (2.6.5). In this more complicated situation, we proceed as follows:

We transform each of these regions $\mathcal{B}_J^{(2)}$ to the unit square, via the rotation (2.6.6) and the quasi–linear transformation (2.6.8). For sake of simplifying our argument, we again take $N_x = N_y = N$, $h_x = h_y = h$, and $m_x = m_y = m = 2\,N+1$, so that the Sinc grid takes the form $(\xi_i\,,\ \eta_j) = (z_i\,,\ z_j)$, where $\{z_j\}_{-N}^{N}$ are the Sinc points on $(0,1)$.

If $\bar{r}_{ij} = \bar{r}(z_i\,,\ z_j) \in \mathcal{B}_J^{(2)}$ and we are integrating over $\mathcal{B}_K^{(2)}$, with $K \neq J$, we simply use repeated Sinc quadrature to approximate each of the $p-1$ integrals

$$P_{JK}(\bar{r}_{ij}) = \int_{\mathcal{B}_K} H(\bar{r}_{ij}, \bar{\rho})d\bar{\rho} \qquad (2.6.42)$$

for $K = 1,\ \ldots,\ \nu_2$, where H is the integral in (2.6.34), after transformation of the variable $\bar{r}$ over (a possibly rotated) $\mathcal{B}_J$ and the variable $\bar{r}'$ over (a possibly rotated) $\mathcal{B}_K$.

We skip here the case when $K = J$, since this case is discussed in detail in §2.5.3, where an explicit separation of variables algorithm, Algorithm 2.5.4, is described for approximation of the integral.

Laplace–Dirichlet Problems in Three Dimensions.

Consider the Laplace–Dirichlet problem

$$\nabla^2 u(\bar{r}) = 0, \quad \bar{r} \in \mathcal{B}^{(3)}$$
$$\tag{2.6.43}$$
$$u(\bar{r}) = g(\bar{r}), \quad \bar{r} \in A = \partial\mathcal{B}^{(3)}$$

for which g is a real–valued function belonging to $\mathbf{X}^{(2)}(S)$. The solution u of (2.6.43) can be represented in the form

$$u(\bar{r}) = \frac{1}{2\pi} \int_S \left\{ \frac{\partial}{\partial n(\bar{r}')} \frac{1}{|\bar{r} - \bar{r}'|} \right\} \mu(\bar{r}')\, dA(\bar{r}'), \qquad (2.6.44)$$

where $\mathbf{n}(\bar{r}')$ is the outward normal to $\partial \mathcal{B}^{(3)}$ at the point $\bar{r}' \in S$, and where μ is a function to be determined.

We assume, of course, that $S = \partial \mathcal{B}^{(3)}$ is the union of ν_3 surface patches S_J discussed at the end of §2.5 above, so that $g \in \mathbf{X}^{(2)}(S_J)$ on each of these patches. Moreover, by proceeding as in [S1, §6.5.4], it follows, with reference to (2.6.44), that if $\mu \in \mathbf{X}^{(2)}((0,1)^2)$, then $u \in \mathbf{X}^{(3)}\left(\mathcal{B}^{(3)}\right)$.

Now, by letting $\bar{r}$ approach a point of smoothness $\bar{\rho}$ of $\partial \mathcal{B}^{(3)}$, we find that μ is given by the solution of the boundary integral equation

$$\mu + K\mu = g, \tag{2.6.45}$$

where

$$(K\mu)(\bar{\rho})$$
$$= -\frac{1}{2\pi} \int_S \left\{ \frac{\partial}{\partial n(\bar{r}')} \frac{1}{|\bar{\rho} - \bar{r}'|} \right\} \mu(\bar{r}') \, dA(\bar{r}'), \tag{2.6.46}$$

This operator K is known to have a simple eigenvalue 1, for which the corresponding eigenfunction is 1. This eigenvalue is the largest eigenvalue in modulus, and since all other eigenvalues are real, all other eigenvalues are less than 1 in modulus [Za]. We are thus able to solve the integral equation (2.6.45) by successive approximation (i.e., using a one dimensional separation of variable procedure, proceeding similarly as for the case of Equation (2.6.24)).

Details of proofs of analyticity and convergence for 3–d Dirichlet problems are similar to those of the two dimensional case, and we omit them (see [S1, §6.5.4]).

Laplace–Neumann Problems in Three Dimensions.

These are problems of the form

$$\nabla^2 u(\bar{r}) = 0, \quad \bar{r} \in \mathcal{B}^{(3)}$$
$$\frac{\partial}{\partial n(\bar{r})} u(\bar{r}) = g(\bar{r}), \quad \bar{r} \in \partial \mathcal{B}^{(3)}. \tag{2.6.47}$$

It is customary, for this case, to start with a representation of the form

$$u(\bar{r}) = \frac{\partial}{\partial n(\bar{r})} \frac{1}{2\pi} \int_S \frac{1}{|\bar{r} - \bar{r}'|} \mu(\bar{r}') \, dA(\bar{r}') \,, \quad \bar{r} \in \mathcal{B}^{(3)} \,, \qquad (2.6.48)$$

v where $S = \partial \mathcal{B}^{(3)}$, $\mathbf{n}(\bar{r})$ is the outward normal to $\partial \mathcal{B}^{(3)}$ at the point $\bar{r} \in S$, and μ is a function to be determined.

The analyticity properties of μ of (2.6.48) and the solution u of (2.6.47)–(2.6.48) as well as the proofs of such properties are the same as those for three dimensional Dirichlet problems discussed above, and we skip them here. However the solution of (2.6.47) using (2.6.48) is more complicated since it is non–unique, and we therefore briefly discuss methods of solving (2.6.48).

By letting $\bar{r} \to \bar{\rho} \in S$ in (2.6.48), we get the integral equation

$$\mu - K^* \mu = g \qquad (2.6.49)$$

where

$$(K^* \mu)(\bar{\rho}) = \frac{1}{2\pi} \int_S \frac{\partial}{\partial \mathbf{n}(\bar{\rho})} \frac{1}{|\bar{\rho} - \bar{r}'|} \mu(\bar{r}') dA(\bar{r}'). \qquad (2.6.50)$$

The operator K^* is the adjoint of the operator K of (2.4.46) above, and it therefore has an eigenvalue 1, which simple, and which is its eigenvalue of largest modulus. Consequently, the equation

$$\mu - K^* \mu = 0 \qquad (2.6.51)$$

has a non–trivial solution, μ_0, and equation (2.6.49) is therefore solvable if and only if (see Theorem 6.2.8 of [S1])

$$\int_S g(\bar{r}) \, \mu_0(\bar{r}) \, dA(\bar{r}) = 0. \qquad (2.6.52)$$

Solving the Homogeneous Equation. The eigensolution μ_0 of (2.6.51) can be obtained from the convergent iteration procedure

$$\left. \begin{aligned} p_0 &= 1 \\ q_{j+1} &= K^* p_j \\ p_{j+1} &= \frac{q_{j+1}}{\|q_{j+1}\|} \end{aligned} \right\} j = 0, \, 1, \, 2, \, \ldots \,, \qquad (2.6.53)$$

where $\| \cdot \|$, can be any $\mathbf{L}^p$ norm. Moreover, (2.6.53) can be readily carried using one dimensional Sinc convolution methods, in which case division by the largest element of $\mathbf{q}$, the discretized version of q is appropriate. We omit the details.

Solving the Non–Homogeneous Equation. One can also obtain a solution to the non-homogeneous equation without computing the solution to the homogeneous equation. This is possible by setting

$$v(\bar{r}) = u(\bar{r}) + \frac{1}{2\pi |\bar{r} - \bar{r}_0|} \int_S \mu(\bar{r}') \, dA(\bar{r}'). \tag{2.6.54}$$

where $\bar{r}_0$ is a point strictly on the exterior of $\mathcal{B}$, and surface $S = \partial\mathcal{B}^{(2)}$. The boundary condition $v(\bar{r}) = g(\bar{r})$ is now satisfied for any given g if $\mu(\bar{r})$ is the solution to the integral equation

$$\mu(\bar{\rho}) + \frac{1}{2\pi |\bar{\rho} - \bar{r}_0|} \int_S \mu(\bar{r}') \, dA(\bar{r}') = (K\,\mu)(\bar{\rho}) + g(\bar{\rho})\,, \bar{\rho} \in S. \tag{2.6.55}$$

This equation can also be solved by iteration involving rectangular matrices. After discretization, one starts with an initial functional form $\mu = \mu^0 \equiv 1$, one forms the matrix $G = g(\bar{\rho}_{ij})$, one evaluates $\kappa = [\kappa_{ij}] = [(K\,\mu)(\bar{\rho}_{ij})]$ by Sinc convolution, one sets $P = [P_{ij}] = G + \kappa$, and finally, one sets $\mu = [\mu^1_{ij}] = [P_{ij}/(1 + 1/(2\pi |\bar{\rho}_{ij} - \bar{r}_0|))]$. Next we take $\mu = \mu^1$ on the right hand side, and so on. This process is also convergent, and can be carried out using one dimensional Sinc operations, that is, using of separation of variables.

Poisson Problems in Three Dimensions.

A particular solution to the problem

$$\nabla^2 u(\bar{r}) = -f(\bar{r})\,, \quad \bar{r} \in \mathcal{B}\,, \tag{2.6.56}$$

is given by

$$u(\bar{r}) = \frac{1}{4\pi} \int_\mathcal{B} \frac{f(\bar{r}')}{|\bar{r} - \bar{r}'|} \, d\bar{r}\,, \quad \bar{r} \in \mathcal{B}. \tag{2.6.57}$$

We assume here, of course that depending on f and the boundary of the three dimensional region $\mathcal{B}^{(3)}$ we can subdivide $\mathcal{B}^{(3)}$ into a finite number, ν_3, of rotations of subregions $\mathcal{B}^{(3)}_J$ of the form $(2.6.11)$, such that the restriction f_J of f to $\mathcal{B}^{(3)}_J$ belongs to $\mathbf{X}^{(3)}\left(\mathcal{B}^{(3)}_J\right)$.

Next, a separation of variables method for evaluating (2.6.57) is described explicitly in §2.5.2 for the case of $\mathcal{B}^{(3)}$ a rectangular region, and in §2.5.4, for the case of $\mathcal{B}^{(3)}$ a region of the form in (2.6.11). In the case when $\mathcal{B}^{(3)}$ is a curvilinear region, we can proceed as follows (similarly to the case of two dimensional Poisson problems).

For sake of simplifying our argument, we set $N_x = N_y = N_z = N$, $h_x = h_y = h_z = h$, and $m = 2N+1$, so that the Sinc grid takes the form $(\xi_i,\,\eta_j,\,\zeta_k) = (z_i,\,z_j,\,z_k)$, where $\{z_j\}_{-N}^{N}$ are Sinc points on $(0,1)$.

If $\bar{r}_{ijk} = \bar{r}(z_i,\,z_j,\,z_k) \in \mathcal{B}_J$ and we are integrating over $\mathcal{B}_K$, with $K \neq J$, we simply use repeated Sinc quadrature to approximate each of the $\nu_3 - 1$ integrals

$$P_{JK}(\bar{r}_{ijk}) = \int_{\mathcal{B}_K} H(\bar{r}_{ijk}, \bar{\rho})d\bar{\rho} \qquad (2.6.58)$$

for $K = 1,\ \ldots,\ \nu_3$, after transformation of the variable $\bar{r}$ over (a possibly rotated) $\mathcal{B}_J$ and the variable $\bar{r}'$ over (a possibly rotated) $\mathcal{B}_K$. The evaluation of (2.6.53) for the case when $K = J$ is discussed in detail in §2.5.4, where an explicit separation of variables algorithm is described for approximation of the integral.

Showing that the result of evaluation of (2.6.58) belongs to $\mathbf{X}^{(3)}$ for the case when $K = J$ is left to Problem 2.5.4, and is omitted.

It thus follows that if $f \in \mathbf{X}^{(3)}\left(\mathcal{B}^{(3)}\right)$, then the function u defined in (2.5.46) also belongs to $\mathbf{X}^{(3)}\left(\mathcal{B}^{(3)}\right)$.

Hyperbolic Problems. As an extension of (2.3.6), we consider problems of the form

$$\frac{1}{c^2}\frac{\partial^2 u(\bar{r}, t)}{\partial t^2} - \nabla^2 u(\bar{r}, t) = f(\bar{r}, t), \qquad (\bar{r}, t) \in \mathcal{B}^{(d)} \times (0, T),$$

$$u(\bar{r}, 0^+) = g(\bar{r}), \qquad \frac{\partial u}{\partial t}(\bar{r}, 0^+) = h(\bar{r}),$$

$$p(\bar{r}, t) \in (\bar{r}, t) \in \partial\mathcal{B}^{(d)} \times (0, T),$$

$$(2.6.59)$$

where d is either 2 or 3, $\mathcal{B}^{(d)}$ is a (not necessarily strict) subset of $\mathbb{R}^d$, and T is either bounded or infinity. and for given functions

f, g, h, and p. The Green's function $\mathcal{G}(\bar{r}, t)$ which we require for the solution to this problem is determined by the solution to the following initial–boundary problem on $\mathbb{R}^d \times (0, \infty)$:

$$\frac{1}{c^2} \frac{\partial^2 \mathcal{G}(\bar{r}, t)}{\partial t^2} - \nabla^2 \mathcal{G}(\bar{r}, t) = \delta(t) \, \delta^d(\bar{r}),$$

(2.6.60)

$$\mathcal{G}(\bar{r}, 0^+) = 0, \qquad \frac{\partial \mathcal{G}}{\partial t}(\bar{r}, 0^+) = 0.$$

A particular solution to the non-homogeneous part of (2.6.59) is given by

$$w(\bar{r}, t) = \int_0^t \int_{\mathcal{B}^{(d)}} \mathcal{G}(\bar{r} - \bar{r}', t - t') \, f(\bar{r}', t') \, d\bar{r}' \, dt' \qquad (2.6.61)$$

Local analyticity of the function w defined by this integral follows from the second of the above Cauchy–Kowalevsky theorems, and indeed, $w \in \mathbf{X}^{(d)}\left(\mathcal{B}^{(d)}\right)$ whenever $f \in \mathbf{X}^{(d)}\left(\mathcal{B}^{(d)}\right)$. This solution can be computed via separation of variables using the Laplace transform given Lemma 2.4.5 for the case of $d = 2$, and by Lemma 2.4.6 for the case of $d = 3$.

If we take the Laplace transform of the function $\mathcal{G}(\bar{r}, \cdot)$ in (2.6.60) with respect to t,

$$\hat{\mathcal{G}}(\bar{r}, \tau) = \int_0^\infty \mathcal{G}(\bar{r}, t) \, \exp(-t/\tau) \, dt, \qquad (2.6.62)$$

and apply the given initial conditions, we obtain the Helmholtz equation,

$$\nabla^2 \hat{\mathcal{G}}(\bar{r}, \tau) - \frac{1}{c^2 \tau^2} \hat{\mathcal{G}}(\bar{r}, \tau) = \delta^d(\bar{r}). \qquad (2.6.63)$$

Alternatively, we could apply a Laplace transform of the homogeneous part of the PDE in (2.6.59) with respect to t to obtain a Helmholtz boundary value problem for each τ.

Our motivation for taking these Laplace transforms is that we can now use the same techniques to study these equations that we used to study Laplace and Poisson problems above. Indeed, in two and three dimensions, the Green's functions that satisfy (2.6.63) are given by

$$\widehat{\mathcal{G}_2}(\bar{\rho}) = \frac{1}{2\pi} K_0 \left(\frac{\log |\bar{\rho}|}{c\tau} \right)$$

$$(2.6.64)$$

$$\widehat{\mathcal{G}_3}(\bar{r}) = \frac{\exp\left(-\frac{|\bar{r}|}{c\tau}\right)}{4\pi |\bar{r}|}.$$

Since sums and products of functions in $\mathbf{X}^{(2)}\left(\mathcal{B}^{(2)}\right)$ again belong to $\mathbf{X}^{(2)}\left(\mathcal{B}^{(2)}\right)$, the analyticity proofs of [S1, §6.5.4] for elliptic PDE extend directly to the solutions Helmholtz PDE.

If we set $\rho = |\bar{\rho}|$, then $\widehat{\mathcal{G}_2}$ has the form

$$\widehat{\mathcal{G}_2}(\bar{\rho}) = A(\rho)\,\log(\rho) + B(\rho),$$

$$(2.6.65)$$

where A and B are entire functions of ρ^2. Thus, the analyticity proofs are almost verbatim the same as those for the above two dimensional elliptic problems.

Similarly, if we set $r = |\bar{r}|$, then $\widehat{\mathcal{G}_3}$ takes the form

$$\widehat{\mathcal{G}_3}(\bar{r}) = C(r)/r + D(r)$$

$$(2.6.66)$$

where C and D are entire functions of r^2.

Sums and products of linear and nonlinear functions in $\mathbf{X}^{(3)}\left(\mathcal{B}^{(3)}\right)$ again belong to $\mathbf{X}^{(3)}\left(\mathcal{B}^{(3)}\right)$, so that the analyticity proofs of [S1, §6.5.4] for elliptic PDE extend directly to the solutions Helmholtz PDE, and we omit them, leaving these proofs to the problems at the end of this section.

Separation of variables follow directly, via the algorithms in §2.5. The example in §4.2.1 illustrates the general approach for solving wave problems over $\mathbb{R}^3 \times (0, T)$.

Parabolic Problems.

Let us consider first, the special parabolic problem

$$U_t(\bar{r}, t) - \varepsilon \nabla^2 U(\bar{r}, t) = f(\bar{r}, t), \quad (\bar{r}, t) \in \mathbb{R}^d \times (0, T),$$

$$(2.6.67)$$

$$U(\bar{r}, 0^+) = u_0(\bar{r}), \quad \bar{r} \in \mathbb{R}^d,$$

which has the solution

$$
\begin{aligned}
U(\bar{r}, t) \;=\; & \int_0^t \int_{\mathbb{R}^d} \mathcal{G}(\bar{r} - \bar{r}', t - t')\, f(\bar{r}', t')\, d\bar{r}'\, dt' \\
& + \int_{\mathbb{R}^d} \mathcal{G}(\bar{r} - \bar{r}', t)\, u_0(\bar{r}')\, d\bar{r}',
\end{aligned}
\tag{2.6.68}
$$

where the Green's function $\mathcal{G}$ is a solution in $\mathbb{R}^d \times (0, \infty)$ of the initial value problem PDE

$$
\begin{aligned}
\mathcal{G}_{t'} - \varepsilon \nabla^2 \mathcal{G} &= \delta(t - t')\, \delta^d(\bar{r} - \bar{r}'), \\
\mathcal{G}(\bar{r}, 0^+) &= 0.
\end{aligned}
\tag{2.6.69}
$$

These free space Green's functions $\mathcal{G}(\bar{r}, t)$ are given explicitly in §2.3.1.

If we take the Laplace transform of the function $\mathcal{G}(\bar{r}, t)$ with respect to t,

$$
\widehat{\mathcal{G}}(\bar{r}, \tau) = \int_0^\infty \mathcal{G}(\bar{r}, t)\, \exp(-t/\tau)\, dt
\tag{2.6.70}
$$

and apply the given initial conditions, we again get a Helmholtz equation,

$$
\nabla^2 \widehat{\mathcal{G}}(\bar{r}, \tau) - \frac{1}{\varepsilon \tau}\, \widehat{\mathcal{G}}(\bar{r}, \tau) = \frac{1}{\varepsilon}\, \delta^d(\bar{r}).
\tag{2.6.71}
$$

The analyticity and separation of variables discussion of this case is similar to that of the wave equation above, and is omitted.

Consider next, solution of the homogeneous problem,

$$
\begin{aligned}
U_t(\bar{r}, t) - \varepsilon \nabla^2 U(\bar{r}, t) &= 0, \quad (\bar{r}, t) \in \mathcal{B}^{(d)} \times (0, T), \\
U(\bar{r}, 0^+) &= u_0(\bar{r}), \quad \bar{r} \in \mathcal{B}^{(d)}, \\
U(\bar{r}, t) &= g(\bar{r}, t), \quad (\bar{r}, t) \in \partial \mathcal{B}^{(d)} \times (0, T).
\end{aligned}
\tag{2.6.72}
$$

In (2.6.72) functions $u_0 \in \mathbf{X}^{(d)}\left(\mathcal{B}^{(d)}\right)$ and $g \in \mathbf{X}^{(d)}(S \times (0, T))$ are given, and assumed to belong to appropriate spaces as defined in §2.6.1.

The solution to (2.6.72) can be expressed in the form

$$U(\bar{r}, t) = V_0(\bar{r}, t) + W(\bar{r}, t),$$ (2.6.73)

where

$$V_0(\bar{r}, t) = \int_{\mathcal{B}^{(d)}} \mathcal{G}(\bar{r} - \bar{r}', t) u_0(\bar{r}') \, d\bar{r}'$$ (2.6.74)

and where W can be expressed in terms of a double layer potential,

$$W(\bar{r}, t) = (2\,\varepsilon) \int_0^t \int_{\partial \mathcal{B}^{(d)}} \frac{\partial}{\partial n} (\bar{r}') \mathcal{G}(\bar{r} - \bar{r}', t - t') \mu(\bar{r}', t') \, dA(\bar{r}') \, dt',$$ (2.6.75)

where $\mathbf{n}(\bar{r}')$ is the outward normal to $\partial \mathcal{B}^{(d)}$ at $\bar{r}'$, and where μ is a function to be determined. By letting $\bar{r} \to \bar{\rho} \in \partial \mathcal{B}^{(d)}$, we get an integral equation for μ,

$$\mu(\bar{\rho}, t) + \int_0^t \int_{\partial \mathcal{B}^{(d)}} \frac{\partial}{\partial n} (\bar{r}') \mathcal{G}(\bar{\rho} - \bar{r}', t - t') \mu(\bar{r}', t') \, dA(\bar{r}') \, dt'$$

$$= -T(\bar{\rho}, t),$$ (2.6.76)

where

$$T(\bar{\rho}, t) = g(\bar{\rho}, t) - \lim_{\bar{r} \to \bar{\rho}} V_0(\bar{r}, t).$$ (2.6.77)

The integral equation (2.6.72) for μ can be solved via iteration. In theory, one constructs a sequence of functions $\left\{ \mu^{(k)} \right\}_{k=0}^{\infty}$, starting with $\mu = \mu^{(0)} = w$ in the integral, evaluating the integral, then subtracting the result from $-w$ to get $\mu^{(1)}$, and so on. This process can also be carried out via Sinc convolution.

Consider first the case of $d = 2$. Recall Lemma 2.4.7, in which an explicit expressions for the Laplace transform $G(u, v, \tau)$ of the two dimensional free–space Green's function $\mathcal{G}(x, y, t)$ and was obtained. In determining a solution via Sinc procedure we first determine the spacial accuracy required for approximation on $\partial \mathcal{B}^{(2)}$, before determining the time accuracy for approximation on $\partial \mathcal{B}^{(2)} \times (0, T)$. This means that we first select the spacial Sinc indefinite integration matrices. We thus fix the eigenvalues (that occupy the positions (u, v)

in the expression for $G(u, v, \tau)$ (this is $\widehat{G^{(2)}}(\bar{\Lambda}, \tau)$ of Theorem 2.4.11) in the process of Sinc convolution) as well as the eigenvectors of these matrices. The corresponding eigenvalues of Sinc time indefinite integration matrices for integration over $(0, T)$ are just T times the corresponding eigenvalues for integration over $(0, 1)$, while the eigenvectors of these matrices are the same as those for integration over $(0, 1)$. It thus follows, upon replacing τ by $T\tau$ in the expression $G(u, v, \tau)$, then keeping u, v, and τ fixed, that $G(u, v, T\tau) \to 0$ as $T \to 0$. Consequently, for the case of $d = 2$, the Sinc approximated integral part of the integral equation (2.6.76) is a contraction for all T sufficiently small, which implies that the sequence $\left\{\mu^{(k)}\right\}_{k=0}^{\infty}$ converges to an approximate solution of (2.6.76). Moreover, this sequence of approximations is obtainable using one dimensional separation of variables Sinc methods. It is readily established, moreover, that the solution μ of (2.6.76) belongs to $\mathbf{X}^{(2)}(\partial\mathcal{B}^{(d)} \times (0, T))$. It thus follows that we are able to also get a uniformly accurate approximation to μ. Finally, having gotten an accurate approximation to μ, we can then get an accurate approximation to W in (2.6.75) using the Sinc convolution procedure.

Similar remarks can be made for the case of $d = 3$.

Nonlinear Problems.

Sinc methods offer several advantages of other methods for solution of linear and nonlinear heat and wave problems when the solution can be expressed as a solution to a convolution–type Volterra IE. In theory, such equations can be solved by successive approximations, a procedure that preserves analyticity. Such a successive approximation procedure can also be carried out explicitly via Sinc methods, based on separation of variables, since we can obtain explicit successive accurate approximation of the nonlinear term. This result follows, directly from Theorem 1.5.13, namely, if $u \in \mathbf{M}$, and if $U_m = \sum_{k=-N}^{N} u(z_k)\omega_k$ denotes an $m = 2N + 1$–term uniformly accurate Sinc basis approximation to u on $(0, 1)$, then there exists a positive constant c that is independent of m, such that

$$\sup_{t \in (0,1)} \left| u(t) - \sum_{k=-N}^{N} u(z_k)\, \omega_k(t) \right| = \mathcal{O}\left(\exp\left(-c\, m^{1/2} \right) \right), \quad (2.6.78)$$

with $z_k = z_k^{(m)}$ denoting the Sinc points on $(0,1)$. Let $F(u(t),t)$ denote the nonlinear part of a differential equation, $L\,u = F$, with L a linear (Laplace, wave, or heat) equation operator, and let $u = u(t) \in [c,d]$, such that both functions $F(\cdot,t)$ for fixed $t \in [0,1]$ and $F(\tau,\cdot)$ for fixed $\tau \in [c,d]$ belong to $\mathbf{M}'$. It then follows that $g \in \mathbf{M}'$, where $g(t) = F(u(t),t)$. Hence, setting $u_k = u(z_k)$, we get

$$\sup_{t \in (0,1)} \left| F(u(t),t) - \sum_{k=-N}^{N} F(u_k, z_k)\,\omega_k(t) \right| = \mathcal{O}\left(\exp\left(-c\,m^{1/2} \right) \right).$$

$$(2.6.79)$$

This type of approximation was used in [SGKOrPa] to solve nonlinear ordinary differential equation initial value problems. It was also used in §3.9 to obtain solutions of nonlinear ordinary differential equations via successive approximations.

This method of approximating solutions to nonlinear equations extends readily to more than one dimension. As one dimensional examples, successive approximation was used in [SBaVa] to solve Burgers' equation, and in §4.3.1 below to solve a nonlinear population balance equation. Successive approximation of this type via Sinc methods, has also been successfully applied to multidimensional problems, illustrated in Chapter 4 of this handbook. (See also [N].)

It is thus quite straight forward to obtain solutions to problems of the form

$$\Delta u = -f((u(\bar{r}), \nabla u(\bar{r})\,\bar{r}) \quad \bar{r} \in \mathcal{B}; \qquad (2.6.80)$$

or

$$\frac{\partial u}{\partial t} - \Delta u = f(u(\bar{r}), \nabla u(\bar{r}), \bar{r}, t), \quad \bar{r}, t) \in \mathcal{B} \times (0, T); \qquad (2.6.81)$$

or

$$\frac{\partial^2 u}{\partial t^2} - \Delta u = f((u(\bar{r}), \nabla u(\bar{r}), \bar{r}, t), \bar{r}, t) \in \mathcal{B} \times (0, T), \qquad (2.6.82)$$

with $\mathcal{B}$ a curvilinear region in $\mathbb{R}^d$, $d = 1, 2, 3$, via successive approximation and Sinc convolution; thus circumventing the use of large

matrices that classical finite element or finite difference methods required important applied problems, e.g., conservation law problems
such as Navier–Stokes equations (see §2.4.5 and 3.4.2), nonlinear
wave equations problems, and nonlinear equations involving material properties.

PROBLEMS FOR SECTION 2.6

2.6.1 Prove that if $g \in \mathbf{M}_{\alpha,\beta,d}(\mathcal{D})$, for some $\alpha \in (0, 1]$, $\beta \in (0, 1]$
and $d \in (0, \pi)$, where φ denotes a conformal map of $\mathcal{D}$ onto $\mathcal{D}_d$,
and if $\gamma \in \mathbf{M}$ denotes an analytic arc, $\gamma : (0, 1) \to \Gamma = \varphi^{-1}(\mathbb{R})$,
then $f = g \circ \gamma \in \mathbf{M}$.

2.6.2 Let $Q = (0, 1) \times (0, 1)$, and consider the integral operation

$$(\kappa f)(\bar{\rho}) = \int \int_Q F(\bar{\rho}, \bar{\sigma}) \log |\bar{\rho} - \bar{\sigma}| \, f(\bar{\sigma}) \, d\bar{\sigma}.$$

with $\bar{\rho} = (x, y)$ and with $\bar{\sigma} = (\xi, \eta)$, $d\bar{\sigma} = d\xi \, d\eta$. Let F, f and
X_α be defined as in [S1, §6.5.4]. Prove that $\kappa f \in X_\alpha$.

2.6.3 Determine a precise definition of a suitable space $\mathbf{X}$, which
enables you to prove if $f = f(z, y)$ is a function of two variables,
if $\mathcal{B}$ is defined as in (2.5.8), if $f \in \mathbf{X}$ with respect to $\mathcal{B}$, if G_2
is defined as in (2.5.31), and if w is defined as in (2.6.43), then
$w \in \mathbf{X}(\mathcal{B})$.

2.6.4 Determine a precise definition of a suitable space $\mathbf{X}$, which
enables you to prove that if $f \in \mathbf{X}(\mathcal{B})$, then the solution u
given by (2.6.52) also belongs go $\mathbf{X}(\mathcal{B})$.

(See [S1, §6.5.4] for help with these problems.)

3

Explicit 1–d Program Solutions via Sinc–Pack

ABSTRACT   We provide Sinc programs written in MATLAB® meant to illustrate and implement the theory of Chapter 1. Sinc programming procedure is demonstrated through fully commented programs solving elementary and advanced one dimensional problems.

3.1   Introduction and Summary

Sinc–Pack is a package of programs in MATLAB®, which implements the algorithms of Chapters 1 and 2. In this chapter we focus on the elementary operations of calculus. We shall see how any operation of calculus on functions f that is defined on an interval or arc Γ is approximated using a simple multiplication of a matrix times a column vector $\mathbf{f}$ whose components are values of f or approximations of such values at "Sinc points" on Γ. The resulting column vector can then be evaluated at any point of Γ through explicit Sinc interpolation.

Let

$$F(x) = \mathcal{O}f(x) \qquad (3.1.1)$$

denote a *calculus operation*, with operator $\mathcal{O}$ acting on a function $f = f(x)$ of one variable, can be any of the operators as given in §1.5. above.

The corresponding *Sinc–Pack operation* approximates F at x as

$$F(x) \approx \mathbf{w}\,\mathbf{O}\,\mathbf{f}\,, \qquad (3.1.2)$$

where $\mathbf{O}$ is the $m \times m$ Sinc–Pack matrix representation of operator $\mathcal{O}$, $\mathbf{w} = \mathbf{w}(x)$ is a row vector of $m = M+N+1$ *Sinc basis functions*,

$$\mathbf{w}(x) = (\omega_{-M}(x),\ \ldots,\ \omega_N(x))\,, \qquad (3.1.3)$$

and

$$\mathbf{f} = (f(z_{-M}),\ \ldots,\ f(z_N))^T,\qquad (3.1.4)$$

is the vector for f evaluated at the *Sinc points* z_k.

Dropping $\mathbf{w} = \mathbf{w}(x)$ in Equation (3.1.2) simplifies to

$$\mathbf{F} = \mathbf{O}\mathbf{f},\qquad (3.1.5)$$

with vector $\mathbf{F} = (F_{-M},\ldots,F_N)^T$. The components F_k of the resulting vector $\mathbf{F}$ satisfy

$$F_k \approx F(z_k)$$

with *error* $\sim e^{-cm^{1/2}}$, provided that M is proportional to N, and f is *analytic* on Γ, with allowable singularities at the end–points of Γ.

Our experience shows that 3 place accuracy of solution is achievable for $m = M + N + 1$ equal to 17 and 5 place accuracy occurs when m is as small as 31. By comparison, the limit with finite elements, and this only for large very large m, is 3 places of least squares accuracy, this reflecting less accuracy at "corners". Often, for hard engineering problems, just one place accuracy is an upper limit on accuracy.

With Sinc, Equations (3.1.2) and (3.1.5) show that the m components of $\mathbf{f}$ is all the input needed to carry out approximating the result of operator $\mathcal{O}$ acting of function $f(x)$. Vector $\mathbf{f}$ is the Sinc vector of orthogonal expansion coefficients with respect to the Sinc basis functions. Equation (3.1.5) then gives the action at Sinc points of operator $\mathcal{O}$ with respect to this basis, while Equation (3.1.2) then interpolates this data back to arbitrary points x simply by taking dot product with basis vector $\mathbf{w}(x)$. However Sinc interpolation is often unnecessary. We will show it can suffice to simply plot values $\{f_k\}$ versus Sinc points $\{z_k\}$. Interpolated values will then all lie within the band of this plot.

Finally we take note that computation of $\mathbf{f}$ is particularly simple via point evaluation of f at Sinc points. Basis function coefficients for other orthogonal expansions come at a higher cost via computing integral type inner products.

Keeping in mind that the $m \times m$ matrices $\mathbf{O}$ of (3.1.5) are typically small, Equations (3.1.2) and (3.1.5) then support Sinc representing a level computational efficiency that stands out among all other methods of numerical analysis.

Let us suppose that one has an approximation problem defined by a calculus expression over some interval or arc Γ. We can then turn to §1.5.3, and obtain one of several useful transformations for approximation on Γ. For example, a map φ, which transforms Γ to real line gives Sinc points $z_j = \varphi^{-1}(j\,h)$, and Sinc weights $w_j = (h/\varphi'(z_j))$. Parameters M, N, and h can be chosen independent of the exact nature of the singularities of F so as to achieve exponential convergence.

3.2   Sinc Interpolation

The theory of Sinc interpolation on an arbitrary interval or arc is treated in §1.5.4. Sinc interpolation bases as defined in Definition 1.5.12 have properties that are convenient and important for applications. These are discussed in the comments preceding the statement of Theorem 1.5.13). In this section we provide MATLAB®. programs which exhibit these properties.

Let us suppose that one has an approximation problem defined by a calculus expression over interval/arc Γ. We can then turn to §1.5.3, and choose, depending on features of the problem at hand, the proper transformation $\varphi : \Gamma \rightarrow \mathbb{R}$ for approximation on Γ. *Sinc points* z_j are given by $\varphi^{-1}(j\,h)$. The associated *Sinc basis vector* **w** has components w_j with expressions found in (1.5.21). They depend on x and parameters M, N, and h. We will see that these parameters can be adjusted independent of the exact nature of the singularities of F so as to achieve uniform approximation.

The goal is to numerically implement interpolation Theorem 1.5.13. **Sinc-Pack** provides programs which are indexed $n = 1, \ldots, 7$ corresponding to the maps of §1.5.3:

1. The subscript 1 corresponding to the identity map of $\mathbb{R} \rightarrow \mathbb{R}$ in Example 1.5.4;

2. The subscript 2 corresponding to the finite interval (a, b), in Example 1.5.5;

3. The subscript 3 corresponding to the interval $(0, \infty)$ which is mapped to $\mathbb{R}$ via the map $\varphi(x) = \log(x)$, in Example 1.5.6;

4. The subscript 4 corresponding to the interval $(0, \infty)$ which is mapped to $\mathbb{R}$ via the map $\varphi(x) = \log(\sinh(x)$, in Example 1.5.7;

5. The subscript 5 corresponding to the map $\varphi(x) = \log(x + \sqrt{1 + x^2})$, which maps $\mathbb{R}$ onto $\mathbb{R}$, in Example 1.5.8;

6. The subscript 6 corresponding to

$$\varphi(x) = \log \left\{ \sinh \left[x + \sqrt{1 + x^2} \right] \right\},$$

which maps $\mathbb{R}$ to $\mathbb{R}$, in Example 1.5.9;

7. The subscript 7 corresponding to

$$\varphi(x) = i(v - u)/2 + \log \left[\left(z - e^{iu} \right) / \left(e^{iv} - z \right) \right]$$

which maps the arc $z = e^{i\theta} : u < \theta < v$ on the unit disc to R, in Example 1.5.10; and

8. Sometimes it is possible to obtain an even more rapid rate of convergence of Sinc methods, via use of a double exponential transformation. Such transformations have been extensively studied by Mori, Sugihara, and their students. Many such transformations are usually compositions of the above formulas. Two such examples are

$$x = \varphi^{-1}(t) = \tanh \left(\tfrac{\pi}{2} \sinh(t) \right),$$

$$x = \varphi^{-1}(t) = \sinh \left(\tfrac{\pi}{2} \sinh(t) \right).$$

which are compositions of the above maps $\varphi(x) = \log((1 + x)/(1 - x))$ and $\varphi(x) = (2/\pi) \log \left(e^x + \sqrt{1 + e^{2x}} \right)$.

3.2.1 Sinc Points Programs

The Sinc points z_k are defined as $\varphi^{-1}(k\,h) = \psi(k\,h)$, where ψ denotes the inverse function of φ. The program Sincpoints*.m, (with $* = 2$) is a MATLAB® program for determining the Sinc points $\varphi^{-1}(k\,h)$, $k = -M, \ldots, N$ on the interval (a, b). The 2 in this program refers to transformation #2 above. The following code computes a row vector of $m = M + N + 1$ Sinc points $\{z_k\}_{-M}^{N}$ for the finite interval $\Gamma = (a, b)$.

```
%
% Sincpoints2.m
%
function [z] = Sincpoints2(a,b,M,N,h)
%
% Here, the formula is
%
%     z(j) = (a + b*e^{j*h})/(1 + e^{j*h})
%
% for j = -M, ..., N .  This program thus
%
% thus computes a vector of m = M+N+1
% values z_j.  See Equation (1.5.1), this
% handbook.
%
% Notice: Our version of Matlab requires that
% components of vectors be stored in the order
% from 1 to m; hence z(-M) here is stored as
% z(1), z(-M+1) as z(2), etc., and similarly,
% z(N) here is stored as z(m).   We thus
% compute these in the manner in which they
% are stored.
%
% Incidentally, this ''storage discrepancy''
% persists for all programs of Sinc-Pack.
%
clear all
m = M+N+1;
T = exp(h);
TT = 1/T^M;
for j = 1:m
    z(j) = (a + b*TT)/(1+TT);
    TT = T*TT;
end
```

For example the MATLAB® statement

"$zz = \text{Sincpoints2}(0, 7, 6, 4, \text{pi}/2)$"

assigns a row vector of m $= 11$ Sinc points to the vector "zz" :

```
>> Sincpoints2(0,7,6,4,pi/2)

ans =

  Columns 1 through 5

    0.0006 0.0027 0.0130 0.0623 0.2900

  Columns 6 through 11

    1.2047 3.5000 5.7953 6.7100 6.9377 6.9870
```

3.2.2 SINC BASIS PROGRAMS

Basis functions $\omega_j(x)$, $j = -M, \dots, N$ are specified in Definition 1.5.12. There are seven MATLAB® programs in Sinc–Pack basis1.m, basis2.m, ..., basis7.m, each relying on basis.m common to them all. The MATLAB® statement W = basis*(x,M,N,h) initiates this program for $* \neq 2, 7$, whereas for $* = 2, 7$, corresponding to $\varphi_*(x)$ mapping a finite interval (a, b), MATLAB® statement
W = basis*(a,b,x,M,N,h) is used.

Example 3.2.1 For example, the programs basis2.m and basis3.m take the form:

```
%
% basis2.m
%
function [val] = basis2(a,b,x,M,N,h)
%
% This program is copyrighted by SINC, LLC
% and belongs to SINC, LLC
%
u = phi2(a,b,x);  % See Example 1.5.5 of Sec. 1.5.3
val = basis(u,M,N,h);
```

where the routine phi2.m is given by the program

```
%
% phi2.m
%
% This program is copyrighted by SINC, LLC
% and belongs to SINC, LLC
%
% Here, phi2 = Sinc map of (a,b) -> R
%
% returns a vector phi2(a,b,x) , where
% phi2(x) = log((x-a)/(b-x)).
%
function y = phi2(a,b,x)
%
y=log((x-a)/(b-x));
```

and with basis3.m given in the program

```
% basis3.m
%
function [val] = basis3(x,M,N,h)
%
% This program is copyrighted by SINC, LLC
% and belongs to SINC, LLC
%
u = phi3(x);  %  See Example 1.5.6 of Sec. 1.5.3
val = basis(u,M,N,h);
```

with phi3.m evaluated by the program

```
%
% phi3.m
%
% This program is copyrighted by SINC, LLC
% and belongs to SINC, LLC
%
% Sinc map #3 of (0,infinity) -> R
%
% returns a vector phi3(x) = log(x)
%
function y = phi3(x)
y=log(x);
```

The common program, **basis.m** takes the form

```
%
% basis.m
%
% This program is copyrighted by SINC, LLC
% and belongs to SINC, LLC
%
function [val] = basis(u,M,N,h)
%
% We omit here the details of the
% computations which are described in
% Theorem 1.5.13 and its proof.
% basis(u,M,N,h) computes a vector of the
% sinc functions omega_j for j = -M, ..., N,
% for the case of the transformation #1 in
% Example 1.5.4 of Section 1.5.3.
%
% Each of the programs basis1.m, ...
% basis7.m calls basis.m, after making
% the transformation v = phi1(u), ...,
% v = phi7(a,b,u).
%
  m = M+N+1;
  v = pi/h*u;
%
% First compute the m sinc functions,
% gamma, called ''val'' here:
%
for k=1:m
    kk = k - M - 1;
    w = kk*pi;
    sarg = v - w;
    if abs(sarg) > eps;
        val(k) = sin(sarg) / sarg;
    else;
        val(k) = 1;
    end;
end;
%
```

```
% Next, compute the left end basis:
%
rho = exp(u);
val1 = 1/(1+rho);
for k=2:m;
    kk = k - M - 1;
    val1 = val1 - (1/(1+exp(kk*h)))*val(k);
end;
%
% Finally, compute the right end basis --
%
valm = rho/(1+rho);
for k=1:m-1;
    kk = k - M - 1;
    valm = valm - (exp(kk*h)/(1+exp(kk*h)))*val(k);
end;
%
val(1) = val1;
val(m) = valm;
%
```

3.2.3 INTERPOLATION AND APPROXIMATION

Example 3.2.2 We wish to approximate the function $x^{2/3}$ on the interval $(0, 7)$. Starting with $m = M + N + 1$ Sinc points z_j and function values $z_j^{2/3}$, $j = -M, \ldots, N$ we form the expression

$$x^{2/3} \approx \sum_{j=-M}^{N} z_j^{2/3}\, \omega_j(x)$$

where the basis functions ω_j are as in Definition 1.5.12. In the MATLAB® program interp_07.m we evaluate and plot each side of this equation at 50 equally spaced points $x_j = 7\,(j - 0.5)/50$, $j = 1, 2, \ldots, 50$. With $M = N = 3$, we are able to achieve a maximum error of about 0.025; whereas doubling M and N the maximum error drops to less than 0.0045. See Figure 3.1

```
%
%  interp_07.m
%
```

```
% This is an interpolation method,
% based on Sinc, for interpolating
% the function x^{2/3} on the
% interval (0,7).  Here we take
% phi(x) = log(x/(7-x)), and we
% first interpolate the function
% x^{2/3} over (0,7) by means of the
% basis in Sec. 1.5.4.
%
clear all;
%
% The following illustrates a usual start
% of any Sinc program.
%
M = 3
N = 3;
m = M+N+1;
c = pi/sqrt(2/3);
h = c/\sqrt(N);
%
% Convergence to zero of the max
% error occurs for any positive
% number c, at a rate
% O(exp(-c' p^{1/2})), with
% p = min(M,N), and c' a positive
% constant that is independent of
% M and N.
%
% This particular value of c is
% close to the ''best'' value; it
% is selected according to the
% discussion of \S 1.5.2.
% See the remarks below,
% regarding the selection of
% M, N, and h in practice.
%
% First compute the Sinc points
% as well as the function values
% at these points, on the interval
```

```
% (0,7).  See Example 1.5.5.
% We remark that we could also
% have computed the Sinc points
% z(j) in the loop which follows
% via use of the Sinc--Pack routine
% Sincpoints2.m, by means of
% the command
% 'z = Sincpoints2(0,7,M,N,h);'
% (see Sec. 1.5.2), or directly,
% as we have done here.
%
T = exp(h);
TT = 1.0/T^M;
for j=1:m
   z(j) = 7*TT/(1+TT);
   y(j) = z(j)^(2/3);
   TT = TT*T;
end
%
yy = y.'; %(yy is now a column vector)
%
% Next, compute the points x(j)
% where we wish to evaluate the
% interpolant, as well as the
% exact function values at these
% points:
%
step = 7/50;
for j = 1:50
   x(j) = (j-0.5)*step;
   fx(j)= x(j)^(2/3);
end
%
% Now, evaluate the interpolant at
% each of the points x(j) -- see
% Definition 1.5.12\, .
% The routine basis.m is used for
% this purpose.
%
```

```
for j=1:50
   val = basis2(0,7,x(j),N,N,h);
   Y(j) = val*yy;
end
%
% Now plot the exact, and computed
% solutions:
%
plot(x,y,'+',x,Y)
%
```

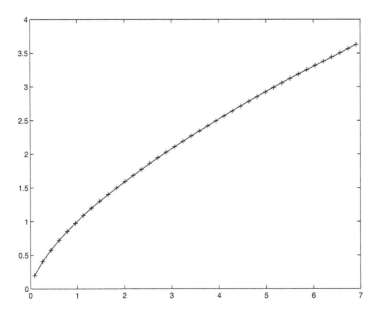

FIGURE 3.1. Seven Point Sinc Approximation to $x^{2/3}$

Remarks.

1. The parameter h used in the above program can be defined in more than one way.

   If the function to be interpolated is known explicitly, then we can determine h theoretically, as described in §1.5.

   To get a good h in the case when f has unknown singularities at end-points of an interval, one can "tweak" the constant c in the formula $h = c/\sqrt{N_1}$, with $N_1 = min(M, N)$. One always gets convergence with any such an h in the limit as $N_1 \to \infty$, and with error of the order of $\exp\left(-b(N1)^{1/2}\right)$, with b a positive number that is independent of N; on the other hand, one can get a more efficient (i.e., with a larger value of b) approximation via inspection of the error. Suppose, for example, we start with $c = 1$, and we plot the error of Sinc approximation, with $h = c/\sqrt{N_1}$. In cases when the exact value of the function is unknown, we would plot the result of smaller values of N_1 against larger, since for this purpose, the results of larger values of N_1 may be taken to be exact. Suppose, then that modulus of the error is larger (resp., smaller) near the end–points of the interval, than in the interior; then by repeating the experiment with a larger (resp., smaller) c, we would decrease (resp., increase) the error at the end–points, and increase (resp., decrease) it near the middle of the interval. Also, if, e.g., the error is larger on the left hand end-point than on the right, then, increasing the parameter M in the above program decreases the error near the left hand end-point.

Remark. Since Sinc methods converge so rapidly, experienced users will rarely need to utilize Steps 1. through 9. It will nearly always suffice to perform a small number of trial solutions for different M, N, and h, and pick off values that achieve a predetermined desired accuracy.

The above adjustments of h could easily be made numerically more specific, as described in Algorithm 3.2.3.

2. For most problems, approximations to the function values

$$(y(1),\ y(2),\ \ldots, y(m)) = f(z_{-M}),\ \ldots,\ f(z_N)$$

are already available, and could be used directly in plotting the solution, i.e., we do not need to evaluate the answer at equi-spaced points via use of the Sinc basis program. To this end, we could, e.g., use the MATLAB® plot command

$$\text{plot}(z, f)$$

3. **Warning:** Plotting the formula (1.5.27) over all of Γ is difficult, since the function φ is infinite at end–points of Γ. For example, for the case of the interval $[0, 1]$, z_{-N} is sufficiently close to 0 to be practically indistinguishable from 0, and z_N is similarly very close to 1, and a plot over $[z_{-N},\ z_N]$ should nearly always be satisfactory (see the properties of Sinc interpolation in the statements preceding Theorem 1.5.13).

The above steps are formalized in the following algorithm.

Algorithm 3.2.3 Uniform Approximation

This algorithm determines M, N and h to achieve a near uniform approximation on arc Γ.

1. Run a given Sinc program simply choosing $M = N$ and $h = 2/\sqrt{N}$, giving approximate solution F_N.

2. Repeat Step 1. with larger values $M' = N'$ approximately equal to $3N/2$, and obtain approximate solution $F_{N'}$.

3. Due to the efficiency of Sinc, $F_{N'}$ will be substantially more accurate than F_N, and for purposes of this algorithm will be considered to be exact. Hence plot difference $|F_{N'} - F_N|$ in order to reveal left and right hand errors E_R and E_L associated with singularities at the end-points of Γ.

4. Determine positive constant γ_L so that near the left end-point of Γ the error E_L roughly equals $e^{-\gamma_L\, N^{1/2}}$.

5. Repeat Step 4. applied to right end-point error E_R to obtain positive constant γ_R.

6. Choose new value M given by

$$M = \left[|\log(E_L)/\log(E_R)|^2\ N \right].$$

7. Run the Sinc program with values N, M as in 6. and $h = 2/\sqrt{N}$ as in Step 1. This will result in left and right end-point errors both close to E given by $|\log(E_L)/\log(E_R)|^2$ associated with positive constant γ equal to $\gamma_L = \gamma_R$.

8. Let E_C denote the error near the center of Γ. Compute positive values c and h given by

$$c = \left| \frac{\log(E)}{\log(E_C)} \right|^{1/2} , \quad h = \frac{c}{\sqrt{N}}$$

.

9. Inputs M, N and h as in 6. and 8., will result in errors of approximately the same at the center of Γ as at both endpoints, characterized by positive constant δ satisfying

$$|E_C| = e^{-\delta/h}.$$

3.2.4 SINGULAR, UNBOUNDED FUNCTIONS

Suppose we have computed approximate values f_{-N}, $\ldots$, f_N, of a function f defined on $[0,1]$, at the Sinc points z_{-N}, $\ldots$, z_N, as defined in §1.5.2, and we know that $f(x) = \mathcal{O}\left(x^{-1/3}\right)$ as $x \to 0$. We may then assume that Sinc approximation of g is accurate on $[0,1]$, where, $g(x) = x^{1/3} f(x)$, provided that f is sufficiently smooth (analytic) on $(0,1)$. Thus, with $g_k = z_k^{1/3} f_k$, and with ω_k as defined in Definition 1.5.12, the approximation

$$f(x) \approx x^{-1/3} \sum_{k=-N}^{N} g_k \, \omega_k(x) \tag{3.2.1}$$

is then accurate in the sense of a relative error, i.e., in the sense that the approximation

$$\frac{f(x)}{x^{-1/3}} = g(x) \approx \sum_{k=-N}^{N} g_k \, \omega_k(x) \tag{3.2.2}$$

is accurate.

PROBLEMS FOR SECTION 3.2

3.2.1 Write a program Sinc_07.m that computes the Sinc points for the interval $(0,7)$, as required in the above program, sinc_1d_interpolation.m.

[Hint: look at the program Sincpoints.m].

3.2.2 Write a program for approximating the function $x^{-1/2}e^{-x}$ on $(0, \infty)$.

3.2.3 Run a variation of the above MATLAB® program, to study what happens to the difference between $x^{2/3}$ and its *Sinc* approximation, e.g., by taking $M = (3/2)N$ as both N increases.

3.3 Approximation of Derivatives

There are several ways of approximating derivatives of functions. Two program classes of formulas in Sinc-Pack are denoted by D*_1.m (with * referring to the *th* transformation in §1.5.3) which is based on differentiating the formula

$$F(x) \approx \sum_{k=-M}^{N} f(z_k)S(k, h) \circ \varphi(x)$$

and Domega*_1.m, which is based on differentiating the Sinc interpolation formulas (1.5.27). A second class is possible for approximating the first derivative on Γ, by simply using the inverse of the indefinite integration matrices of §1.5.8. A third class of formulas is derived in §1.7, and is based on differentiating a composite polynomial which interpolates a function at the Sinc points. This latter class is preferable particularly if such derivatives exist and are required in neighborhoods of finite end-points of an arc Γ.

Example 3.3.1 *Derivative Approximation via Sinc–Pack.* In the program which follows, we approximate the derivative of e^x on $(0, 1)$ via two procedures, the program Domega2_1.m, which is based on differentiating the approximation (1.5.27) (see Figure 3.2), and Dpoly2_1.m, which is based on the polynomial method of §1.7, for the interval $(0, 1)$ (see Figure 3.3).

We have thus found that the program Domega2_1.m has computes very poor approximations near the end-points of $(0, 1)$, as the reader can observe by running the program, and we have thus not bothered including this plot. However, although the results are bad in the neighborhood of the end-points, it is provable that this procedure does in fact produce results which converge to the derivative as $N \to \infty$ on any closed subinterval of $(0, 1)$. We have therefore plotted the results on a subset of the set of Sinc points.

We have also included a plot of the results using the program Dpoly2_1.m based on the method derived in §1.7. These latter results are accurate, even in the neighborhood of the end-points of $(0, 1)$.

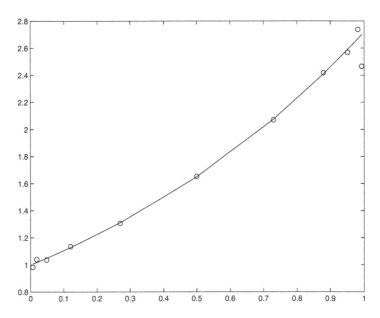

FIGURE 3.2. Restricted plot of e^x and Derivative

```
%
% der_test33.m
%
% This program is copyrighted by SINC, LLC
% and belongs to SINC, LLC
%
% This program is an example for approximating the
% derivative of exp(x) on (0\,,1).   See Sec. 3.3
%
clear all
%
N=10;
m = 2*N+1;
h=pi/sqrt(N);
z = Sincpoints2(0,1,N,N,h);
%
% Exact and derivative values of ff
```

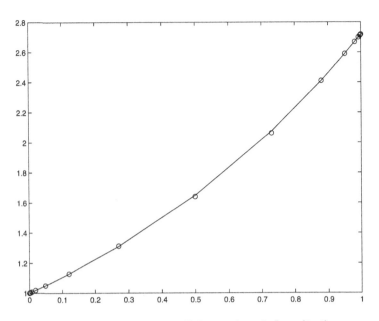

FIGURE 3.3. "Polynomial" Derivative of e^x on $(0,1)$

```
%
for j=1:m
     fff(j) = exp(z(j));
end
%
ff = fff.'; % Now a column vector
clear fff
%
% Values of derivative at Sinc pts.
%
fd = Domega2_1(0,1,N,N,h)*ff;
%
figure(1)
plot(z,fd,'o',z,ff,'-')
pause
% title('Computed & Exact Deriative of e^x on (0\,,1)')
% print -dps2 All_Sp.01.ps
%
figure(2)
for k=1:m-10
```

```
    fdd(k) = fd(k+5);
    fff(k) = ff(k+5);
    zz(k) = z(k+5);
end
%
plot(zz,fdd,'o',zz,fff,'-')
pause
% title('')
print -dps2 restr_der.ps
%
fp = Dpoly2_1(0,1,N,N,h)*ff;
%
figure(3)
plot(z,fp,'o',z,ff,'-')
pause
% title('Computed and Exact "Poly." Deriv. of e^x')
print -dps2 poly_der.ps
```

Remarks:

1. If an end–point a of the interval Γ is finite, then the derivative of the Sinc interpolant is unbounded at a. However, if Sinc interpolation is accurate, then, as this example shows, the derivative of this interpolant is nevertheless accurate on strict closed subintervals on the interior of Γ.

2. Approximation via differentiating a composite polynomial that interpolates a function f at the Sinc points of Γ is much better than the approximation based on differentiating the Sinc interpolant in the case when f is smooth in the neighborhood of the end–points of Γ.

PROBLEMS FOR SECTION 3.3

3.3.1 Verify the second of the above remarks, by finding the maximum difference on Γ between $(1/\varphi'(x)) \{x^{2/3}\}'$ and $(1/\varphi'(x))$ times the derivative of the interpolant

$$\sum_{k=-3}^{3} z_k^{2/3}\, \omega_k(x).$$

3.3.2 Compare the method of approximating derivatives used in this section with using inverses of indefinite integration matrices.

3.3.3 In the routine D22_1.m of the program der_test01.m, the term $h/\varphi'(z_j)$ was computed by the formula

$$\frac{h}{\varphi'(z_j)} = \frac{h\,e^{jh}}{(1+e^{jh})^2}.$$

Why is this formula numerically better than the theoretically same formula,

$$\frac{h}{\varphi'(z_j)} = h\,z_j(1-z_j)\,?$$

3.4 Sinc Quadrature

The theory of Sinc quadrature is covered in §1.5.7. In particular, a program based on Algorithm 1.5.18 to approximate a one dimensional integral over any interval by means of one of the transformations of Table 1.1 and the transformations of the integral as given in equations (1.5.45)–(1.5.47). The MATLAB® program below is based on this algorithm. It integrates a function (that the user inputs), and it outputs a value of the resulting integral to within an arbitrary error ε. This algorithm is based on the property of Sinc approximation, namely, if h is replaced by $h/2$, then every second Sinc point is the same as that for the original value of h, i.e., $z_{2k}(h/2) = z_k(h)$, whereas the number of significant figures of accuracy doubles – see, e.g., Eq. (1.3.9). We now give more details about this simple algorithm, which could be made more efficient, but with added complications – see [ShSiS].

Example 3.4.1 Suppose, for example, we wish to obtain an approximation to the integral

$$I = \int_{-\infty}^{\infty} \frac{dt}{(1+t^2)^{1/3}\,\sqrt{(t-1)^2+2}} \tag{3.4.1}$$

to within an accuracy of 'Eps'. In this case, the integrand has algebraic decay at $\pm\infty$, which suggests the use of the transformation #5 given in Ex. 1.5.8. We would thus use the transformation $t = (x - 1/x)/2$ given in Table 4.1, to get the integral

$$I = \int_0^\infty \frac{x^{-2/3}\,(1 + x^2)^{1/3}\,dx}{\sqrt{(x-1)^4 + 4\,(x-1)^2 + 16\,(x-1) + 8}}\,, \qquad (3.4.2)$$

We can now apply Algorithm 1.5.18 to approximate this integral. The following program, quadd1.m is a MATLAB® version of Algorithm 15.18. See also the similar MATLAB® program, quad1.m.

```
%
% quadd1.m
%
% Quadrature algorithm based on
% Algorithm 1.5.18, for
% approximating an integral
% over (0,infinity)
%
clear all
format long e
disp(' ')
Eps = input('Enter the accuracy, Eps:  ')
Esq = sqrt(Eps);
h = 2;
e = exp(h);
q = e;
%
% This routine is written for a function ff --
% see ff.m.
%
% The user should write a function routine, e.g.,
% create a file user.m in which he defines a
% his/her function, e.g., 'user(x)'.  See, for
% example, the routine fun_ex_1_3.m.  He should
% then proceed to modify the line ''y = ...' in
% ff.m, changing it to 'y = user(x)'.
%
T = ff(1);
r1 = 1;
while(r1 > Eps)
```

```
      p = 1/q;
      r = p*ff(p) + q*ff(q);
      T = T+r;
      q = e*q;
      r1 = abs(r);
end
%
T = h*T;
%
r3 = 1;
while(r3 > Esq)
    M = 0;
    sre = sqrt(e);
    q = sre;
    r2 = 1;
    points = 0;
    while(r2 > Eps)
        p = 1/q;
        r = p*ff(p) + q*ff(q);
        M = M+r;
        q = e*q;
        points = points+1;
        r2 = abs(r);
    end
    %
    M = h*M;
    rr3 = abs(T-M);
    T = (T+M)/2;
    e = sqrt(e);
    h = h/2;
    points = 2*points;\small
{

    r3 = rr3;
end
%
% Output
%
fprintf('# of evaluation points      = %d\n', points);
```

```
% fprintf('# of function evaluations    = %d\n', nff);
fprintf('Integral Value to Within Eps = %.16e\n', T);
```

PROBLEMS FOR SECTION 3.4

3.4.1 Approximate each of the following integrals to within a error of 10^{-8} :

(a) $\int_0^\infty t^{-1/2}/(1+t)\,dt$;

(b) $\int_0^\infty t^{-1/2}\,e^{-t}\,\sin(t)\,dt$;

(c) $\int_0^1 t^{-1/2}\,(1-t)^{-2/3}\,dt$;

(d) $\int_{-\infty}^\infty \left\{(1+t^2)\,(3+t^2+t^4)\right\}^{-1/2}\,dt$;

(e) $\int_{-\infty}^\infty \left(\sqrt{t^6+1}+t^3\right)e^{-t}/\left(1+e^{-t}\right)\,dt$.

3.4.2 (a) Write a MATLAB® version of Algorithm 1.5.18.

(b) Either modify the above MATLAB® code, or use the algorithm of the (a)–Part of this problem, to devise an algorithm to evaluate the Gamma function,

$$\Gamma(x) = \int_0^\infty t^{x-1}\,e^{-t}\,dt$$

that is accurate to 8 significant figures for all positive x.

3.4.3 (a) Given that $0 < \Re b < \Re c$, write a modification of the above program to evaluate the integral

$$F(x) = \int_0^1 t^{b-1}\,(1-t)^{c-b-1}\,(1-xt)^{-a}\,dt\,,$$

to within an error of 10^{-6} at all of the points $x = 0$, 0.1, 0.2 , ..., 0.9. How is this integral related to the usual hypergeometric function? (b) For what range of values of exponent a can you also evaluate this integral to the same accuracy when $x = 1$?

3.5 Sinc Indefinite Integration

We illustrate here, the Sinc approximation of the indefinite integral

$$F(x) = \int_x^\infty f(t)\, dt$$

(3.5.1)

$$f(t) = \frac{\log(t)}{t^{1/3}\sqrt{1 + \exp(t)}}.$$

by means of the following MATLAB® program. We use the Sinc transformation #4, since the $F(x) - F(0) = \mathcal{O}\left(x^{2/3}\right)$ as $x \to 0$, and $F(x) = \mathcal{O}\left(\exp(-x/2)\right)$ as $x \to \infty$.

The last 5 lines of this program employ the known Sinc point identity $z_{2k}(h/2) = z_k(h)$. Of course, the "for loop" in these lines takes account of our change of numbering for MATLAB®, i.e., from 1 to $2N + 1$ instead of from $-N$ to N, and from 1 to $8N + 1$ instead of from $-4N$ to $4N$. The $8N + 1$ "solution" is taken to be the "exact" one, since it is not possible to obtain an explicit expression for $F(x)$. See Figs. 3.4 and 3.5.

```
%
% indef_ex_1.m
%
% This program is copyrighted by SINC, LLC
% and belongs to SINC, LLC
%
% Here we approximate the integral
%
%   F(x) = int_x^\infty f(t) dt
%
% where
%
% f(t) = t^{-1/3}\,\log(t)/sqrt{1 + exp(t)}
%
% We can use the Transformation \#3
% of Sec 4.3.4, i.e., phi(x) = log(x).
%
% Also, using an approximation sum_{-N}^N,
% we can use h = pi(3/(2N))^{1/2} -- See
```

```
% the discussion of Example 4.3.1.
%
clear all
%
N =10;
m = 2*N+1;
h = pi*sqrt(1/N);
z = Sincpoints4(N,N,h);
for j=1:m
t = z(j);
ff(j) =  t^(-1/3)*log(t)/sqrt(1 + exp(t));
end
ff = ff.';
FF = Int4_m(N,N,h)*ff;
%
M = 32;
hh = pi*sqrt(1/M);
mm = 2*M+1;
y = Sincpoints4(M,M,hh);
for j=1:mm
t = y(j);
fg(j) = t^(-1/3)*log(t)/sqrt(1 + exp(t));
end
fg = fg';
FG = Int4_m(M,M,hh)*fg;
%
step = 6/100;
for j=1:600
    t = (j-0.5)*step;
    w(j) = t;
    u = phi4(t);
    valz = basis(u,N,N,h);
    valy = basis(u,M,M,hh);
    Fz(j) = valz*FF;
    Fy(j) = valy*FG;
end
figure(1)
plot(w,Fz,'.',w,Fy,'-');
pause
```

```
title('Exact ${-}$ and Approximate ${.}$ Integrations')
pause
print -dps2 ex_approx_indefint.ps
figure(2)
plot(w,Fy-Fz);
pause
title('Error of Indefinite Integration')
pause
print -dps2 err_indefint.ps
```

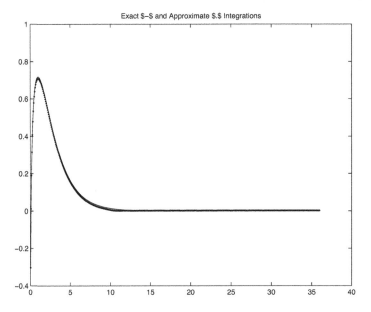

FIGURE 3.4. Plots of Exact and Approximate Indefinite Integration

PROBLEMS FOR SECTION 3.5

3.5.1 Approximate each of the following integrals to within a uniform accuracy of 10^{-8}:

(a) $\displaystyle\int_x^\infty t^{-1/2}/(1+t)\,dt\,, \quad x \in (0,\infty);$

(b) $\displaystyle\int_x^\infty t^{-1/2}\,e^{-t}\,\sin(t)\,dt\,, \quad x \in (0,\infty);$

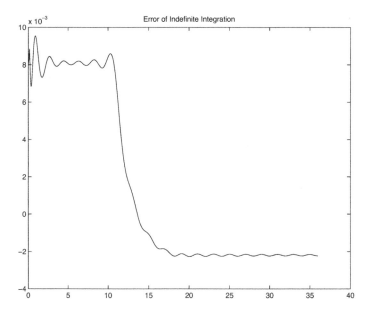

FIGURE 3.5. Plots of Exact and Approximate Indefinite Integration

(c) $\displaystyle\int_0^x t^{-1/2}\,(1-t)^{-2/3}\,dt\,, \quad x \in (0,1);$

(d) $\displaystyle\int_{-\infty}^x \left\{(1+t^2)\,(3+t^2+t^4)\right\}^{-1/2}\,dt\,, \quad x \in (-\infty,\infty);$

(e) $\displaystyle\int_x^\infty \left(\sqrt{t^2+1}+t\right)\,e^{-t}\big/\left(1+e^{-t}\right)\,dt\,, \quad x \in (-\infty,\infty).$

3.5.2 (a) Use the above algorithm to compute an accurate approximation to the incomplete Gamma function

$$\gamma(x) = \int_x^\infty t^{a-1}\,e^{-t}\,dt$$

for given $a > 0$ on an arbitrary closed subinterval of $(0, \infty]$.

(b) Discuss what happens in the following three cases, in a neighborhood of $x = 0$: (i) The case when $a > 0$; (ii) The case when $a = 0$; and (iii) the case when $a < 0$. For example, are the computed results multiplied by x^a accurate (implying that the computed results for $\gamma(x)$ are accurate at the Sinc points, in the sense of a relative error)?

3.6   Sinc Indefinite Convolution

The indefinite convolution procedure is described in detail in §1.5.9.
Let us illustrate the procedure to obtain an approximate solution of
the integrals

$$F_1(x) = \int_0^x (x-t)^{-1/2} t^{-1/2} \, dt \,,$$

$$F_2(x) = \int_x^1 (t-x)^{-1/2} t^{-1/2} \, dt.$$

$$(3.6.1)$$

Here F_1 and F_2 may be explicitly expressed as

$$F_1(x) = \begin{cases} 0 & \text{if} \quad x = 0; \\ \pi & \text{if} \quad x > 0, \end{cases}$$

$$F_2(x) = \log\left(\frac{1 + \sqrt{1-x}}{1 - \sqrt{1-x}} \right).$$

The function F_1 is the Heaviside function, which is difficult to approximate, since Sinc methods produce continuous approximations.
The second function, F_2, also contains a difficulty, in that it is unbounded at $x = 0$. We provide a complete MATLAB® program (except for setting up the $I^{(-1)}$ matrix, which is given in the MATLAB® routine Im.m) for computing $F_1(x)$ and $F_2(x)$. Notice the role of the Laplace transform, which is the same for both integrals. See Figures 3.6 and 3.7.

```
%
%    ex_abel.m
%
% Example of Application of Sinc Convolution
%
% Here we wish to find approximations to two
% integrals:
%
% Exact1(x)
% = \int_0^x (x-t)^{-1/2} t^{-1/2} dt
% = 0 if x = 0 and = \pi if x > 0
%
% and
%
```

```
%   Exact2(x)
%   = \int_x^1 (t-x)^{-1/2} t^{-1/2} dt
%
%   = \log((1 + (1-x)^{1/2})/(1 - (1-x)^{1/2})).
%
% if 0 < x (< =) 1
%
M = 16;
N = 8;
m = M+N+1;
Ei = Im(m); % This is the m x m marix  I^{-1}.
h = sqrt(2/N);
%
% Computation of sinc-points, weights,
% g(x) = x^{-1/2}, and exact solution at Sinc
% points.
%
z = Sincpoints2(0,1,M,N,h);
%
% The function x^{-1/2} evaluated at the Sinc
%   points.
%
for j=1:m
    gg(j) = 1/sqrt(z(j));
%
% sq is the function (1-x)^{1/2} evaluated at
% the Sinc points.
%
    sq = sqrt(1 - z(j));
%
% The exact value of the first convolution.
%
    exact1(j) = pi;
%
% The exact value of the second convolution.
%
    exact2(j) = log((1+sq)/(1-sq));
end
%
```

```
g = gg.';
exact1 = exact1.';
exact2 = exact2.';
%
% The Sinc Indefinite integration matrix for
% integration from 0 to x.
%
A = Int2_p(0,1,M,N,h);
%
% The Sinc Indefinite integration matrix for
% integration from x to 1.
%
B = Int2_m(0,1,M,N,h);
%
% Eigenvector and eigenvalue matrix
% decomposition of A .
%
[X S] = eig(A);
% S
% Eigenvector and eigenvalue matrix
% decomposition of B .
%
[Y T] = eig(B);
Xi = inv(X);
Yi = inv(Y);
%
% The Laplace transform of f(t) = t^{-1/2} is
%
%   F(s) = (pi*s)^{1/2}.
%
for j=1:m
  ss(j) = sqrt(pi*S(j,j));
end
%
st = ss.';
%
% The next matrix, S, is a diagonal matrix, with
% j^{th} diagonal entry s(j).
%
```

```
S = diag(st);
AA = X * S * Xi; % This AA is now  F(A).
BB = Y * S * Yi; % This BB is now  F(B).
p = real(AA*g);
%
% This is the Sinc approximation of
% exact1(z(i)) at the Sinc points.
%
q = real(BB*g);
%
% This is the Sinc approximation of
% exact2(z(i)) at the Sinc points.
plot(z,exact1,'o',z,p)
pause;
plot(z,exact2,'o',z,q)
%
```

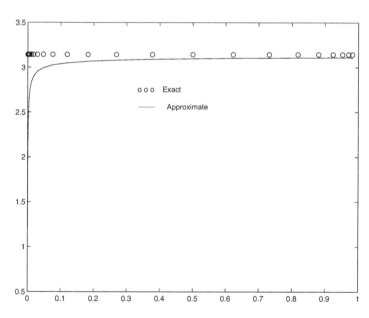

FIGURE 3.6. Exact and Sinc of $\int_0^x (t-x)^{-1/2} t^{-1/2} dt$

The examples of the above MATLAB® program are somewhat extreme; indeed, it is seems difficult to predict via theory that Sinc convolution actually converges for these examples. On the other hand,

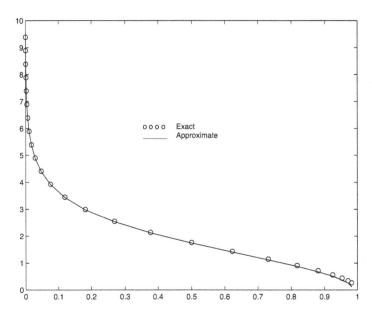

FIGURE 3.7. Exact and Sinc of $\int_x^1 (t-x)^{-1/2} t^{-1/2} dt$

the user can try the method on examples, such as

$$\int_0^x (x-t)^{-2/3} e^{-t} \, dt, \quad x \in (0, \infty),$$

for which one can readily prove uniform convergence on $[0, \infty]$.

PROBLEMS FOR SECTION 3.6

3.6.1 (a) Given that $0 < \Re b < \Re c$, alter the above *Sinc* convolution
algorithm so that it will compute an accurate approximation
on $(0, 1)$ of the function

$$F(x) = \int_0^x (x-t)^{c-b-1} t^{b-1} (1-t)^{-a} \, dt.$$

(b) How is this function related to the standard hypergeometric
function?

3.6.2 With reference to Figs. 3.8 and 3.9, try to explain the reason
for the relatively large error in the result of the Sinc convolution
approximation of the integral $\int_0^x (t-x)^{-1/2} t^{-1/2} \, dt$, compared

to the relatively small error in the approximation of $\int_x^1 (t - x)^{-1/2} t^{-1/2} \, dt$.

3.6.3 With reference to Fig. 3.9 , how would you remove the obvious error between the exact F and its Sinc approximation in the neighborhood of the point $x = 1$?

3.7 Laplace Transform Inversion

Assume that we are given the ordinary Laplace transform of f , i.e., $g(s) = \int_0^\infty \exp(-s\,t)\,f(t)\,dt$. For example suppose that we are given the Laplace transform,

$$g(s) = \frac{(\pi/2)^{1/2}}{(1 + s^2)^{1/2}\,(s + (1 + s^2)^{1/2})^{1/2}} , \qquad (3.7.1)$$

and we want to determine $f(t)$ on the interval $(0, \infty)$. Actually, we have here selected the Laplace transform $g(s)$ of the function $f(t) = t^{-1/2}\sin(t)$, in order to be able to compare our computed with the exact value of f. According to the procedure of §1.5.10, we first set

$$G(s) = g(1/s)/s = \frac{(\pi/2)^{1/2}\,s^{1/2}}{(1 + s^2)^{1/2}\,(s + (1 + s^2)^{1/2})} , \qquad (3.7.2)$$

We initialize, by picking M , N and h, and we then determine a Sinc matrix for indefinite integration from 0 to x on the interval $(0, \infty)$ based on the transformation #4 of §1.5.3 which is given by the MATLAB® program, $A = \texttt{Int4\_p.m}$, so that f is given at the $m = M + N + 1$ Sinc points of the interval $(0, \infty)$ by the vector $\mathbf{f} = G(A)\,\mathbf{1}$, where $\mathbf{1}$ is the column vector of order m with all of its entries a "1". The evaluation of $\mathbf{f}$ requires the matrices of eigenvalues S and eigenvectors X of A, enabling us to set up $G^{[2]} = S^{-1}\,g(S^{-1})$, and then use the routine $\texttt{Lap\_inv.m}$ to then get $\mathbf{f}$ (called 'ffa' in the program). We also need the Sinc points of the interval $(0, \infty)$ for this M , N and h, which are given by the MATLAB® routine $\texttt{Sincpoints4.m}$. The computational details, including a comparison with the exact solution, are given in the Sinc–Pack program $\texttt{ex\_lap\_inv4.m}$ which we cite here, verbatim. A plot of the exact and approximate solution are also given in Figures 3.8 and 3.9 below, and the error of approximation is given in Figure 3.10.

```
%
% ex_lap_inv4.m
%
% This program is copyrighted by SINC, LLC
% an belongs to SINC, LLC
%
% This is an example of Laplace
% transform inversion, via
% Sinc--Pack.  See Sec. 1.5.10.
%
% Here we start with the given
% Laplace transform,
%
% g(s) = hat(f)(s)
% = int_0^infty f(t) exp(-st) dt
% =  (pi/2)^{1/2}/w(s)/sqrt(s + w(s))
%
% with
%
% w(s) = (1 + s^2)^{1/2},
%
% and we wish to determine a function
% f(t) on (0,infinity) which gives rise
% to this transform.  We will also
% compare with the exact solution
%
% f(t) = t^{-1/2} sin(t).
%
% A "usual" start:
%
clear all
%
M = 15;
N = 20;
m = M+N+1;
h = pi/sqrt(N);
%
% Sincpoints and indefinite
% integration matrix for the
```

```
% interval (0,infinity) based on the
% transformation #4 of Sec. 1.5.3
%
z = Sincpoints4(M,N,h);
A = Int4_p(M,N,h);
%
f_approx = real(lap_inv_g(A));
%
for j=1:m
    t = z(j);
    f_exact(j) = t^(-1/2)*sin(t);
end
%
plot(z,f_exact,'-',z,f_approx,'o')
%
pause
%
%figure(1)
title('Lapace Transform Inversions')
xlabel('Sinc Points')
ylabel('Inversions')
print -dps2 Lap_inv.ps
%
pause
% Plotting over the interval (0,4 pi):
step = 4*pi/100;
for j=1:100
    xx(j) = (j-1/2)*step;
    u = phi4(xx(j));
    v = basis(u,M,N,h);
    ffa(j) = v*f_approx;
    ffe(j) = v*f_exact.';
end
%
plot(xx,ffe,'-',xx,ffa,'.')
pause
% figure(2)
title('Sinc Interpolated Laplace Transform Inversion')
xlabel('x-Values')
```

```
ylabel('Inversions')
print -dps2 Lap_inv_interp.ps
pause
plot(xx,ffe-ffa,'-')
pause
% figure(3)
title('Sinc Interpolated Error')
xlabel('x-Values')
ylabel('Error')
print -dps2 Lap_inv_err.ps
```

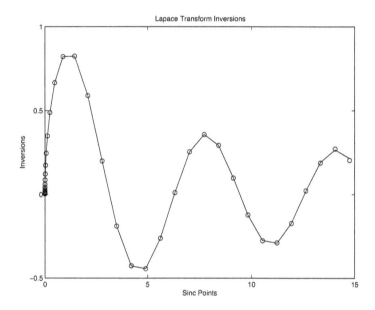

FIGURE 3.8. Exact and Approximate Laplace Transform Inversion of $\text{Lap\_tr}\{t^{-1/2}\sin(t)\}$

PROBLEMS FOR SECTION 3.7

Let a and b denote numbers with positive real parts, and let c denote an arbitrary positive number. Use the formula of Problem 1.5.14 for f given the ordinary Laplace transform $\hat{f}$ to determine accurate approximations to functions whose Laplace

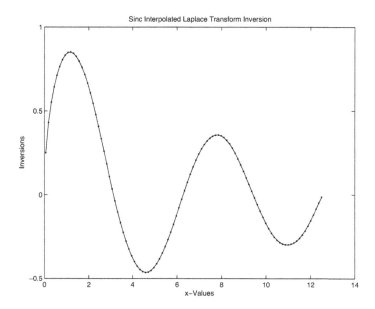

FIGURE 3.9. Sinc Interpolated Laplace Transform Inversion

transforms are

(a)

$$\sqrt{\frac{\pi}{s}} \exp\left(-2\sqrt{s\,b}\right),$$

and

(b)

$$\sqrt{\frac{\pi}{b}} \exp\left(-2\sqrt{s\,b}\right).$$

[Answers: (a) $t^{-1/2} \exp(-b/t)$, (b) $t^{-3/2} \exp(-b/t)$.] [Hint: The solution to the (b)-Part of this problem is unbounded at $t = 0$. Hence, try to get an inversion for $\exp(-2\sqrt{(s+c)\,b})$ for a small positive number c.]

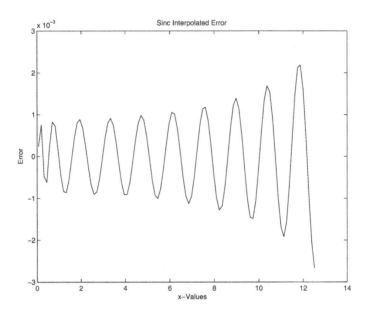

FIGURE 3.10. Sinc Interpolated Error of Inversion

3.8 Hilbert and Cauchy Transforms

Formulas for approximating Cauchy and Hilbert transforms are given in Theorem 1.5.25. We recommend that the user study the `basis.m` program in Sinc–Pack, as a guide for writing a MATLAB® program to approximate these formulas. For example, the following is an approximation for the Hilbert transform $\mathcal{S}F(z_\ell)$ at a Sinc point z_ℓ. Following such evaluations for z_ℓ, $\ell = -N$, ... , N, we can use the `basis.m` routine to approximate $\mathcal{S}F(x)$ for any $x \in \Gamma$.

$$(\mathcal{S}F)(z_\ell) \;\; \equiv \;\; \frac{P.V.}{\pi i} \int_\Gamma \frac{F(t)}{t - z_\ell} dt$$

$$\approx \;\; \frac{h}{\pi i} \sum_{k \neq \ell} \frac{F(z_k)}{\phi'(z_k)} \frac{1 - (-1)^{k-\ell}}{z_k - z_\ell}. \tag{3.8.1}$$

We can approximate the sum on the right hand side of (3.7.1) to within a *relative error* of δ by means of Algorithm 3.8.1.

Algorithm 3.8.1   Evaluation of $(\mathcal{S}F)(z_\ell)$ to Within δ Relative Error

$$k = \ell - 1$$
$$S_1 = 2F(z_k)/[\phi'(z_k)(z_k - z_\ell)]$$
$$u = v = 1, \quad w = |S_1|$$
$$(A) \; k = k - 2$$
$$T = 2F(z_k)/[\phi'(z_k)(z_k - z_\ell)]$$
$$S_1 = S_1 + T$$
$$u = v, \; v = w, \; w = |T|$$
$$(u + v + w)/|S_1| : \delta$$
$$(\geq) \Longrightarrow (A)$$
$$(<)$$

$$k = \ell + 1$$
$$S_2 = 2F(z_k)/[\phi'(z_k)(z_k - z_\ell)]$$
$$u = v = 1, \quad w = |S_2|$$
$$(B) \; k = k + 2$$
$$T = 2F(z_k)/[\phi'(z_k)(z_k - z_\ell)]$$
$$S_2 = S_2 + T$$
$$u = v, \; v = w, \; w = |T|$$
$$(u + v + w)/|S_2| : \delta$$
$$(\geq) \Longrightarrow (B)$$
$$(<)$$

Yamamoto's Hilbert Transforms. Sinc–Pack contains several programs for computation of the Hilbert transform via use of Yamamoto's formula given in Theorem 1.5.25, Hilb_yam.m, Hilb_yam1.m, Hilb_yam2.m, and so on, up to Hilb_yam7.m, where the integers 1 to 7 correspond to the seven explicit transformations given in §1.5.3. Each of the formulas Hilb_yam1.m, Hilb_yam2.m, and so on, up to Hilb_yam7.m calls Hilb_yam.m. We illustrate here, the approximation of the Hilbert transform

$$g(x) = (\mathcal{S}f)(x) = \frac{1}{\pi i} \int_0^1 \frac{f(t)}{t - x} \, dt, \qquad (3.8.2)$$

where $f(t) = t(1 - t)$, for which we require both programs Hilb_yam2.m as well as Hilb_yam.m. The exact solution is $g(x) = 1/(i\pi)(1/2 - x + (x - x^2)\log(x/(1 - x)))$. See Figures 3.11 and 3.12.

```
%
% ex1_hilb_yam.m
%
% This program is copyrighted by SINC, LLC
% and belongs to SINC, LLC
%
% We compute here the Hilbert transform of
%
% t (1-t)   taken over (0\,,1)
%
% via Yamamoto's method given in  Sec. 1.5.
% The exact solution is
```

```
%
% g(x) = (i\,\sqrt{pi})(1/2-x + f(x) log(x/(1-x))).
%
clear all
M = 10;
N = 10;
h = pi/sqrt(N);
m = M+N+1;
%
% Note, if Gamma = (0\,,1), then z_j(1-z_j)
%
% = exp(jh)/(1 + exp(jh))^2.
%
z = Sincpoints2(0,1,M,N,h);
%
TT = exp(-M*h);
Q = exp(h);
for j=1:m
    k = j - M - 1;
    f(j) = TT/(1 + TT)^2;
    TT = Q*TT;
    ge(j) = 1/pi*(1/2 - z(j) - f(j)*k*h);
end
%
ff = f.';
g = ge.';
%
C = Hilb_yam2(0,1,M,N,h);
gg =- imag(C*ff);
%
% Plots
%
figure(1)
plot(z,g,'-',z,gg,'o');
pause
print -dps2 Exact_and_Yamamoto_Approx.ps
pause
figure(2)
plot(z,g-gg)
```

```
pause
print -dps2 Exact_minus_Yamamoto_Approx.ps
```

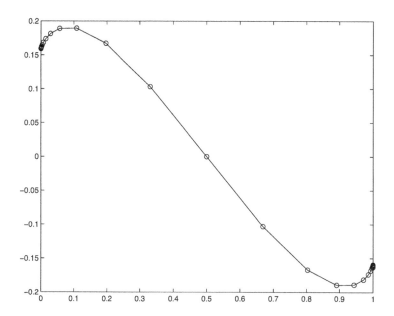

FIGURE 3.11. Approximate via Yamamoto Method and Exact Hilbert Transforms of $x(1-x)$

PROBLEMS FOR SECTION 3.8

3.8.1 Let $0 < \alpha \le 1$, let $g \in \mathbf{L}_{\alpha,d}(\varphi)$, with $\varphi(z) = \log(z)$, take $g(z) = f(z) - f(0)/(1+z)$, take $z_k = e^{kh}$, let N denote a positive integer, and let $h = (\pi d/(\alpha N))^{1/2}$. Prove that there exists a constant, c, independent of N, such that

$$\sup_{x \in (0,\infty)} \left| \mathcal{S}f(x) - f(0) \frac{\log(1/x)}{\pi i(1+x)} - \frac{1}{i} \sum_{j=-N}^{N} g(z_j) t_j(x) \right|$$

$$\le c N^{1/2} e^{-(\pi d \alpha N)^{1/2}}.$$

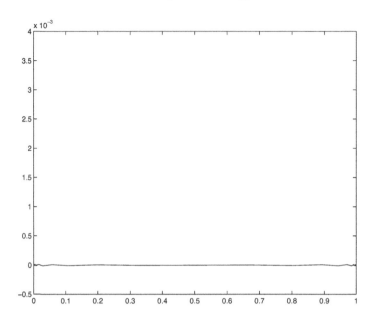

FIGURE 3.12. Exact Minus Yamamoto Approximation

3.9   Sinc Solution of ODE

In this section we illustrate the Sinc solution of ordinary differential equation (ODE) initial value problems of the form

$$y' = f(x, y), \quad x \in (0, T), \quad y(0) = y_0, \qquad (3.9.1)$$

and also, the Sinc solution of a second order ODE problem related to Airy's equation.

The DE are in all cases first transformed into a Volterra integral equation (IE). This IE is first converted to a system of algebraic equations, at first, via Sinc indefinite integration, and later, via use of wavelet indefinite integration. The resulting system of algebraic equations is then solved. For the case of linear equations, the resulting system of algebraic equations is linear, and can be solved directly, even for the case when $T = \infty$. For the case of nonlinear equations, the system is solved iteratively, a process which can be shown to converge for all sufficiently small T (in (3.9.1)), since for a finite interval $(0, T)$, the norm of the Sinc indefinite integration matrix is proportional to T.

3.9.1   Nonlinear ODE–IVP on $(0, T)$ via Picard

Let us illustrate a Sinc method of approximate solution via use of Picard iteration, for the problem

$$y' = 1 + y^2, \quad 0 < x < T, \quad y(0) = 0. \qquad (3.9.2)$$

This problem has the exact solution $y = \tan(x)$. The indefinite integration matrix A^+ used here is defined as in §1.5.8.

We somewhat arbitrarily took $P = 0.6$ in the program ode_ivp_nl.m below. In this case we got convergence to the correct solution. See Figure 3.13. Note, however, that many more points are required to achieve some accuracy for the case of the method based on wavelet approximation given below in §3.10, since the wavelet method tacitly assumes that the solution is periodic with period P on $\mathbb{R}$.

```
%
% ode_ivp_nl1.m
%
% Here we solve the following
%   problem:
%
% y' = f(x,y)
%
% with f(x,y) = "ode_f1(x,y)"
%             = 1+y^2.
%
% The problem is solved on (0,T),
% subject to y(0) = 0, via use
% of Sinc collocation via use
% of Sinc indefinite integration
% on (0,T), combined with Picard
% iteration.
%
% In the program, the
% ode_f1_exact(x) = tan(x), which is
% the exact solution.  Here, T=1,
% although the user is encouraged
% to use other values of T.
%
clear all
```

```
N = 10;
M = 10;
m = M+N+1;
h = 2.7/sqrt(N);
T = 0.6;
y0 = 0;
x = Sincpoints2(0,T,M,N,h);
A = Int2_p(0,T,M,N,h);
Er = 5;
for j=1:m
    v(j) = 0;
end
%
yy = v.';
% pause
while(Er > .00001)
     y = yy;
     for k=1:m
         W(k) = y0 + ode_f1(x(k),y(k));
     end
%
     Z = W.';
%
     yy = A*Z;
     Er = norm(y - yy);
end
%
for j=1:m
   ex(j)= ode_f1_exact(x(j));
end
%
TT = ex.';
plot(x,TT,'o',x,yy)
%
```

3.9.2 LINEAR ODE–IVP ON $(0, T)$ VIA PICARD

Let us next illustrate a Sinc approximate solution via use of Picard
iteration, for the linear problem

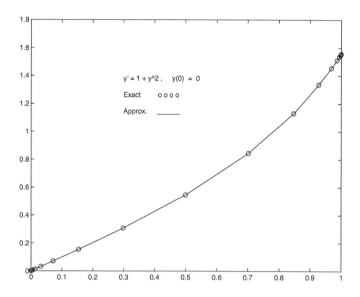

FIGURE 3.13. Sinc-Picard Solution of $y\prime = 1 + y^2$, $y(0) = 0$

$$y' = -y/(1+x^2), \quad 0 < x < T, \quad y(0) = 1. \qquad (3.9.3)$$

This problem has the exact solution $y = \exp(-\tan^{-1}(x))$. We take $T = 0.5$ in the program ode_ivp_li1.m below. See Figure 3.14. Indeed, best accuracies result if we take $T < 0.5$. For $T > 0.5$, we experience a paradoxical phenomenon: our discretized approximation converges to the incorrect result. when using this program we thus recommend repeating the computation, either via use of larger (or smaller) values of M and N, or via use of a different value of T, in order to check the results.

```
%
% ode_ivp_li1.m
%
% Here we solve the problem
%
% y' = f(x,y); y(0) = y_0 on (0,T),
%
% with f(x,y) = ode_f2(x,y)
%               = - y/(1+x^2),
```

```
% If y0=1, then
%
% ode_f2_exact(x) = exp(arctan(x)).
%
% This problem is solved on
% (0,T) for T = 0.5.  Again Picard
% iteration converges for all (0,T)
% and while the exact Picard
% converges to the exact answer, our
% collocated Picard does so only if
% T < 0.5.  It is suggested that
% whenever the user does a Picard
% she/h will do another run with
% larger values of M and N, to see
% if the two computations agree.  If not,
% then she/he should use a smaller
% interval.
%
clear all
%
M=10;
N=10;
m = M+N+1;
h = 2.7/sqrt(N);
y0 = 1;
T = 0.5;
z = Sincpoints2(0,T,M,N,h);
A = Int2_p(0,T,M,N,h);
Er = 5;
%
for j=1:m
yy0(j) = y0;
end
yy = yy0.';
%
while(Er > .00001)
      y = yy;
      for k=1:m
          w(k) =  ode_f2(z(k),y(k));
```

```
      end
%
      Z = w.';
%
      yy = yy0.'+A*Z;
      Er = norm(y - yy);
end
%
for j=1:m
   ex2(j)= ode_f2_exact(z(j));
end
%
TT = ex2.';
plot(z,TT,'o',z,yy)
%
```

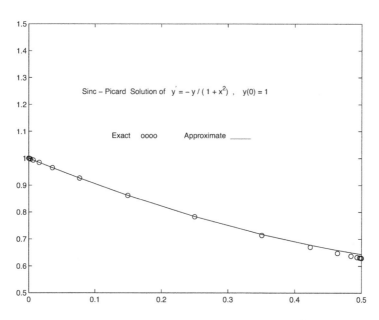

FIGURE 3.14. Sinc-Picard Solution of $y\prime = -y/(1+x^2)$, $y(0) = 1$

3.9.3 LINEAR ODE–IVP ON $(0, T)$ VIA DIRECT SOLUTION

We again solve the same problem as in the previous example. To this end, we discretize it in the same manner, but instead of solving the

problem via use of Picard iteration, we solve the resulting system
of linear equations directly, a procedure which works for all linear
ODE–IVP problems (indeed, even for a system of equations, which
can be similarly collocated and solved). That is, we again solve

$$y' = -y/(1 + x^2), \quad 0 < x < T, \quad y(0) = 1. \tag{3.9.4}$$

We solve this problem on $(0, \infty)$, via solution of the system of linear
equations obtained via Sinc collocation. See Figure 3.15.

```
%
%    ode_ivp_li2.m
%
% 1. The problem
%
% y' = f(x,y); y(0) = y_0; 0 < x < T.
%
% e.g., if f(x,y) = ode_f3(x,y)
%                 = - y/(1+x^2)
%
% subject to y(0) = 1, so that the
% exact solution is ode_f3_exact(x)
% = exp(- tan^{-1}(x)).
%
% This is the same problem as was
% solved over (0,T) in the previous
% subsection.  However, here we solve
% it over (0,infinity), via direct
% solution of the collocated system
% system of linear algebraic
% equations.
%
% It is suggested that the user also
% run this same routine, but with
% Sincpoints3(M,N,h) replaced with
% Sincpoints4(M,N,h), and with
% Int3_p(M,N,h) replaced with
% Int4_p(M,N,h).
%
% This direct solution procedure
```

```
% works for any interval (0,T), and it
% is suggested that the user try this,
% but she/he then needs to use the
% routines Sincpoints2(0,T,M,N,h) and
% Int2_p(0,T,M,N,h).
%
clear all
%
M=15;
N=15;
h = 1/sqrt(N);
%
m = M+N+1;
%
z = Sincpoints3(M,N,h);
A = Int3_p(M,N,h);
%
for j=1:m
    d(j) = 1/(1+z(j)^2);
    ex3(j)= ode_f3_exact(z(j));
end
D = diag(d);
B = eye(m)+A*D;
Er = 5;
y0=1;
%
for j=1:m
yy0(j) = y0;
end
zz = yy0.';
sol = B\zz;
so = sol.';
%
plot(z,ex3,'o',z,so)
%
```

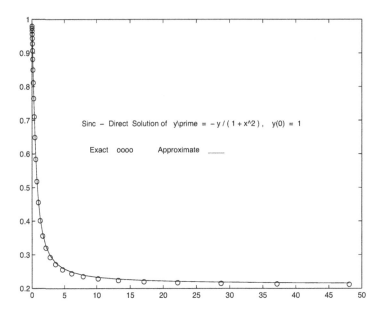

FIGURE 3.15. Sinc-Direct Solution of $y\prime = -y/(1+x^2)$, $y(0) = 1$

3.9.4 SECOND-ORDER EQUATIONS

An important procedure for obtaining asymptotic estimates of ODE is the transformation due to Liouville and Green (see [O]). In making this transformation on the DE,

$$y_{xx} - g(x)\,y(x) = 0\,,\tag{3.9.5}$$

one sets $\zeta = \int^x g^{1/2}(t)\,dt$, $\quad y = cg^{-1/4}(x)\exp(\pm\zeta)\,Y$, with c a constant, to reduce (3.9.5) to the DE $Y_{\zeta\zeta} \pm 2Y_\zeta = -fY$. The function f is "small" and absolutely integrable over $(X\,,\,\infty)$, $X > 0$ for many applications. For example, if $g(x) = x$, then we have *Airy's equation*, and the function $y(x) = Ai(x)$ satisfies (3.9.5), and for this function we get the equation

$$Y_{\zeta\zeta} - 2Y_\zeta = -fY\,,\qquad f(\zeta) = \frac{5}{36\,\zeta^2}\,.\tag{3.9.6}$$

We try an approximation $Y = 1 + \eta$, so that the "error" $\eta = Y - 1$ then satisfies the DE

$$\eta_{\zeta\zeta} - 2\eta = -f\,\eta - f\,,\tag{3.9.7}$$

with f as above. The solution to this equation which vanishes at ∞ can then be shown to satisfy the integral equation

$$\eta(\zeta) = -\frac{1}{2} \int_{\zeta}^{\infty} (1 - \exp(2\zeta - 2t)) \{f(t)\eta(t) + f(t)\} \, dt. \quad (3.9.8)$$

Suppose, for example, starting with Airy's equation,

$$y_{xx} - x\,y = 0\,, \quad (3.9.9)$$

we set $\zeta = (2/3)\,x^{3/2}$, $y = x^{-1/4}\,e^{-\zeta}\,Y$, use an approximation $Y \approx Z = 1$, and then set $\eta = Y - Z$, we get the differential equation

$$\eta_{\zeta\zeta} - 2\,\eta_{\zeta} = -\frac{5}{36\,\zeta^2}\,\eta - \frac{5}{36\,\zeta^2}\,, \quad (3.9.10)$$

We can then compute η on the interval (b, ∞) for $b > 0$ via the program which follows. But, in order to conveniently use the formulas of Sinc–Pack which are written for the interval $(0, \infty)$, we first replace ζ in the functions $f(\zeta)$ defined above by $b + \zeta$, to arrive at the same equation (3.9.10), but with f(t) replaced by $f(t+b)$, and we can now solve this final equation over $(0, \infty)$, via use of Sinc convolution.

In our solution, we have selected the transformation #4 of §1.5.3, since the solution to (3.9.10) is expected reach its limiting value algebraically at b, and exponentially at ∞. For our Sinc convolution procedure we also need the Laplace transform (see §1.6),

$$G(s) = \int_{0}^{\infty} \frac{1 - e^{-2t}}{2}\,e^{-t/s}\,dt = \frac{s^2}{1 + 2\,s}. \quad (3.9.11)$$

After solving for η we add 1 to η, to get $Y(b + \zeta)$, then recall the Airy function identity $Ai(x) = x^{-1/4}\,Y(\zeta)$, as well as the identity (10.4.14) of [Na], to deduce that, with $\xi = a + \zeta$,

$$1 + \eta(\zeta) = 2\,\pi^{1/2}\,(3\xi/2)^{1/6}\,e^{\xi}\,Ai\left((3\xi/2)^{2/3}\right). \quad (3.9.12)$$

where Ai denotes the Airy function.

In the program which follows, we have (somewhat arbitrarily) taken $b = 0.004$, to enable us to get compute the Airy function over (b', ∞), with $b' \approx .033$. We have thus presented 4 plots:

1. In Figure 3.16 below we have plotted both sides of (3.9.12):
   the left hand side computed via Sinc convolution, and the right
   hand side, computed via use of MATLAB® routines;

2. In Figure 3.17, we have plotted the difference between these
   two functions of the previous figure;

```
%
% ode_airy.m
%
% This program is copyrighted by SINC, LLC
% and belongs to SINC, LLC
%
%  Here we solve the problem
%
% v" - 2 v' = f v + R , on (b,infty), b > 0
%
% where for our program, f = R = -5/(36 t^2),
%
% via collocation based on Sinc convolution,
% and direct solution of the collocated
% system of algebraic equations corresponding
% to the integral equation
%
% v(z) = int_z^\infty
%
% (1/2) (1 - exp(2z - 2t)) {f(t) v(t) + R(t)} dt
%
% In this case, the exact 1 + v(z) equals
%
%  2 pi^{1/2} x^(1/4) e^z Ai(x) , where
%
%  x = (3/2 z)^(2/3)
%
% The Laplace transform" F(s) of (1/2)(1-e^(-t))
% is s^2/(1+2s).
%
% of the convolution kernel is used.  Sinc
% convolution usually requires computation of the
% eigenvalues and eigenvectors of the indefinite
```

```
% integration matrix A (see below) to set up the
% matrix F(A).  However, since F(s) is a simple
% rational in s, and B = F(A) = A*A*(I+2A)^(-1)
% can thus be easily computed.
%
clear all
%
% Initializations
%
M=15;
N=10;
h = 4/sqrt(N);
m = M+N+1;
%
b = .004; % Solution sought over (b,infinity)
%
% Sinc points and indifinite integration
% matrix over (0,infinity)
%
z = Sincpoints3(M,N,h);
A = Int3_m(M,N,h);
%
% Entries for functions in the IE; 'eta_ex' is
% the exact 1 + v
%
for j=1:m
    t = z(j)+b;
    p = (3/2*t)^(1/6); %See (3.9.12)
    f(j) = -5/(36*t^2);
    eta_ex(j) = 2*pi^(1/2)*p*exp(t)*airy(p^4);
end
%
% The matrix B = F(A), and collocation of the IE
%
B = A*inv(eye(m) + 2*A)*A;
D = diag(f);
C = eye(m)-B*D;
R = B*f.';
%
```

```
% The ''solution'' v -> 1 + v
%
eta = C\R;
v_apr = eta + 1;
%
%%%%%%%%%%%%%  PLOTS  %%%%%%%%%%%%%
%
% First, plot eta against z
%
plot(z,eta)
pause
%
% Next, plot v against z
plot(z,v_apr)
pause
%s
% Plot against Sinc points
%
plot(z,v_ex,'o',z,v_apr,'-')
%
pause
%
plot(z,v_apr-v_ex)
%
pause
%
% Plot of computed (yy) and exact Ai(x)
% against equi-spaced points x = xx in (aa, bb).
% Here we use Sinc interpolation and Eq. 10.4.59
% of [N] to get the approximate (yy) values,
%
aa = (3/2*(z(1)+b))^(2/3); % left boundary
bb = 2; % right boundary
step = (bb-aa)/100;
for j=1:100
    x = aa + (j-0.5)*step;
    xx(j) = x;
    zz = 2/3*x^(1.5) - b;
    uu = phi3(zz);
```

```
      ww = basis(uu,M,N,h);
      yy(j) = (ww*v_apr)/2/pi^(1/2)/x^(1/4)/exp(2/3*x^(3/2));
      ai(j) = airy(x);
end
%
figure(1)
plot(xx,yy,'x',xx,ai,'-')
pause
title('Exact & Approximate Airy')
print -dps2 airy_ex_apr.ps
%
pause
figure(2)
% Plot of error, yy - ai
pause
title('Difference, exact-computed Airy')
plot(xx,yy-ai)
print -dps2 airy_err.ps
%
```

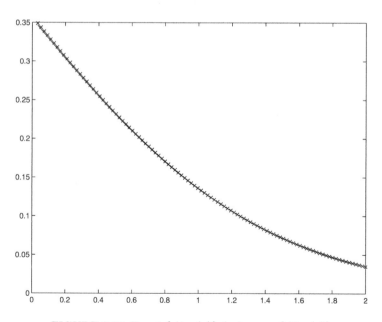

FIGURE 3.16. Exact $\{Airy(x)\}$ & Approx $\{Airy(x)\}$

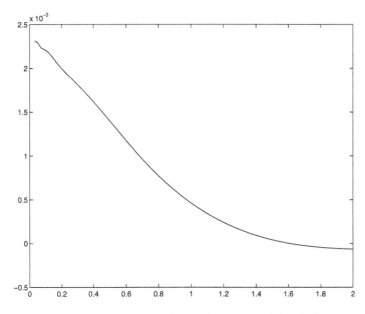

FIGURE 3.17. Exact $\{Airy(x)\}$ - Approx $\{Airy(x)\}$

3.9.5 WIENER–HOPF EQUATIONS

We illustrate here the approximate solution of the Wiener–Hopf problem

$$f(x) + \int_0^\infty e^{-|x-t|} f(t)\, dt = 2(1+x)e^{-x} + (1/3)e^{-2x}, \qquad x \in (0, \infty),$$
$$(3.9.13)$$

whose exact solution is

$$f(x) = 2\, e^{-x} - e^{-2x}. \qquad (3.9.14)$$

The theory of Wiener–Hopf methods is discussed in §1.4, and a method of solution is also described in §6.8 of [S1]. The present method, however, is by far the simplest and most efficient.

The program which follows solves this problem via use of Sinc convolution. See Figure 3.18. We remark that a program similar to the one below could be applied to solve a Wiener–Hopf problem with a more difficult kernel, such as, e.g.,

$$k(t) = \begin{cases} t^{-2/3} \exp(-2t), & \text{if } t > 0 \\ (-t)^{-3/4} \exp(3t), & \text{if } t < 0. \end{cases} \qquad (3.9.15)$$

Additionally, the present method also works for equations over finite intervals.

```
%
% wiener_hopf.m
%
% Here we solve the Wiener-Hopf problem
%
%  f(x) + int_0^infty k(x-t) f(t) dt = g(x)
%
% where k(t) = e^{-|t|} ,   and
% g(x) = 2 (1+x) e^{-x} + (1/3) e^{-2x}.  The
% exact solution is  f(x) = 2 e^{-x} - e^{-2x}.
%
% Initializations
%
M = 13;
N = 10;
m = M+N+1;
h = 3.5/sqrt(m);
z = Sincpoints4(M,N,h);
%
% Indefinite integration matrices
%
A = Int4_p(M,N,h);
B = Int4_m(M,N,h);
%
% Computing the right hand side  g  and the
% exact solution
%
for j=1:m
    gg(j) = 2*(1+z(j))*exp(-z(j)) - 1/3*exp(-2*z(j));
    f_e(j) = 2*exp(-z(j)) - exp(-2*z(j));
end
%
g = gg.';
f_exact = f_e.';
```

```
%
% Evaluation of the Laplace transforms of the Kernel
% and collocation of the IE. See \S 1.9.
%
C = A*inv(eye(m)+A);
D = B*inv(eye(m)+B);
%
E = eye(m) + C + D;
%
f_approx = E\g;
%
plot(z,f_exact,'o',z,f_approx)
%
pause
%
plot(z,f_approx-f_exact)
%
```

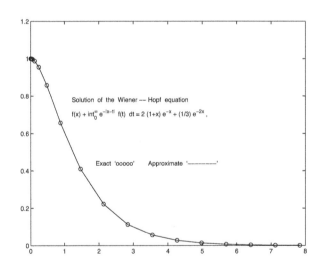

FIGURE 3.18. Exact and Approximate Wiener-Hopf Solutions

3.10   Wavelet Examples

In this section we illustrate two types of wavelet approximations via the procedures described in §1.4.2 and §1.4.5. The first involves

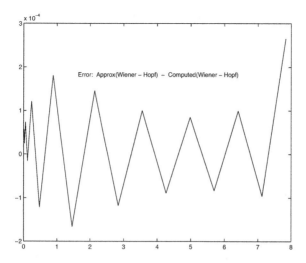

FIGURE 3.19. Error of Wiener-Hopf Solution

the approximation of two functions on $[0, 1]$, a non–periodic and
a periodic one, while the second involves a wavelet solution of an
ordinary differential equation initial value problem.

3.10.1   Wavelet Approximations

We illustrate here two examples of wavelet interpolations.
The quality of approximation of these two functions via the four
classes of wavelet routines

> wavelet_basis_even_half.m,
> wavelet_basis_even_int.m,
> wavelet_basis_odd_half.m, and
> wavelet_basis_odd_int.m

listed in the Sinc–Pack package is similar, and we thus illustrate only
the case of wavelet_basis_odd_half.m.

Example 3.10.1 We illustrate here wavelet approximations on the
interval $[0, 1]$ of the two functions, $\exp(x)$ and $1/\sqrt{1 - 0.5 \cos(2\pi x)}$.
Notice that when considered as a periodic function of period 1
on the real line $\mathbb{R}$, the function $\exp(x)$ is discontinuous, whereas
$1/\sqrt{1 - 0.5 \cos(2\pi x)}$ is smooth on all of $\mathbb{R}$. We use the following
MATLAB® program to produce our results.

```
%
% wavelet_main_odd_half.m
%
% This program is copyrighted by SINC, LLC
% and belongs to SINC, LLC
%
% This is a main program for testing the
% periodic basis functions defined in
% the routine wavelet_basis_odd_half.m .
%
% Here we interpolate the function exp(x)
% on [0,1], which is discontinuous at the
% integers, when considered as a periodic
% function of period 1 on the real line.
%
clear all
%;
M = 21;
P = 1;
h = P/M;
%
```

```
for j=1:M
    t = (j-1/2)*h;
    z(j) = t;
    f1(j) = exp(t);
    f2(j) = 1/sqrt(1 - 0.5*cos(2*pi*t));
end
%
g1_exact = f1.';
g2_exact = f2.';
%
step = P/300;
for j = 1:300
    y = (j-0.5)*step;
    x(j) = y;
    vec = wavelet_basis_odd_half(P,M,y);
    g1_approx(j) = vec*g1_exact;
    g2_approx(j) = vec*g2_exact;
 end
%
plot(z,g1_exact,'o',x,g1_approx,'-');
pause
figure(1)
title('Wavelet Interp. of $e^x$ at 21 Half Interval
Values')
pause
print -dps2 wavelet_basis_odd_half_exp.ps
pause
plot(z,g2_exact,'o',x,g2_approx,'-');
pause
figure(2)
title('Wavelet Interp. of $(1-(1/2)cos(2\pi x))^{-1/2}$
at 21 Half Interval Values')
pause
print -dps2 wavelet_basis_odd_half_cos_root.ps
%
```

Example 3.10.2 We next illustrate the approximation of the func-

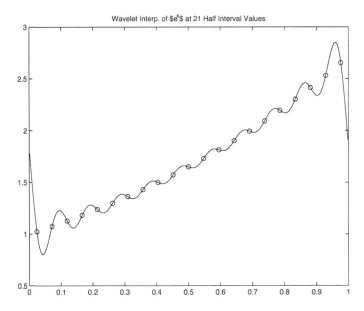

FIGURE 3.20. Exact and Wavelet–Approx. of e^x

tion $\sinh(z)$ on $[0, \pi]$, using variations of the basis `wavelet_basis_even_half` essentially based on Corollary 1.4.5. However, we have already demonstrated a "complete trigonometric basis" approximation in the previous example, and so, instead of using the $2N$–point formula (1.4.18), we shall use the N–point cosine polynomial formula (1.4.20) and the N–point sine polynomial formula (1.4.21).

Also included in this example is approximation via use of Corollary 1.4.5 (ii), but now we do not interpolate at the points $x_k = (2\,k-1)\pi/(2\,N)$; instead we interpolate at the points $z_k = (\pi/2)(1-\cos(x_k))$, which amounts to Chebyshev polynomial interpolation of $\sinh(z)$ on $(0, \pi)$ by a polynomial of degree $N - 1$. (Note here, that the zeros y_k of the Chebyshev polynomial $T_N(y)$ defined on the interval $[-1, 1]$ are just the points $y_k = \cos((2\,k - 1)\,\pi)/(2\,N))$.) The function $\sinh((\pi/2)(1 - \cos(x)))$ is a smooth, even, periodic function with period $2\,\pi$ defined on the whole real line, so that the right hand side of the expression

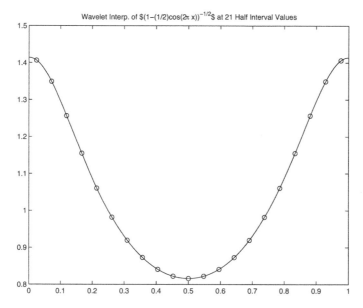

FIGURE 3.21. Exact and Wavelet–Approx. of $(1 - 0.5\,\cos(2\,\pi\,x))^{-1/2}$

$$\sinh(z) \;\; = \;\; \sum_{k=1}^{N} \sinh(z_k)\,\left(w_k(z) + w_{2\,N-k}(z)\right)$$

$$w_k(z) \;\; = \;\; D_e\!\left(N, \left(\cos^{-1}\left(1 - \frac{2\,z}{\pi}\right)\right)\right). \tag{3.10.1}$$

is, in fact, a Chebyshev polynomial interpolation, which is then an accurate approximation of $\sinh(z)$ on $[0, \pi]$.

```
%
% wavelet_main_even_half.m
%
% This program is copyrighted by SINC, LLC
% and belongs to SINC, LLC
%
% This is a main program for testing out
% the periodic basis functions defined
% in the routine wavelet_basis_even_half.m
% defined in Corollary 1.4.4.
%
```

```
% We interpolate here the function sinh(x)
% on [0,pi].
%
% Note that the interpolation of the even function
% f(x) = sinh(|x|) on [-pi,pi] via a linear combination
% of the basis {e^{ikx}: k = -N,..., N} is equivalent
% to interpolation on [0,pi] via the basis
% {cos(kx): k = 0, ..., N}.  As a result we end up
% with a cosine polynomial approximation on the real
% line R of the periodic function f, where f(x + 2 k pi)
% = f(x) for all integers k.  This function is
% continuous and of class Lip_1 on R, and results of
% approximation theory tell us that our uniform error
% of approximation of this function is of the order of
% N^{-1}.
%
% Similarly, the interpolation of the function
% g(x) = sinh(x) on [-pi,pi] via a linear  combination
% of the basis {e^{ikx}: k = -N, ..., N} is equivalent
% to interpolation of g(x) on [0,pi] via the basis
% {sin(kx): k=1, ..., N}.  This function g which is
% defined on R via the formula g(x + 2 pi k) = g(x)
% for all x on R, where k is any integer, is
% discontinuous on R, and our sine polynomial
% thus has very large errors in the neighborhoods of
% the points (2 k - 1) pi .
%
% On the other hand, the function sinh(z) =
% sinh((pi/2)(1 + cos(z))) is analytic in the
% strip {z : |Im(z)| < d} of the complex z-plane,
% for all d > 0, and it is moreover periodic with
% period 2 pi on the real line.  The interpolation
% of this even function of x via a linear combination
% of the basis {e^{ikx}: k = -N,..., N} is
% equivalent  to interpolation via the basis
% {cos(kx): k = 0, ..., N} and this will be an
% accurate interpolation.  This latter interpolation
% is just the formula (3.10.1) above, or
% equivalently, Chebyshev polynomial interpolation
```

```
% via the basis {T_k(1-2z/pi)}, where the T_k(w)
% denote the usual Chebyshev polynomials of degree
% k, defined by w = cos(u) and T_k(w) = cos(ku).
%
clear all
%;
N=8;
M = 2*N;
P = 2*pi;
h = P/M;
%
for j=1:N
    t = (j-1/2)*h;
    u = pi/2*(1 - cos(t));
    fx_ex(j) = sinh(t);
    fz_ex(j) = sinh(u);
end
%
f_exact = fx_ex.';
fz_exact = fz_ex.';
%
step = P/2/300;
for j = 1:300
    t = (j-0.5)*step;
    tz = pi*(1-cos(t))/2;
    tzz = acos(1-2*t/pi);
    xx(j) = t;
    zz(j) = tz;
    fxx(j) = sinh(t);
    cosvec = wavelet_basis_even_half_cos(P,N,t);
    sinvec = wavelet_basis_even_half_sin(P,N,t);
    chebvec = wavelet_basis_even_half_cos(P,N,tzz);
    fcos_approx(j) = cosvec*f_exact;
    fsin_approx(j) = sinvec*f_exact;
    cheb_approx(j) = chebvec*fz_exact;
  end
%
figure(1)
plot(xx,fxx,'.',xx,fcos_approx,'-');
```

```
pause
% title('Cos_wavelet Interp. of sinh(x) at 8 Half
Interval Values')
print -dps2 wavelet_basis_even_half_cos_sinh.ps
pause
figure(2)
plot(xx,fxx,'.',xx,fsin_approx,'-');
 pause
% title('Sin_wavelet Interp. of sinh(x) at 8 Half
Interval Values')
print -dps2 wavelet_basis_even_half_sin_sinh.ps
 pause
 figure(3)
plot(xx,fxx - cheb_approx,'-');
pause
% title('Error of Cheb Poly Interp. of sinh(x) at 8 Vals')
print -dps2 wavelet_basis_cheb_poly_error.ps
%
```

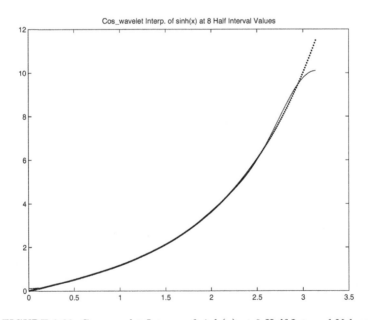

FIGURE 3.22. Cos_wavelet Interp. of sinh(x) at 8 Half Interval Values

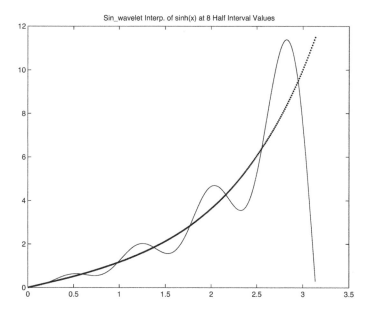

FIGURE 3.23. Sin wavelet Interp. of $\sinh(x)$ at 8 Half Interval Values

3.10.2 WAVELET SOL'N OF A NONLINEAR ODE VIA PICARD

Let us illustrate a method of approximate solution of a nonlinear ordinary differential equation initial value problem via use of wavelet approximation and Picard iteration, for the problem

$$y' = 1 + y^2, \quad 0 < x < P, \quad y(0) = \tan(0.2). \tag{3.10.2}$$

This problem was also solved via Sinc methods in §3.9.1 above. As above, we first convert the problem to a Volterra integral problem. But instead of collocation via Sinc methods, we now collocate via use of the wavelet bases of §1.4.1. This problem has the exact solution $y = \tan(x + 0.2)$. For the routine which follows, wavelet_ode_ivp_nl1.m, we used the wavelet indefinite integration matrix routine waveletintmat_e_int.m, for which the entries of the indefinite integration matrix are defined in (1.4.103).

We somewhat arbitrarily took $T = 0.6$ in the program ode_ivp_nl.m below. In this case we get convergence to the correct solution. Note, however, that many more points are required to achieve some accuracy for the case of the method based on wavelet approximation,

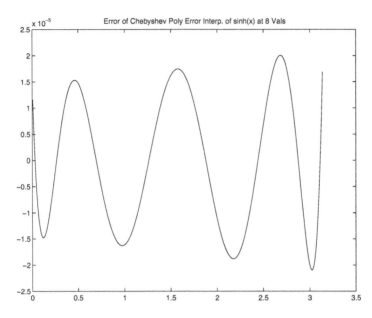

FIGURE 3.24. Chebyshev Poly Interp. Error of sinh at 8 Values

since the method tacitly assumes that the solution is periodic on $\mathbb{R}$, with period $T = 0.6$. See Figures 3.25 and 3.26.

```
%
% wavelet_ode_ivp_nl1.m
%
% This program is copyrighted by SINC, LLC
% and belongs to SINC, LLC
%
% Here we solve the following
%   problem:
%
% y' = f(x,y); y(0) = tan(0.2), with
%
%   f(x,y) = "ode_f1"(x,y) = 1+y^2.
%
% The problem is solved on (0,0.6),
% subject to y(0) = tan(0.2), using
% Sinc collocation and Sinc-wavelet
% indefinite integration combined
% with Picard iteration, over the
```

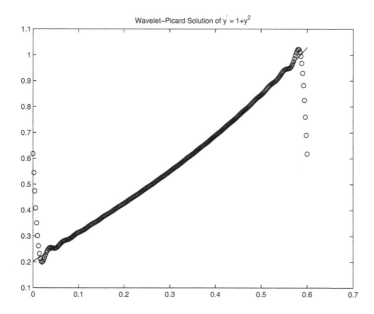

FIGURE 3.25. Wavelet-Picard Solution of $y' = 1 + y^2$, $y(0) = \tan(.2)$

```
% interval $(0,P) = (0,0.6)$.
%
% In the program, the routine
% ode_f1_exact(x) somputes the exact
% solution, tan(x+0.2).
%
clear all
T = 0.6;
N = 20;
M = 2*N+1;
h = P/M;
y0 = tan(0.2);
for j = 1:M;
    x(j) = (j-1)*h;
    y00(j) = y0;
end
A = waveletintmat_e_int(N,P);
Er = 5;
%
yy = y00.';
```

```
y = y00.';
%
while(Er > .00001)
      for k = 1:M
          W(k) =  ode_f1(x(k),y(k));
      end
%
      y = y00.' + A*W.';
      Err = norm(y - yy);
      yy = y;
      Er = Err
end
%
% Plots at j*P/300
%
step = T/300;
for j=1:301
      t = (j-1)*step;
      ex(j) = ode_f1_exact(t);
      xx(j) = t;
      w = wavelet_basis_even_int(P,N,t);
      Y(j) = w*yy;
end
%
figure(1)
plot(xx,ex,'-',xx,Y,'o')
%
pause
%
% title('Wavelet-Picard Solution of y^\prime = 1+y^2')
%
pause
print -dps2 wavelet_nl_ode.ps
%
figure(2)
plot(xx,Y-ex)
pause
title('Wavelet-Picard Error of y^\prime = 1+y^2')
pause
```

```
print -dps2 wavelet_nl_ode_error.ps
```

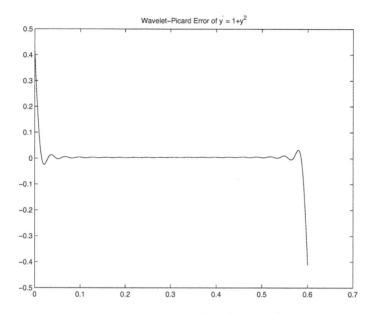

FIGURE 3.26. Wavelet-Picard Error of $y' = 1 + y^2$, $y(0) = \tan(.2)$

4

Explicit Program Solutions of PDE via Sinc–Pack

ABSTRACT We illustrate the theory of Chapter 2 by solving a full range of partial differential equations with programs from **Sinc–Pack**. A number of these PDE correspond to research level problems of interest in theory and in industry.

4.1 Introduction and Summary

Separation of Variables is the unifying theme of this chapter, wherein multidimensional problems over curvilinear regions can be reduced to solving a relatively small number of corresponding one dimensional problems. In §4.2 solutions are analytic continuations, hence harmonic continuations, of Sinc functions. In other sections we obtain particular solutions of non–homogeneous PDE through Sinc evaluation of linear and nonlinear convolution integrals over rectangular and curvilinear regions. The solution to the PDE including boundary conditions is then obtained using the *boundary integral equation* (BIE) method. Convergence will seen to be exponential and absolute error is uniform throughout the entire region.

4.2 Elliptic PDE

We illustrate here various methods of solving Laplace and Poisson problems in two and three dimensions.

4.2.1 HARMONIC SINC APPROXIMATION

Let $\mathcal{B}$ denote a simply connected and bounded domain in $\mathbb{R}^2$, and let

$$\begin{aligned}
\nabla^2 u &= 0, & z \in \mathcal{B} \\
u &= g, & z \in \partial\mathcal{B}
\end{aligned} \qquad (4.2.1)$$

315

with

$$\partial \mathcal{B} = \bigcup_{i=1}^{L} \Gamma_i, \tag{4.2.2}$$

where the Γ_i are analytic arcs (see Definition 2.6.2). We now apply the results of §1.5.13 and §2.6.2 to obtain an approximate solution to a Dirichlet problem on $\mathcal{B}$. Let us assume that for each $i = 1, 2, \cdots, L$, there exists a function $\varphi_i : \Gamma_i \to \mathbb{R}$, with φ_i satisfying Definition 1.5.2, We assume, in addition, that $\psi_i = \varphi_i^{-1} : \mathbb{R} \to \Gamma_i$, and furthermore, that the endpoints a_i and b_i of Γ_i satisfy the equations $a_{i+1} = b_i$, $i = 1, 2, \cdots, L$, and $a_{L+1} = a_1$. For $h > 0$, and $k \in \mathbb{Z}$, select the Sinc points $z_{i,k} = \psi_i(kh)$. In the same in manner as the region $\mathcal{D}^+$ was defined for the function φ in §1.5.13 for $i = 1, 2, \ldots, L$, we let

$$\mathcal{D}_i^+ = \{\psi_i(u + iv) : u \in \mathbb{R}, v \geq 0, \ \psi(u = iv) \ defined\}.$$

Let us assume furthermore, that the closure of $\mathcal{B}$ belongs to each $\mathcal{D}_i^+$, i.e., that $\overline{\mathcal{B}} \subseteq \mathcal{D}_i^+$ for $i = 1, 2, \ldots, L$. Let $\delta_{i,j}$ be defined for φ_i just as δ_j was defined for φ in (1.5.107). We then form the approximation

$$u(z) \approx u_N(z) = \sum_{j=1}^{L} \sum_{\ell=-N}^{N} v_{j,\ell}\, \delta_{j,\ell}(z), \tag{4.2.3}$$

where the constants $\{v_{i\ell}\}$ are to be determined. Equating $u_N(z) = g(z)$ for $z = z_{i,k}$, we get a linear system of equations of the form

$$\begin{bmatrix} I & B^{12} & \cdots & B^{1L} \\ B^{21} & I & \cdots & B^{2L} \\ \vdots & \vdots & \ddots & \vdots \\ B^{L1} & B^{L2} & \cdots & I \end{bmatrix} \begin{bmatrix} \mathbf{v}^1 \\ \mathbf{v}^2 \\ \vdots \\ \mathbf{v}^L \end{bmatrix} = \begin{bmatrix} \mathbf{g}^1 \\ \mathbf{g}^2 \\ \vdots \\ \mathbf{g}^L \end{bmatrix}, \tag{4.2.4}$$

where $\mathbf{v}^i = (v_{i,-N}, \cdots, v_{i,N})^T$, where I is the unit matrix of order $2N + 1$, where the $(k, \ell)^{th}$ entry of the matrix $B^{i,j}$ is $\delta_{j,\ell}(\psi(z_{i,k}))$, and where $\mathbf{g}^i = (g(z_{i,-N}), \ g(z_{i,-N+1}), \ \ldots, \ g(z_{i,N}))^T$.

We remark here that the function g is allowed to be discontinuous at the the the points $a_1, a_2, \ldots, a_L$, although we shall assume that on Γ_i, we have $g \in \mathbf{M}_{\alpha,d}(\varphi_i)$. Moreover, it will usually occur, that for all (i, j), and $i \neq j$, the maps φ_i are not only defined on every Γ_j, but satisfy on Γ_j the condition $\varphi_i/\varphi_j' \in \mathbf{L}_{\alpha,d}(\varphi_j)$. In this case, selecting $h = [\pi d/(\alpha N)]^{1/2}$, it follows that for each $i = 1,\, 2,\, \ldots,\, L$, we have

$$\sup_{z \in \Gamma_i} \left| g(z) - \sum_{j=1}^{L} \sum_{\ell=-N}^{N} v_{j,\ell}\, \delta_{j,\ell}(z) \right| = \mathcal{O}(\varepsilon_N),$$

where ε_N is defined as in (1.3.27). Since the difference between the solution u and the approximation u_N on the right hand side of (4.1.3) is also harmonic in $\mathcal{B}$, it follows by the maximum principle for harmonic functions, which states that a harmonic function takes on its maximum and minimum values on the boundary of $\mathcal{B}$, that u_N approximates g on $\mathcal{B}$ to within a uniform error bound of $CN^{1/2}e^{-(\pi d\alpha N)^{1/2}}$, where C is independent of N. Finally, if different values $N = N_i$ are selected on each Γ_i, then the off diagonal blocks of matrices $B^{i,j}$ will be rectangular, i.e., no longer square.

It may be shown that the system of equations (4.1.4) is always non–singular and also well conditioned whenever the region $\mathcal{B}$ is not unduly complicated – see the discussion following Equation (2.6.46). We emphasize again, that the boundary function g is allowed to have discontinuities at the junctions of the Γ_i, and also, that re–entrant corners pose no problems provided that the logarithmic mapping functions are properly evaluated. Indeed, we can allow "overlapping" re–entrant corners, at which Γ_i and Γ_{i+1} intersect at an interior angle greater than 2π. We remark here, that given any complex number of the form $z = x + iy = r\,e^{i\theta}$, with $-\pi < \theta \le \pi$, the MATLAB® statement $\log(z)$ returns the complex number $\log|r| + i\arg(z)$, with $\arg(z) = \theta$. With the continuous extension of a function $\varphi(z) = \log((z-a)/(b-z))$ from a line segment Γ connecting a and b on which it is real valued to the line segment connecting another arc Γ' we must warn the user to exercise care in the evaluations of the functions $\delta_i(z_{j,k})$ for $i \ne j$. For non-convex regions, the maps φ_j require the evaluation of logarithms outside their usual principal value range $(-\pi, \pi]$ and thus require their correct analytic continuation values.

We emphasize this procedure produces a solution that is uniformly accurate in $\mathcal{B}$, even though the function u may have discontinuities at the corner points of $\partial\mathcal{B}$.

Example 4.2.1 To be specific, let us solve the problem

$$\nabla^2 u = 0 \quad \text{in} \quad \mathcal{B}$$

$$u = g \quad \text{on} \quad \partial \mathcal{B},$$

$$\text{(4.2.5)}$$

where $\mathcal{B}$ is the "Pac–Man" region, $\mathcal{B} = \{(r, \theta) : 0 < r < 1, \ \pi/4 < \theta < 7\pi/4\}$, with $r = |x + iy|$, and $\theta = \arg(x, y)$. Setting $w = e^{i\pi/4}$, we may describe the boundary of $\mathcal{B}$, which consists of 3 arcs, in an *oriented manner* as follows:

$$
\begin{aligned}
\Gamma_1 &= \{z = x + iy \in \mathbb{C} : z = wt, \quad 0 < t < 1\}, \\
\Gamma_2 &= \{z = e^{it} \in \mathbb{C} : \pi/4 < t < 7\pi/4\}, \quad \text{and} \\
\Gamma_3 &= \{z \in \mathbb{C} : z = w^3(1 - t), \quad 0 < t < 1\}.
\end{aligned}
\qquad \text{(4.2.6)}
$$

Finally, as our boundary condition, we take

$$g(z) = Arg(z) + \Re(w - z)^{2/3} \qquad \text{(4.2.7)}$$

Notice that this function g is discontinuous at $z = 0$, and has a singularity at the point $z = 1$. We may furthermore note that $g(x + iy)$ is harmonic in $\mathcal{B}$, and so our problem has the exact solution $u(x, y) = g(z)$, where we now use the same expression (4.2.7) to evaluate $g(z) = g(x + iy)$ in the interior of $\mathcal{B}$.

We can analytically continue our approximation into the interior of $\mathcal{B}$ by means of

$$u(z) \approx u_N(z) = \sum_{\ell=1}^{3} \sum_{k=-N}^{N} c_k^\ell \delta_k^\ell(z), \qquad \text{(4.2.8)}$$

which is harmonic in $\mathcal{B}$. Setting $b = w^3$ and $t = e^{\ell h}/(1 + e^{\ell h})$ we can then define Sinc points z_ℓ^j by (see the transformation # 2 in §1.5.3)

$$z_\ell^1 = wt, \quad z_\ell^2 = b(1 + bt)/(t + b), \quad z_\ell^3 = b^2(1 - t), \qquad \text{(4.2.9)}$$

and vectors $\mathbf{g}^\ell = \left(g_{-N}^\ell, \ldots, g_N^\ell\right)^T$ where, with Γ_ℓ parametrized as above, we have $g_j^\ell = g(z_j^\ell)$. Then, we similarly define $\mathbf{u}^\ell$, and similarly define the $(k\,\ell)^{th}$ element of B^{ij} by $\delta_{j,\ell}(\psi(z_k^i))$. The interested user should pay particular attention to the way the entries of the matrices B^{ij} are evaluated in the MATLAB® program below, due to the presence of the re-entrant corner on the boundary of $\mathcal{B}$.

The system of equations for the vectors $\mathbf{c}^\ell = \left(c^\ell_{-N}, \ldots, c^\ell_N\right)^T$, then takes the form:

$$
\begin{bmatrix}
\mathbf{I} & B^{12} & B^{13} \\
B^{21} & \mathbf{I} & B^{23} \\
B^{31} & B^{32} & \mathbf{I}
\end{bmatrix}
\begin{bmatrix}
\mathbf{c}^1 \\
\mathbf{c}^2 \\
\mathbf{c}^3
\end{bmatrix}
=
\begin{bmatrix}
\mathbf{g}^1 \\
\mathbf{g}^2 \\
\mathbf{g}^3
\end{bmatrix}.
\tag{4.2.10}
$$

Once we have solved (4.2.10) for the constants c^ℓ_j, we can use (4.1.8) to compute u at points (t_j, t_k) in the interior of $\mathcal{B}$. Here, too, we must take care regarding the evaluation of the sum (4.3.8) above, due to the presence of the re-entrant corner.

This procedure was effectively used in [Nr, SCK, SSC] and in §4.2.2 and §4.2.5 below.

In the remaining examples of this section we illustrate the application of the *Sinc convolution* procedures to obtain approximate solutions of elliptic, parabolic, and hyperbolic differential equations. We also illustrate the solution of a PDE problem over a curvilinear region.

The elliptic problems we shall solve by Sinc convolution in this section include a Poisson–Dirichlet problem over $\mathbb{R}^2$, a Poisson-Dirichlet problem over a square, a Poisson-Dirichlet problem over $\mathbb{R}^3$, and a Poisson–Dirichlet problem over the unit disk, a bounded curvilinear region.

4.2.2   A Poisson-Dirichlet Problem over $\mathbb{R}^2$

We illustrate here some explicit algorithms for the solution of a simple PDE problem, namely,

$$\nabla^2 u(x,y) = -g(x,y), \quad (x,y) \in \mathbb{R}^2, \qquad (4.2.11)$$

where

$$g(x,y) = \frac{8 - 16\,r^2}{(1+r^2)^4}, \quad r = \sqrt{x^2 + y^2}. \qquad (4.2.12)$$

The solution u of this problem is $u = 1/(1+r^2)^2$, and is given by the convolution integral

$$u(x,y) = \int\!\!\int_{\mathbb{R}^2} \mathcal{G}(x - \xi, y - \eta)\, g(\xi, \eta)\, d\xi\, d\eta, \qquad (4.2.13)$$

where $\mathcal{G}$ is the two dimensional Green's function,

$$\mathcal{G}(x,y) = \frac{1}{2\pi} \log \frac{1}{\sqrt{x^2 + y^2}}. \qquad (4.2.14)$$

The following is a detailed MATLAB® program for the solution to this problem. In it, the initial parameters, Mj, Nj, hj, for $j = 1, 2$ are selected based on our assumed form for the solution to this problem over $\mathbb{R}^2$, namely,

$$u(x,y) \approx \sum_{j=-M1}^{N1} \sum_{k=-M2}^{N2} U_{j,k} S(j,h1) \circ \varphi_5(x)\, S(k,h2) \circ \varphi_5(y). \quad (4.2.15)$$

The solution is illustrated in Figures 4.1 and 4.2.

```
%
% pois_R_2.m
%
% This program is copyrighted by SINC, LLC
% and belongs to SINC, LLC
%
% parameters
%
clear all
M1 = 15;
```

```
N1 = M1;
m1 = M1+N1+1;
h1 = 0.6/sqrt(N1);
M2 = M1;
N2 = M1;
m2 = m1;
h2 = 0.6/sqrt(N1);
%
% Sinc points
%
z1 = Sincpoints5(M1,N1,h1);
%
z2 = z1;
%
% Evaluate the function g, for solution
% u = 1/(1+x^2+y^2)^2
%
for j=1:m1
   for k=1:m2
      r2 = z1(j)^2+z2(k)^2;
      g(j,k) = (8-16*r2)/(1+r2)^4;
   end
end
%
clear r2
%
% Evaluate the exact solution, U_ex,
% at the Sinc points
for j=1:m1
    for k=1:m2
        U_ex(j,k) = 1/(1+z1(j)^2 + z2(k)^2)^2;
    end
end
%
% Indefinite integration matrices
%
AA = Int5_p(M1,N1,h1);
BB = Int5_m(M1,N1,h1);
%
```

```
% Eigenvalues
%
[X S] = eig(AA);
[Y T] = eig(BB);
% clear A
% clear B
s = diag(S);
%
% Evaluate the 2d Laplace transform
%
for j=1:m1
   for k=1:m2
         G(j,k) = lap_tr_poi2(s(j),s(k));
   end
end
%
%
% Now, solve the PDE Lap U = - g
%
U = zeros(m1,m2);
for n=1:4
   if n == 1
     A=X; B=X;
   elseif n == 2
     A=X; B=Y;
   elseif n == 3
     A=Y; B=X;
   else
     A=Y; B=Y;
   end
   U = U + conv2_vec(A,m1,B,m2,G,g);
end
%
UU = real(U);
clear U
surf(z1,z2,UU);
pause
print -dps2 pois_R_2.ps
W = UU(:,N1+1);
```

```
plot(z1,W)
print -dps2 ex_sec_pois_R_2.ps
pause
surf(z1,z2,U_ex)
pause
surf(z1,z2,UU-U_ex)
```

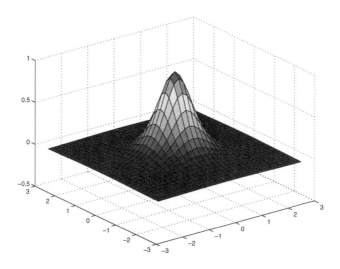

FIGURE 4.1. Sol'n to $\nabla^2 u = (8 - 16 * r2)/(1 + r2)^4$

4.2.3   A POISSON–DIRICHLET PROBLEM OVER A SQUARE

In this first example, we illustrate the use of the Sinc convolution method in combination with the BIE method for finding a solution u on region B which fulfills a non-homogeneous partial differential equation plus boundary conditions. In this example, the boundary conditions are Dirichlet boundary conditions.

The Poisson – Dirichlet problem over the square $(-1, 1) \times (-1, 1)$ *in 2–d takes the form*

$$u_{xx} + u_{yy} = -f(x, y), \quad (x, y) \in B \equiv (-1, 1) \times (-1, 1)$$

$$\text{(4.2.16)}$$

$$u = g \quad \text{on } \partial B.$$

We now solve this problem using the Green's function

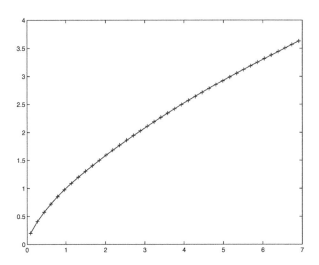

FIGURE 4.2. X–Section of Computed u

$$G(x, y) = \frac{1}{2\pi} \log \frac{1}{\sqrt{x^2 + y^2}} \qquad (4.2.17)$$

Let U be a particular solution to the non-homogeneous equation, given by

$$U(x, y) = \int\int_B G(x - \xi, y - \eta) \, f(\xi, \eta) \, d\xi \, d\eta. \qquad (4.2.18)$$

Next, let p be a solution to the homogeneous equation problem

$$p_{xx} + p_{yy} = 0, \quad (x, y) \in B \equiv (-1, 1) \times (-1, 1)$$
$$\qquad (4.2.19)$$
$$p = g - U \quad \text{on } \partial B.$$

We then solve for p via the procedure of §2.5.4. Indeed, p is a harmonic function, as the real part of a function F which is analytic in the region $D = \{z = x + i\,y : (x, y) \in B\}$, and as shown in §2.5.4, we can set

$$F(z) = \frac{1}{\pi i} \int_{\partial D} \frac{r(t)}{t - z} \, dt, \quad z \in D, \qquad (4.2.20)$$

where r is a real valued function.

Upon letting $z \to \zeta \in \partial D$, with ζ not a corner point of ∂D, and taking real parts, we get the integral equation

$$r(\zeta) + (K\,r)(\zeta) = g(\zeta), \tag{4.2.21}$$

with $K\,r = \Re \mathcal{S}\,r$.

It may be shown that this integral equation always has a unique solution, which can be obtained via iteration after Sinc collocation as described in §2.5.4. However, since ∂B consists of only 4 arcs, it is also efficient to solve the resulting system of equations directly.

We first parametrize $\partial D = \cup_{j=1}^{4}\Gamma_j$, in an oriented fashion, in complex variable notation, as follows:

$$
\begin{array}{rcl}
\Gamma_1 & = & \{z = z^1(t) = 1 + i\,t, -1 \le t \le 1\} \\
\Gamma_2 & = & \{z = z^2(t) = i - t, -1 \le t \le 1\} \\
\Gamma_3 & = & \{z = z^3(t) = -1 - i\,t, -1 \le t \le 1\} \\
\Gamma_4 & = & \{z = z^4(t) = -i + t, -1 \le t \le 1\}
\end{array}
\tag{4.2.22}
$$

Next, setting $r|_{\Gamma_j} = r^j$, and $g|_{\Gamma_j} = g^j$, we see that the real part of the Hilbert transform of r^1 taken over Γ_1 is zero when $\zeta \in \Gamma_1$; whereas for $\zeta \in \Gamma_1$, each of the terms $\Re \int_{\Gamma_j} r^j(t)/(t - \zeta)\,dt$ may be collocated for $j = 2, 3, 4$ using Sinc quadrature, since the singularities are not on the interior of the interval of integration. Indeed, these singularities are on the exterior of the assumed region of analyticity of the integrand $r^j/(t-\zeta)$ (see Theorem 1.5.16). For $\zeta = 1+i\tau \in \Gamma_1$, the explicit forms $\Re \mathcal{S}\,r^j$ taken over Γ_j for $j = 2,3,4$ are

$$\text{Over } \Gamma_2 : J^{12}\,r^2(\tau) = \frac{1 - \tau}{\pi} \int_{-1}^{1} \frac{r^2(t)}{(1 + t)^2 + (1 - \tau)^2}\,dt$$

$$\text{Over } \Gamma_3 : J^{13}\,r^3(\tau) = \frac{2}{\pi} \int_{-1}^{1} \frac{r^3(t)}{4 + (t + \tau)^2}\,dt \tag{4.2.23}$$

$$\text{Over } \Gamma_4 :: J^{14}\,r^4(\tau) = \frac{1 + \tau}{\pi} \int_{-1}^{1} \frac{r^4(t)}{(1 - t)^2 + (1 + \tau)^2}\,dt.$$

Our parametrization of ∂D enables us to replace the above integral equation (4.2.21) by a system of 4 integral equations over $(-1, 1)$ of the form

$$
\begin{bmatrix}
I & J^{12} & J^{13} & J^{14} \\
J^{21} & I & J^{23} & J^{24} \\
J^{31} & J^{32} & I & J^{34} \\
J^{41} & J^{42} & J^{43} & I
\end{bmatrix}
\begin{bmatrix}
r^1 \\ r^2 \\ r^3 \\ r^4
\end{bmatrix}
=
\begin{bmatrix}
g^1 \\ g^2 \\ g^3 \\ g^4
\end{bmatrix}
\tag{4.2.24}
$$

Now, it is readily checked, as follows by symmetry that $J^{23} = J^{34} = J^{41} = J^{12}$, $J^{24} = J^{31} = J^{42} = J^{13}$, and $J^{21} = J^{32} = J^{43} = J^{14}$, so that we only need to collocate three of the integrals. Collocation thus leads us to the block system of equations

$$
\begin{bmatrix}
I & A & B & C \\
C & I & A & B \\
B & C & I & A \\
A & B & C & I
\end{bmatrix}
\begin{bmatrix}
\mathbf{r}^1 \\ \mathbf{r}^2 \\ \mathbf{r}^3 \\ \mathbf{r}^4
\end{bmatrix}
=
\begin{bmatrix}
\mathbf{g}^1 \\ \mathbf{g}^2 \\ \mathbf{g}^3 \\ \mathbf{g}^4
\end{bmatrix}
\tag{4.2.25}
$$

where $\mathbf{r}^j = \left(r^j_{-N}, \ldots, r^j_N \right)^T$ and similarly for $\mathbf{g}^j = \left(g^j_{-N}, \ldots, g^j_N \right)^T$, with $g^j_\ell = g^j(z_\ell)$, z_ℓ denoting a Sinc point of $(-1,1)$. Thus,

$$
A = [A_{jk}], \quad A_{jk} = \frac{1 - z_j}{\pi} \frac{w_k}{(1 - z_j)^2 + (1 + z_k)^2}
$$

$$
B = [B_{jk}],, \quad B_{jk} = \frac{2}{\pi} \frac{w_k}{4 + (z_j + z_k)^2}
\tag{4.2.26}
$$

$$
C = [C_{jk}], \quad C_{jk} = \frac{1 + z_j}{\pi} \frac{w_k}{(1 + z_j)^2 + (1 - z_k)^2},
$$

with $w_k = h/\varphi'(z_k)$, and where $\varphi(z) = \log((1 + z)/(1 - z))$. We can then evaluate our approximation to p in the interior of B. We could do this either using the basis given in §1.5.12 or equivalently, using the harmonic Sinc basis of §1.5.13. We have chosen the latter procedure in the program which follows, by approximating p at the exact Sinc points in B for which we obtained our above Sinc convolution approximation to U.

Finally, we can approximate the solution u to our original problem, by forming the expression $u = U + p$.

```
%
% lap_square.m
```

```
%
%  Program for Solution of (4.2.16)
%
% Solving Poisson problem
%
% nabla u = -f in D = (-1,1) x (-1,1)
%
% with f(x,y) = -2,
%
% subject to
%
%  u(x,y) = x^2 on the boundary of D.
%
% Our procedure is based on use of
% coupled SINC and BIE.
%
clear all
N = input('Enter number of Sinc points:   ')
h = 2.5/sqrt(N);
m = 2*N+1;
format long e
%
% Sinc points and weights over (-1,1)
%
z = Sincpoints2(-1,1,N,N,h);
w = weights2(-1,1,N,N,h);
W = diag(w);
%
% Construction and diagonalization of
% indefinite integration matrices
%
A1 = Int2_p(-1,1,N,N,h);
B1 = Int2_m(-1,1,N,N,h);
[X S] = eig(A1);
Xi = inv(X);
[Y S] = eig(B1);
%
% Note: A1 and B1 have the same
% eigenvalue matrix.
```

```
%
Yi = inv(Y);
%
s = diag(S);
%
% I. The particular solution
%
% Computing Laplace transform of the
% Green's function at the eigenvalues
% of A1 and B1 -- see Lemma 2.4.2
%
gamma = 0.5772156649015328606065;
%
for j = 1:m
    p=s(j);
    for k = 1:m
        q=s(k);
        G(j,k) = 1/(1/p^2 + ...
1/q^2)*(-1/4 +1/(2*pi)*(q/p*(gamma-log(q))+ ...
            p/q*(gamma-log(p))));
    end
end
%
% Computing Right Hand Side --  matrix --
% f = [f(z_j\,,z_k)], with f(x,y) = 2.
%
%
for j = 1:m
  for k = 1:m
   f(j,k) = 2;
  end
end

%
% Computing the Sinc Convolution of the
% Green's Function  G  and f by Sinc
% convolution, based on Algorithm 2.5.1.
%
% We thus get a particular solution of the
```

```
% non-homogeneous PDE.
%
U = X*(G.*(Xi*f*Xi.'))*X.';
U = U + Y*(G.*(Yi*f*Xi.'))*X.';
U = U + X*(G.*(Xi*f*Yi.'))*Y.';
U = U + Y*(G.*(Yi*f*Yi.'))*Y.';
UR = real(U);
%
% Plot of UR
%
figure(1)
surf(z,z,UR)
print -dps2 dirich_pois_partic.ps
pause
%
% II.  Solving the Homogenous Problem
%
% We next need to compute a solution of the
% homeogenous PDE to satisfy the boundary
% conditions.  This is done via BIE, i.e., via
% the procedure described above.
%
% Constructing right-hand side vectors
%
for j = 1:m
    g2(j) =  z(j)^2 - UR(j,m);
    g4(j) =  z(j)^2 - UR(j,1);
  end
%
for j=1:m
    g1(j) = z(m)^2 - UR(m,j);
    g3(j) = z(1)^2 - UR(1,j);
end
%
% Setting up the blocks A, B, C
%
for j = 1:m
    for k = 1:m
        A1(j,k) = (1-z(j))/pi/((1-z(j))^2 + (1+z(k))^2);
```

```
            A2(j,k) = 2/pi/(4 + (z(j) + z(k))^2);
            A3(j,k) = (1+z(j))/pi/((1+z(j))^2 + (1-z(k))^2);
        end
end
%
A1 = A1*W;
A2 = A2*W;
A3 = A3*W;
D = eye(m);
%
% Now use the above created matrices and vector
% to create the 'big system' of equations.
%
A = [D A1 A2 A3; A3 D A1 A2; ...
        A2 A3 D A1;  A1 A2 A3 D];
g = [g1 g2 g3 g4].';
%
% Solving linear equations
%
r = A\g;
%
% Approximate solution via Harmonic Sinc
%
    r1 = r(1:m);
    r2 = r(m+1:m+m);
    r3 = r(m+m+1:m+m+m);
    r4 = r(m+m+m+1:m+m+m+m);
%
% Solution computed on Sinc grid
%
p = zeros(m,m);
%
for j=1:m
    for k=1:m
        zz = z(j) + i*z(k);
        u1 = phi2(1-i,1+i,zz);
        u2 = phi2(1+i,-1+i,zz);
        u3 = phi2(-1+i,-1-i,zz);
        u4 = phi2(-1-i,1-i,zz);
```

```
            p1 = basis_harm(u1,N,N,h);
            p2 = basis_harm(u2,N,N,h);
            p3 = basis_harm(u3,N,N,h);
            p4 = basis_harm(u4,N,N,h);
            p(j,k) = p1*r1 + p2*r2 + p3*r3 + p4*r4;
        end
end
% Approximte solution
%
 v = p+UR;
%
% Plots of Homogenous & Complete Sols
%
figure(2)
surf(z,z,p);
print -dps2 dirich_lap.ps
pause
%
figure(3)
surf(z,z,v);
print -dps2 dirich_tot.ps
pause
%
approx = diag(v);
exact = z.^2;
figure(4)
plot(z, approx, 'o', z, exact, '-');
xlabel('Sinc points')
ylabel('Approximate, "o" and Exact, "-"')
 print -dps2 X-Sec_lap_poi_sq.ps
```

The particular solution (4.2.18) is given in Figure 4.3; the solution
p which solves the problem (4.2.19) subject to $u = x^2 - U(x,y)$ on
∂B is given in Figure 4.4, and the solution u of the problem (4.2.16)
with $g(x,y) = x^2$ is given in Figure 4.5.

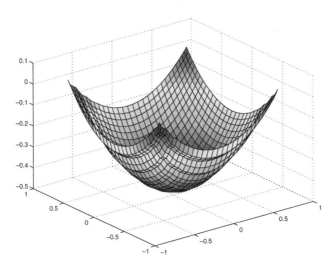

FIGURE 4.3. Particular Solution to $\nabla^2 U = -2$ on $[-1,1]^2$

4.2.4 NEUMANN TO A DIRICHLET PROBLEM ON LEMNISCATE

Consider obtaining an approximate solution to the Neumann problem

$$u_{xx} + u_{yy} = 0, \quad (x,y) \in B$$

$$\frac{\partial u}{\partial n} = p \quad \text{on } \partial B\,, \tag{4.2.27}$$

where $\partial\,/(\partial n)$ denotes differentiation in the direction of the unit outward normal $\mathbf{n}$ on ∂B. If $\int_{\partial B} p\,ds = 0$, then the conditions of the Fredholm alternative are satisfied. Consequently, a function u which solves (4.2.27) is not unique, i.e., if u_1 is a solution to (4.2.27) then, with c an arbitrary constant, $u_2 = u_1 + c$ also solves (4.2.27). Now, if v is a conjugate harmonic function of u, such that $F = u + i\,v$ is analytic in $D = \{z \in \mathbb{C} : z = x + i\,y,\ (x,y) \in B\}$, then (see Equations (2.6.28) and (2.6.29))

$$\frac{\partial u}{\partial n} = -\frac{\partial v}{\partial t} = p\,, \tag{4.2.28}$$

where $\partial\,/(\partial t)$ denotes differentiation in the direction of the unit tangent, $\mathbf{t}$ on ∂B. We can thus produce $q = -\int p\,ds$ via Sinc indefinite integration. But since v is a conjugate of u, we have

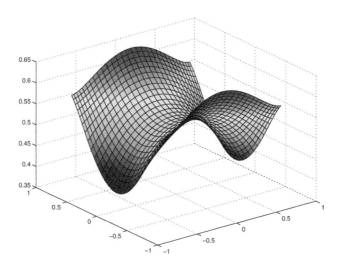

FIGURE 4.4. Particular Solution to $\nabla^2 p = 0$ on Square

$u(\zeta) = (\mathcal{S}\,v)(\zeta)$, $\quad \zeta \in \partial B$, where $\mathcal{S}$ denotes the Hilbert transform operator, defined as in (2.6.22). That is, if g is defined on ∂B by $g = \mathcal{S}\,q$, then $g(\zeta) = u(\zeta)$, $\zeta \in \partial B$, with u a solution to (4.2.27).

After constructing g, we can proceed as outlined in §2.6 to find the function u which solves the PDE problem (4.2.27) via solution of a Dirichlet problem.

We now illustrate with a MATLAB® program the construction of the function g for the case when B is the lemniscate

$$B = \{(\rho,\theta) : 0 < \rho < 1 - \cos(\theta),\ 0 < \theta < 2\,\pi\}. \qquad (4.2.29)$$

In polar coordinates the boundary of B is given by

$$\partial B = \{(r,\theta) : r = 1 - \cos(\theta),\ 0 < \theta < 2\,\pi\}. \qquad (4.2.30)$$

Clearly, this parametrization of ∂B defines it as an analytic arc on $(0, 2\,\pi)$.

To test our algorithm, we take as the solution to (4.2.27) the harmonic function

$$u(x, y) = \Re(x + i\,y)^3 = x^3 - 3\,x\,y^2. \qquad (4.2.31)$$

We then obtain the following expressions for $(x, y) \in \partial B$:

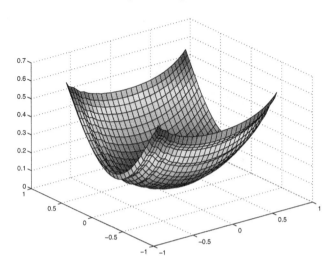

FIGURE 4.5. Solution $u = U + p$ to $\nabla^2 u = -2$ on Square

$$
\begin{aligned}
r &= 2\sin^2(\theta/2)\,, \\
x &= r\cos(\theta)\,, \quad y = r\sin(\theta)\,, \\
\mathbf{n} &= (\sin(3\,\theta/2)\,, -\cos(3\,\theta/2))\,, \\
\nabla u &= (u_x\,, u_y) = (3(x^2 - y^2)\,, 6\,x\,y) \\
p &= \frac{\partial u}{\partial n} = (\nabla u)\cdot\mathbf{n}\,.
\end{aligned}
\tag{4.2.32}
$$

Performing indefinite integration of p, we get

$$
q(\phi) = -\int_{-\pi}^{\phi} p(\theta)\,ds = -\int_{-\pi}^{\phi} p(\theta)\,\frac{ds}{d\theta}\,d\theta.
\tag{4.2.33}
$$

Finally, by performing the Hilbert transform on the function q we get g,

$$
\begin{aligned}
g(\theta) &= \frac{P.V}{\pi i}\int_{\partial B}\frac{q\,dz}{z - \zeta} \\
&= \frac{P.V.}{\pi i}\int_{-\pi}^{\pi}\frac{q(\phi)}{\phi - \theta}\frac{\phi - \theta}{z - \zeta}\frac{d}{d\phi}\left(2\sin^2(\phi/2)\,e^{i\,\phi}\right)d\phi.
\end{aligned}
\tag{4.2.34}
$$

Thus, with $\bar{r} = (x, y)$, $r = |\bar{r}|$, a function u which solves the Neumann problem (4.2.27) on the lemniscate (4.2.29) is then also given by the solution to the Dirichlet problem

$$\nabla^2 u(\bar{r}) = 0, \quad \bar{r} \in B,$$

$$u(r \, \cos(\theta), r \, \sin(\theta)) = g(\theta), \bar{r} \in \partial B. \tag{4.2.35}$$

The MATLAB® program which follows illustrates the computation for $g(\theta)$ at the Sinc points on the interval $(0, 2\pi)$.

```
%
% lemniscate.m
%
% Determining q defined as in (4.2.33).
%
clear all
%
N = 15;
m = 2*N+1;
h = .55/N^(1/2);
TT = 1/exp(N*h);
T = exp(h);
for j=1:m
     t(j) = 2*pi*TT/(1+TT);
     TT = TT*T;
end
A = Int2_p(0,2*pi,N,N,h);
C = Hilb_yam2(0,2*pi,N,N,h);
%
for j=1:N+1
     s2(j) = sin(t(j)/2);
     r(j) = 2*s2(j)^2;
     x(j) = r(j)*cos(t(j));
     y(j) = r(j)*sin(t(j));
     z(j) = x(j)+i*y(j);
     Normzdot(j) = 2*s2(j);
     unitnormalx(j) = sin(3*t(j)/2);
     unitnormaly(j) = cos(3*t(j)/2);
     u(j) = x(j)^3-3*x(j)*y(j)^2;
     ux(j) = 3*(x(j)^2-y(j)^2);
     uy(j) = 6*x(j)*y(j);
     gexact(j) = u(j);
```

```
    p(j) = ux(j)*unitnormalx(j) ...
        + uy(j)*unitnormaly(j);
    p_integrand(j) = - Normzdot(j)*p(j);
end
%
for k=1:N
    j = m+1-k;
    s2(j) = s2(k);
    r(j) = r(k);
    x(j) = x(k);
    y(j) = -y(k);
    z(j) = x(j)+i*y(j);
    Normzdot(j) = 2*s2(j);
    unitnormalx(j) = unitnormalx(k);
    unitnormaly(j) = - unitnormaly(k);
    u(j) = u(k);
    ux(j) = ux(k);
    uy(j) = -uy(k);
    gexact(j) = gexact(k);
    p(j) = p(k);
    p_integrand(j) = p_integrand(k);
end
%
pp = p_integrand.';
q = A*pp;
%
for j=1:m
    for k=1:m
        if abs(j-k) > 0;
            part1 = ...
            (t(k) - t(j))/(z(k) - z(j));
            part2 = ...
            2*s2(k)*(unitnormaly(j) ...
                + i*unitnormalx(j));
            phi_theta_matrix(j,k) = (part1*part2);
          else;
            phi_theta_matrix(j,k) = 1;
        end;
    end;
```

```
end;
%
Hilb_yam_matrix = C.*phi_theta_matrix;
% Hilb_yam_matrix = real(C.*phi_theta_matrix);
%
gr = Hilb_yam_matrix*q;
gimag = imag(gr.');
%
figure(1)
plot(t,gimag,'-',t,gexact,'o');
print -dps2 Dir_bc_exact_appr.ps
pause
figure(2)
plot(t,gimag-gexact);
print -dps2 Dir_bc_exact_appr_err.ps
```

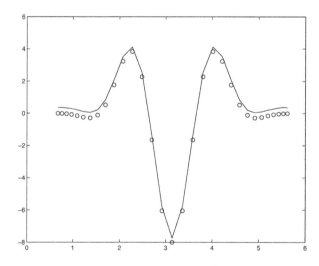

FIGURE 4.6. Solution $u = U + p$ to $\nabla^2 u = -2$ on Square

Note that our computed values in Figure 4.6 are slightly larger by
a constant than the exact ones throughout the interval $(0, 2\pi)$, resulting from the fact that the solution to the Neumann problem is
not unique. Note also the large error near the ends of the interval.
In part, this is due to the inaccuracy of the Sinc points t_j near 2π,
and also, due to the fact that MATLAB® computes eigenvalues and

eigenvectors to within an absolute, rather than a relative error.

An approximate solution to a harmonic function u which solves the problem (4.2.27) can now be obtained via solution of (4.2.35). The simplest procedure for accomplishing this is to use the analytic continuation procedure of §1.5.13. To this end, we would take for φ in (1.5.107) the function $\varphi(t) = \log(t/(2\pi - t))$. Then, instead of using (1.5.109), we would set

$$u_m(z) = \sum_{k=-N}^{N} c_k w_k(z) \frac{\delta_k(t)}{t - t_k} \frac{t - t_k}{z - z_k} 2 \sin(t_k/2) \exp(2\,i\,t_k).$$

$$(4.2.36)$$

with t_k the Sinc points of the map φ, and with $z_k = 2\sin^2(t_k/2)\exp(i\,t_k)$. The complex numbers t and z are related simply by the equation $z = 2\sin^2(t/2)\exp(i\,t)$. The functions $w_k(z)$ are given in terms of the functions δ_k defined as in (1.5.109) by

$$w_k(z) = \frac{\delta_k(t)}{t - t_k} \frac{t - t_k}{z - z_k} 2 \sin(t_k/2) \exp(2\,i\,t_k), \qquad (4.2.37)$$

in which we take t to be a function of z, so that as $z \to z_j$ then $t \to t_j$, and $w_k(z) \to \delta_{j,k}$ with $\delta_{j,k}$ now denoting the Kronecker delta. Hence, if $\mathbf{W}$ denotes the $m \times m$ matrix with $(j,k)^{th}$ entry $w_k(z_j)$, and if $\mathbf{g}$ denotes the column vector of order m with entry $g(t_j)$, where g is defined as in (4.2.35), then the vector $\mathbf{c}$ with j^{th} entry c_j is given by the solution of the non-singular system of equations $\mathbf{W}\,\mathbf{c} = \mathbf{g}$.

4.2.5　A Poisson Problem over a Curvilinear Region in $\mathbb{R}^2$

Let $\mathcal{D}$ denote the unit disk, and let C_+ (resp., C_-) denote the part of $\partial\mathcal{D}$ on the upper (resp., lower) half plane. We illustrate here a MATLAB® programs for the solution to the problem

$$u_{xx} + u_{yy} = -4, \quad (x, y) \in \mathcal{D}$$

$$(4.2.38)$$

$$u(x, y) = \begin{cases} 1 & \text{if } (x, y) \in C_+ \\ -1 & \text{if } (x, y) \in C_-. \end{cases}$$

The particular solution, $U(x, y) = 1 - x^2 - y^2$, to his problem is obtained via Sinc convolution, using the Green's function representation,

$$U(x, y) = \int\int_{\mathcal{D}} \mathcal{G}(x - \xi, y - \eta)\, f(\xi, \eta)\, d\xi\, d\eta = (1 - x^2 - y^2), \quad (4.2.39)$$

where, as above, $\mathcal{G}$ denotes the free space Green's function,

$$\mathcal{G}(x, y) = \frac{1}{2\pi} \log \frac{1}{\sqrt{x^2 + y^2}}, \qquad (4.2.40)$$

and where $f(x, y) = 4$.

We initially obtain an approximate solution to this non-homogeneous problem based on the curvilinear region algorithm in §2.5.3 in lap_poi_disc.m. We then give plots of the our computed solution, the exact solution, and the difference between the computed and the exact one. This solution is illustrated in Figure 4.7.

Following these plots, lap_harm_disc.m computes the solution to

$$v_{xx} + v_{yy} = 0, \quad (x, y) \in \mathcal{D},$$

$$v = \begin{cases} b_1 \equiv 1 & \text{if } (x, y) \in C_+ \\ b_2 \equiv -1 & \text{if } (x, y) \in C_-, \end{cases} \qquad (4.2.41)$$

using the harmonic Sinc approximation procedure described in §1.5.3, the implementation of which is illustrated in §4.2.1. This solution is illustrated below in Figure 4.8.

Indeed, it is possible to show that the exact solution is just

$$v(x, y) = \frac{2}{\pi} \arctan\left(\frac{2y}{1 - x^2 - y^2} \right), \qquad (4.2.42)$$

so that we can again compare computed and exact solutions. These solutions as well as the error, are plotted following lap_harm_disc.m which computes only the solution in the upper half plane, since the solutions in the upper and lower half planes are negatives of each other.

As already stated, the MATLAB® program for the solution to the convolution integral problem follows. It uses the two dimensional convolution algorithm derived in §2.4.3. Notice also, that we have computed a solution that is uniformly accurate in D in spite of discontinuous boundary conditions.

```
%
% lap_poi_disc.m
%
% We solve here the problem
%
% nabla u = -ff in D,
%
% with D the unit disc, and with
% ff = 4;
%
% The method of solution is to use
% the free space Green's function
%
% {\cal G}(x,y) = (4 pi)^{-1} log(x^2 + y^2)
%
% to produce a particular solution
%
% U(x,y)
% = \int\int_D {\cal G}(x-p,y-q) f(p,q) dp dq
%
% via use of the Sinc convolution
% algorithm convnl2_vec.m.
%
clear all
M1 = input('Enter Sinc point number, M1:   ')
N1 = input('Enter Sinc point number, N1:   ')
M2 = input('Enter Sinc point number, M2:   ')
N2 = input('Enter Sinc point number, N2:   ')
m1 = M1+N1+1;
m2 = M2+N2+1;
h1 = 0.45/sqrt(N1);
h2 = 0.45/sqrt(N2);
format long e
%
% We provide here the explicit
% computation of the zeros
% z_k = (e^{kh}-1)/(e^{kh}+1) for
% (-1,1) using the formula
% 1-z_k^2 = 4e^{kh}/(1+e^{kh})^2
```

```
% instead of computing (1-z_k^2) after z_k
% is computed,, since this latter method
% is not a good procedure to use in the
% case when z_k is close to 1.
%
al = 2;
TT = exp(h1);
TM1 = 1/TT^M1;
for j=1:m1
   T2 = sqrt(TM1);
   den = 1+TM1;
   bet(j) = 4*T2/den;
   z1(j) = TM1/den;
   w1(j) = h1*TM1/den^2;
   TM1 = TM1*TT;
end
%
TT = exp(h2);
TM2 = 1/TT^M2;
for k=1:m2
    den = 1 + TM2;
    z2(k) = TM2/den;
    w2(j) = h2*TM2/den^2;
    TM2 = TM2*TT;
end
% Computing the right hand side
% vector ff
%
for ii = 1:m1
  for j = 1:m2
   ff(ii,j) = 4;
  end
end
%
% Construction and diagonalization
% of indefinite integration matrices
% Aj and Bj, j=1,2 <=> x,y
%
A1 = Int2_p(0,1,M1,N1,h1);
```

```
B1 = Int2_m(0,1,M1,N1,h1);
A2 = Int2_p(0,1,M2,N2,h2);
B2 = Int2_m(0,1,M2,N2,h2);
%
[X1 S] = eig(A1);
[Y1 S] = eig(B1);
[X2 T] = eig(A2);
[Y2 T] = eig(B2);
clear A1
clear B1
clear A2
clear B2
%
s = diag(S);
clear S
sig = diag(T);
clear T
%
% Now, the Laplace transform, as
% given in Step #6 of Algorithm 2.1.2
% and using the routine lap_tr_poi2.m.
%
gamma = 0.5772156649015328606065;
%
for ii=1:m1
    p = al*s(ii);
    for k=1:m2
        for j=1:m1
            q = bet(j)*sig(k);
            F(ii,j,k) = 1/(1/p^2 + 1/q^2)*(-1/4 + ...
            1/(2*pi)*(q/p*(gamma-log(q)) + ...
            p/q*(gamma-log(p))));
        end
    end
end
U = zeros(m1,m2);
for n=1:4
    if n == 1
    A = X1; B = X2;
```

```
      elseif n == 2
      A = X1; B = Y2;
      elseif n == 3
        A = Y1; B = X2;
      else
        A = Y1;  B = Y2;
      end
U = U + convnl2_vec(A,m1,B,m2,F,ff);
end
%
U = real(U);
%
surf(z2,z1,U);
pause
%
% Now, compare with the exact solution,
% U_ex(x,y) = 1-x^2-y^2.  But this must first
% be approximated on the square:
%
for j=1:m1
    for k=1:m2
        x = -1 + al*z1(j);
        y = -bet(j)/2+bet(j)*z2(k);
        U_ex(j,k) = 1 - x^2 -y^2;
    end
end
%
surf(z2,z1,U_ex)
pause
surf(z2,z1,U - U_ex)
pause
%
% L1 and L2 norms of error
%
Er1 = 0;
Er2 = 0;
for j=1:m1
    for k=1:m2
        eee = abs(U(j,k) - U_ex(j,k));
```

```
        Er1 = Er1 + w1(j)*w2(k)*eee;
        Er2 = Er2 + w1(j)*w2(j)*eee^2;
    end
end
%
L1error = Er1 % L1 norm
L2error = sqrt(Er2) % L2 norm
```

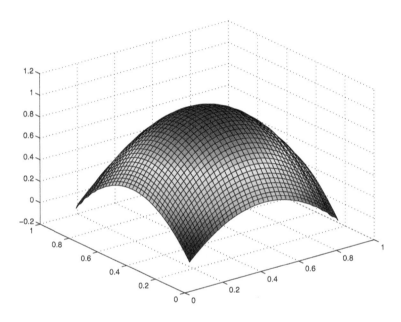

FIGURE 4.7. L H Side of (4.2.39) Computed via Algorithm 2.5.4

This next program illustrates the computation of the solution to (4.2.38).

```
%
% lap_harm_disc.m
%
% We solve here the problem
%
% nabla u = 0   in D
%
% with D the unit disc, and
```

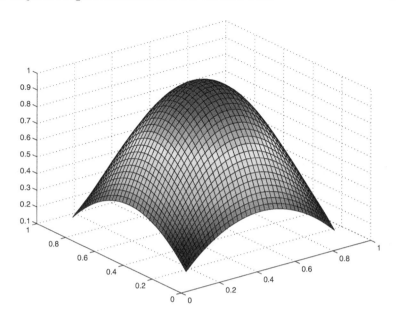

FIGURE 4.8. Exact Solution of (4.2.39)

```
% subject to the boundary conditins
%
% u = 1 on C_+, u = -1 on C_-
%
% where C_+ (resp., C_-) denotes
% the boundary of D in the upper
% (resp., lower) half of the
% complex plane.
%
% The exact solution to this problem
% is
%
% u =  2/pi*tan^{-1}(2y/(1-x^2-y^2)).
%
% The method of solution of via use
% of the harmonic Sinc basis defined
% in Sec. 1.5.13, the implementation
% of which is illustrated in Sec. 4.1.1.
% We thus firs solve a 2 x 2 block
% system of equations
```

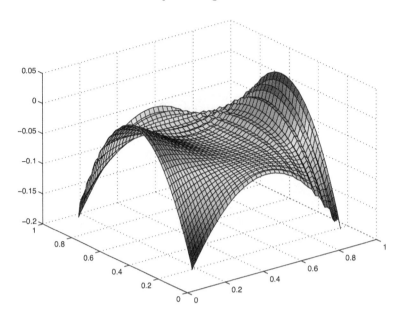

FIGURE 4.9. Difference Between Exact and Computed (4.2.39)

```
%
%     [ I  B ] c1 = '1'
%     [ C  I ] c2 = '-1'
%
% we determine the vector (c1 , c2)^T,
% and we use the expression
%
% U = sum_j [c1_j del1,j(z) + c2_j del2(j,z)
%
% to compute the solution in the
% interior of D.
%
% First compute the vectors c1 & c2:
clear all
N = 20;
h = 3/sqrt(N);
m = 2*N+1;
z1 = Sincpoints7(0,pi,N,N,h);
z2 = Sincpoints7(pi,2*pi,N,N,h);
B = zeros(m,m);
```

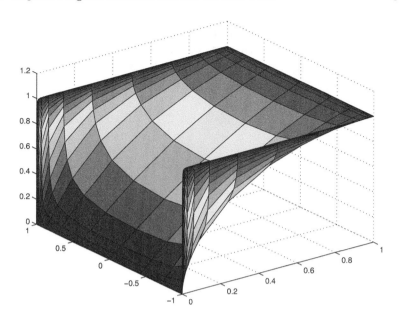

FIGURE 4.10. Sinc-Harmonic Solution to (4.2.41)

```
C = B;
E = eye(m);
%
for j=1:m
    C(j,:) = basis_harm7(pi,2*pi,z1(j),N,N,h);
    b1(j) = 1;
    B(j,:) = basis_harm7(0,pi,z2(j),N,N,h);
    b2(j) = -1;
end
%
Mat = [E B;C E];
b = [b1 b2].';
%
sol = Mat\b;
%
c1 = real(sol(1:m));
c2 = real(sol(m+1:2*m));
%
% Next compute the solution in the
% interior of D.  However, we will plot
```

```
% only the part in the upper half plane,
% since the suolution in the lower half
% is just the negative (with respect to
% the vertical coordinate) of that in the
% lower half plane
%
h2 = 0.9;
T = exp(h2);
TT = 1/T^N;
for j=1:m
    rt = sqrt(TT);
    den = 1 + TT;
    x(j) = (TT-1)/den;
    r(j) = 2*rt/den;
    TT = T*TT;
end
%
T = exp(h2);
TT = 1/T^N;
for k=1:m
    y(k) = TT/(1+TT);
    TT = T*TT;
end
%
U = zeros(m,m);
%
for j=1:m
    for k=1:m
        xx = x(j);
        yy = r(j)*y(k);
        zz = xx + i*yy;
        v1 = basis_harm7(0,pi,zz,N,N,h);
        v2 = basis_harm7(pi,2*pi,zz,N,N,h);
        U(j,k) = v1*c1 + v2*c2;
        u_exact(j,k) = 2/pi*atan(2*yy/(1-xx^2-yy^2));
    end
end
%
% Now, for some plots:
```

```
%
figure(1)
surf(y,x,U);
pause
figure(2)
surf(y,x,u_exact);
pause
figure(3)
surf(y,x,U - u_exact);
```

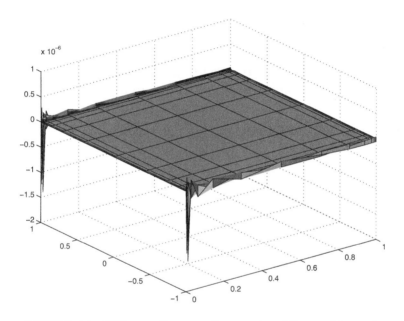

FIGURE 4.11. Difference Between Computed and Exact of (4.2.41)

4.2.6 A Poisson Problem over $\mathbb{R}^3$

MATLAB® program pois_R_3.m provides a Sinc solution to the Poisson problem

$$u_{xx} + u_{yy} + u_{zz} = -g(x, y, z),$$

$$(x, y, z) \in B = \mathbb{R}^3,$$

$$(4.2.43)$$

with

$$g(x, y, z) = (4\,r^2 - 6)\,\exp(-r^2),$$

with $r = \sqrt{x^2 + y^2 + z^2}$, so that the solution to (4.2.43) is just $u(x, y, z) = \exp(-r^2)$. Although this function is only a function of r, no advantage was taking of this fact by our method of solution. Also, since the region is infinite, we did not need to apply any boundary conditions.

Our Sinc method of solution convolution is based on the Green's function representation

$$u(x, y, z) = \int \int \int_B \mathcal{G}(x - \xi, y - \eta, z - \zeta)\, g(\xi, \eta, \zeta)\, d\xi\, d\eta\, d\zeta, \quad (4.2.44)$$

with

$$\mathcal{G}(x, y, z) = \frac{1}{4\,\pi\,\sqrt{x^2 + y^2 + z^2}}. \qquad (4.2.45)$$

In order to apply *Sinc convolution* we split the multi-dimensional convolution over the region B into the 8 indefinite convolution integrals,

$$\iiint_{\mathbb{R}^3} =$$

$$\int_{-\infty}^{x}\int_{-\infty}^{y}\int_{-\infty}^{z} + \int_{x}^{\infty}\int_{-\infty}^{y}\int_{-\infty}^{z} + \int_{x}^{\infty}\int_{y}^{\infty}\int_{-\infty}^{z}$$

$$+ \int_{x}^{\infty}\int_{y}^{\infty}\int_{z}^{\infty} + \int_{-\infty}^{x}\int_{y}^{\infty}\int_{z}^{\infty} + \int_{-\infty}^{x}\int_{-\infty}^{y}\int_{z}^{\infty}$$

$$+ \int_{x}^{\infty}\int_{-\infty}^{y}\int_{z}^{\infty} + \int_{-\infty}^{x}\int_{y}^{\infty}\int_{-\infty}^{z},$$

$$(4.2.46)$$

each of these being evaluated via the Sinc convolution algorithm, and then summed to produce a particular solution to (4.2.43).

MATLAB® program pois_R_3.m clearly points out the remarkable simplicity of the *Sinc convolution* algorithm. The main program loops over the eight indefinite convolutions, setting up the needed matrices, and then calling the conv3_vec subroutine to apply the matrices in the same order for each indefinite convolution. A plot of the solution is given in Figure 4.2.12.

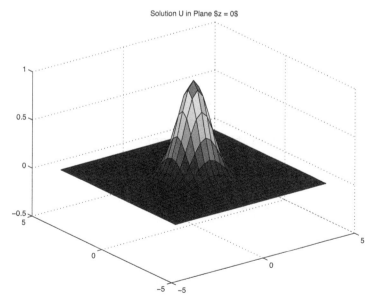

FIGURE 4.12. Solution of $\Delta u = (6 - 4r^2)\,e^{-r^2}$ in $\mathbb{R}^3$

```
%
% pois_R_3.m
%
% This is a Sinc convolution solution to the problem
%
% \nabla u = - (6 - 4 r^2) exp(-r^2)
%
% via Sinc methods -- see Sec. 3.1.2 of this handbook.
%
% Parameters
%
clear all
%
M1 = 14;
N1 = M1;
m1 = M1+N1+1;
h1 = 1.1/sqrt(N1);
M2 = M1;
N2 = M1;
m2 = m1;
h2 = h1;
M3 = M1;
N3 = M1;
m3 = m1;
h3 = h1;
%
% Sinc points.  The''1'' in Sincpoints1
% refers to the tranformation #1 given
% in Example 1.5.3 of this handbook.
%
z1 = Sincpoints1(M1,N1,h1);
z2 = Sincpoints1(M2,N2,h2);
z3 = Sincpoints1(M3,N3,h3);
%
% Next we proceed to approximate the
% solution U according to the convolution
% algorithm, Algorithm 2.5.2 of this
% handbook.
%
```

```
% Evaluate the function g
%
for j=1:m1
   for k=1:m2
      for l=1:m3
         r2 = z1(j)^2+z2(k)^2+z3(l)^2;
         g(j,k,l) = - (4*r2-6)*exp(-r2);
      end
   end
end
%
clear r2
%
% Indefinite integration matrices.  These
% were also computed in previous examples.
%
A1 = h1*Im(m1);
B1 = A1.';
A2 = h2*Im(m2);
B2 = A2.';
A3 = h3*Im(m3);
B3 = A3.';
%
% Eigenvalues and eigenvectors
%
[X1 S1] = eig(A1);
[Y1 T] = eig(B1);
s1 = diag(S1);
%
[X2 S2] = eig(A2);
[Y2 T] = eig(B2);
s2 = diag(S2);
%
[X3 S3] = eig(A3);
[Y3 T] = eig(B3);
s3 = diag(S3);
%
% Evaluate the 3d Laplace transform,
% as given in Theorem 2.4.4 of this handbook.
```

```
%
for j=1:m1
   for k=1:m2
      for l=1:m3
         G(j,k,l) = lap_tr_poi3(s1(j),s2(k),s3(l));
      end
   end
end
%
%
% Now, solve the PDE Lap U = - g
% via repeated (8 times) use of the convolution
% algorithm, Algorithm 2.5.2 of this handbook.
%
U = zeros(m1,m2,m3);
for n=1:8
   if n == 1
      A=X1; B=X2; C=X3;
   elseif n == 2
      A=X1; B=X2; C=Y3;
   elseif n == 3
      A=X1; B=Y2; C=X3;
   elseif n == 4
      A=X1; B=Y2; C=Y3;
   elseif n == 5
      A=Y1; B=X2; C=X3;
   elseif n == 6
      A=Y1; B=X2; C=Y3;
   elseif n == 7
      A=Y1; B=Y2; C=X3;
   else
      A=Y1; B=Y2; C=Y3;
   end
   U  = U + conv3_vec(A,m1,B,m2,C,m3,G,g);
end
%
% We now do some surface, and cross-section plots.
% The cross-section plots are checks, to see if
% one gets the same solution through each
```

```
% plane through the origin, for symmetric
% problems.
%
UU = real(U);
clear U
V1 = UU(:,:,M3+1);
surf(z1,z2,V1);
title('Solution U in Plane $z=0$')
% print -dps2 pois_3_d_appr.ps
pause
for j=1:m1
   for l=1:m3
      V2(j,l) = UU(j,M2+1,l);
   end
end
surf(z2,z3,V2);
pause
for k=1:m2
   for l=1:m3
      V3(k,l) = UU(M1+1,k,l);
   end
end
surf(z2,z3,V3);
pause
for j = 1:m1
   exact(j) = exp(-z1(j)^2);
end
%
% plot(z1,UU(:,M2+1,M3+1),z1,exact,'o')
% title('Error Plot in Planes $y = z = 0$')
% print -dps2 gauss_err.ps
% pause
% plot(z3,V2(M1+1,:))
% pause
% plot(z2,V3(:,M3+5))
% pause
clear V1
clear V2
clear V3
```

4.3   Hyperbolic PDE

We illustrate here solutions of wave equations over regions of the
form $B \times (0, T)$, where B is a region in $\mathbb{R}^3$, and $(0, T)$ is a time
interval, and where either B or T (or both) may be unbounded.

4.3.1   SOLVING A WAVE EQUATION OVER $\mathbb{R}^3 \times (0, T)$

The program wave_3p1.m contains the details of the Sinc solution to
the wave equation problem

$$\frac{1}{c^2} \frac{\partial^2 u(\bar{r}, t)}{\partial t^2} - \nabla^2 u(\bar{r}, t) = \left(7 - \frac{3}{ct} + \frac{3}{4\,(ct)^2} - 4\,r^2 \right) t^{3/2} \exp(-r^2 - c\,t),$$

$$\bar{r} = (x, y, z), \quad (\bar{r}, t) \in B = \mathbb{R}^3 \times (0, \infty)$$

(4.3.1)

The exact solution is

$$u(\bar{r}, t) = t^{3/2} \exp(-|\bar{r}|^2 - c\,t)$$

(4.3.2)

The program wave_3p1.m calls two important programs from Sinc–
Pack, namely, conv4_vec.m, which enables four dimensional convolution, and also, lap_tr4_wave.m, which evaluates the four dimensional
transform of the Green's function, as given in Theorem 2.4.10.

We include wave_3p1.m here, for sake of completeness. Note, in this
program, the initializations are similar to those in previous programs.
The dimension of the problem (4.3.1) is large here, and for this reason, it was necessary to specify four sets of Sinc points, and eight
indefinite integration matrices.

In the program below we took 21 Sinc points in the x, y, and z
directions, and 15 time points, thus we computed an approximate
solution requiring only $21^3 \times 15 = 138,915$ points.

```
%
% wave_3p1.m
%
% This program is copyrighted by SINC, LLC
% an belongs to SINC, LLC
```

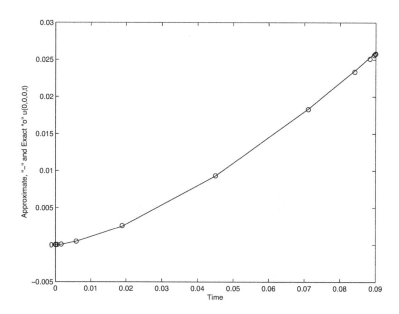

FIGURE 4.13. Solution $u(0,0,0,t)$ to (4.3.1)

```
%
% This is the MAIN routine for computation of
% the wave problem c^{-2} u_{tt} = Delta u + f
% on R^3 x (0,\infty), via Sinc convolution.
% The exact solution is
%
% u(r,t) = t^(3/2)*exp(-r^2 - c*t).
%
% Initializations:
%
clear all
%
tic
c=1/2;
%
TT = 0.1;
%
p1=10;
p2=10;
p3=10;
```

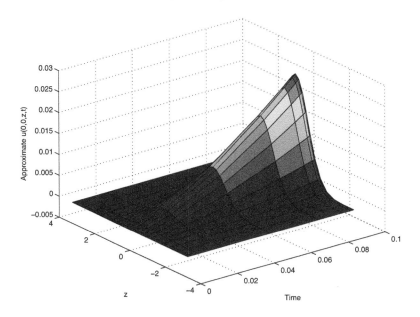

FIGURE 4.14. Solution $u(0, 0, z, t)$ to (4.3.1)

```
p4=7;
%
M1 = p1;
N1 = p1;
m1 = M1+N1+1;
h1 = .6/sqrt(N1);
xx = Sincpoints5(M1,N1,h1);
[W1,W2,R] = Int_both5(M1,N1,h1);
rrr = diag(R);
M2 = p2;
N2 = p2;
m2 = M2+N2+1;
h2 = .6/sqrt(N2);
[X1,X2,S] = Int_both5(M2,N2,h2);
sss = diag(S);
yy = Sincpoints5(M2,N2,h2);
M3 = p3;
N3 = p3;
m3 = M3+N3+1;
h3 = .6/sqrt(N3);
```

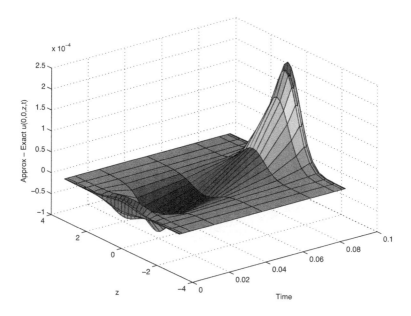

FIGURE 4.15. z–Time Error Plot for PDE (4.3.1)

```
zz = Sincpoints5(M3,N3,h3);
[Y1,Y2,T] = Int_both5(M3,N3,h3);
ttt = diag(T);
M4 = p4;
N4 = p4;
m4 = M4+N4+1;
h4 = 3.5/sqrt(N4);
tt = Sincpoints2(0,TT,M4,N4,h4);
A4 = Int2_p(0,TT,M4,N4,h4);
[D U] = eig(A4);
uuu = diag(U);
clear R
clear S
clear T
clear U
%
% Setting up the right hand side
% of c^{-2}u - Delta u = f
% f (called ff here) and the
% exact solution, ex.
```

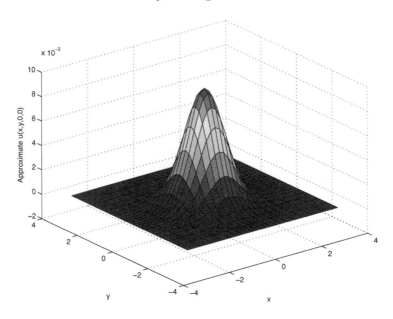

FIGURE 4.16. Solution $u(x, y, 0, 0)$ to (4.3.1)

```
%
for ii=1:m1
    x = xx(ii);
    for j=1:m2
        y = yy(j);
        for k=1:m3
            z = zz(k);
            r2 = x^2+y^2+z^2;
            f1 = exp(-r2);
            for l=1:m4
            t = tt(l);
            ct = 1/c/t;
            th = sqrt(t);
            ee = th^3*exp(-r2-c*t);
            fe = (7-3*ct+3/4*ct^2-4*r2);
            ff(ii,j,k,l) = fe*ee;
            ex(ii,j,k,l)= ee;
        end
    end
end
```

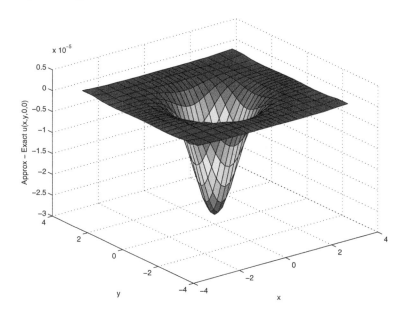

FIGURE 4.17. Spacial Error Plot for PDE (4.3.1)

```
%
% Setting up of the Laplace transform
% of the Green's function, as described
% in Theorem 2.4.10 of this handbook.
%
for ii=1:m1
    u=rrr(ii);
    for j=1:m2
        v = sss(j);
        for k=1:m3
            w = ttt(k);
            for l=1:m4
                t = uuu(l);
                G(ii,j,k,l) = lap_tr_wave4(u,v,w,t,c);
            end
        end
    end
end
%
% Now, solve the PDE
```

```
%
U = zeros(m1,m2,m3,m4);
%
for n=1:8
   if n == 1
    A = W1; B = X1; C = Y1;
   elseif n == 2
    A = W1; B = X1; C = Y2;
   elseif n == 3
    A = W1; B = X2; C = Y1;
   elseif n == 4
    A = W1; B = X2; C = Y2;
   elseif n == 5
    A = W2; B = X1; C = Y1;
    elseif n == 6
    A = W2; B = X1; C = Y2;
   elseif n == 7
    A = W2; B = X2; C = Y1;
   else
    A = W2; B = X2; C = Y2;
   end
   U = U + conv4_vec(A,m1,B,m2,C,m3,D,m4,G,ff);
end
%
U = real(U);
%
clear ff
clear G
%
for l=1:m4
    for k=1:m3
        VR1(k,l) = U(M1+1,M2+1,k,l);
        ER1(k,l) = ex(M1+1,M2+1,k,l);
    end
    ER3(l) = U(M1+1,M2+1,M3+1,l);
    EX3(l) = ex(M1+1,M2+1,M3+1,l);
end
%
toc
```

```
figure(1)
plot(tt,EX3,'o',tt,ER3,'-')
xlabel('Time')
ylabel('Approximate, "-" and Exact "o" u(0,0,0,t)')
print -dps2 Exact.vs.Appr_3d_wave_t_plot.ps
pause
%
figure(2)
surf(tt,zz,VR1)
xlabel('Time')
ylabel('z')
zlabel('Approximate u(0,0,z,t)')
print -dps2 Appr_3d_wave_zt_plot.ps
pause
%
figure(3)
surf(tt,zz,VR1-ER1)
xlabel('Time')
ylabel('z')
zlabel('Approx - Exact u(0,0,z,t)')
print -dps2 Approx_Exact_3d_wave_zt_plot.ps
pause
%
for ii=1:m1
   for j=1:m3
      VR2(ii,j) = U(ii,j,M3+1,M4+1);
      ER2(ii,j) = ex(ii,j,M3+1,M4+1);
   end
end
%
figure(4)
surf(yy,xx,VR2)
xlabel('x')
ylabel('y')
zlabel('Approximate u(x,y,0,0)')
print -dps2 Appr_3d_wave_xy_plot.ps
pause
%
figure(5)
```

```
surf(yy,xx,VR2-ER2)
xlabel('x')
ylabel('y')
zlabel('Approx - Exact u(x,y,0,0)')
print -dps2 Approx_Exact_3d_wave_xy_plot.ps
```

4.3.2   SOLVING HELMHOLTZ EQUATION

We present here a method of solution to the Helmholtz equation
(2.3.11) derived from the wave equation (2.3.5). However, instead
of applying Sinc methods directly to (2.4.58), we refer the reader
to the analysis in §2.4.4 leading to Equations (2.4.61) and its two
dimensional Fourier transform, (2.4.64). We shall also assume here
that $V = (0, a) \times (-b, b) \times (-c, c)$. We only summarize the method
here; details will appear elsewhere [SRA].

We begin by taking the $2 - d$ Fourier transform of the integral
equation (2.4.61),

$$\widetilde{f^{sc}}(x, \bar{\Lambda}) = -\kappa^2 \int_0^a \frac{e^{-\mu|x-\xi|}}{2\mu} \left(\gamma \widetilde{f^{sc}}(\xi, \bar{\Lambda}) + f^{in}(\xi) \, \widetilde{\gamma}(\xi, \bar{\Lambda}) \right) d\xi, \tag{4.3.3}$$

and then proceed to solve for $\widetilde{f^{sc}}(x \, \bar{\Lambda})$ in order to obtain $f^{sc}(\bar{r})$.

Here $f^{sc} = f - f^{in}$ is the scattered field, f is the total field which
satisfies (2.4.55), and $f^{in}(\bar{r}) = e^{-\kappa x}$ is the input field. Also, $\bar{\Lambda}$ is the
$2 - d$ Fourier transform variable, when taking the Fourier transform
of (2.4.58) with respect to (y, z), and $\kappa = 1/(c_0 \, s)$ with c_0 the speed
of sound in the medium surrounding the support of γ, namely $V =
(0, a) \times (-b, b) \times (-c, c)$. Finally we let $\mu = \sqrt{\kappa^2 + \Lambda^2}$.

Test problems are easy to formulate. Starting with Equation (2.4.61)
for the scattered field, and starting with any function f^{sc}, and taking
as our input field the function $e^{-\kappa x}$, we can readily solve for the
potential

$$\gamma = \frac{\nabla^2 f^{sc} - \kappa^2 f^{sc}}{\kappa^2 \left(f^{sc} + f^{in} \right)}. \tag{4.3.4}$$

We thus take as our solution the function $f^{sc}(\bar{r}) = e^{-r^2 - \kappa x}$, and
obtain

$$\gamma = \frac{4(y^2 + z^2 - 1)\, e^{-r^2}}{\kappa^2(1 + e^{-r^2})}. \tag{4.3.5}$$

We can evaluate the right hand side of (4.3.3) for each fixed $\bar{\Lambda}$ as a function of x using indefinite convolution. To this end, we shall require the Laplace transform,

$$-\kappa^2 \int_0^\infty e^{-\mu x - x/\sigma}\, dx = -\frac{\kappa^2}{2\mu}\frac{\sigma}{1 + \sigma\mu}. \tag{4.3.6}$$

A sequence of functions $\{f_n^{sc}\}_{n=-\infty}^\infty$ can then be obtained using successive approximation based on the equation

$$\widetilde{f_{n+1}^{sc}}(x, \bar{\Lambda}) = -\kappa^2 \int_0^a \frac{e^{-\mu|x-\xi|}}{2\mu}\left(\gamma\,\widetilde{f_n^{sc}}(\xi, \bar{\Lambda}) + f^{in}(\xi)\,\widetilde{\gamma}(\xi, \bar{\Lambda})\right) d\xi, \tag{4.3.7}$$

starting with $f_0^{sc} = \gamma$. This iteration can also be carried out using Sinc approximations, and it can moreover be shown that the resulting sequence of Sinc approximations converges.

4.4   Parabolic PDE

We illustrate here the Sinc solution to two problems:

1. Two separate population density problems over (i) $(0, \infty) \times (0, \infty)$ and (ii) over $(0, \infty) \times (0, T)$; and

2. A Navier–Stokes problem over $R^3 \times (0, T)$.

4.4.1   A Nonlinear Population Density Problem

We consider the nonlinear integro–differential equation problem

$$F_t(x, t) = A(x, t) - B(x, t), \quad (x, t) \in \mathbb{R}^+ \times (0, T),$$

$$\tag{4.4.1}$$

$$F(x, 0+) = F_0(x), \quad x \in \mathbb{R}^+.$$

with $\mathbb{R}^+ = (0, \infty)$.

In (4.4.1),

$$A(x,t) \;=\; \int_x^\infty G(x,u)\, b(u)\, F_x(u,t)\, du$$

$$B(x,t) \;=\; \mu(t) \int_0^x F_x(v,t) \int_{x-v}^\infty \frac{a(v,w)}{w}\, F_x(w,t)\, dw\, dv \,,$$

$$(4.4.2)$$

where the initial condition function F_0, and functions G, b, μ and a are assumed to be given.

This problem was discussed, for example, in [RM, VrPMS], and a solution of this problem for an engineering application has been successfully carried out in [SRi] using programs similar to the ones presented here.

No solutions to this problem seems to exist in the literature.

At the outset the above PDE is transformed into the integral equation problem,

$$F(x,t) = F_0(x) + U(x,t) - V(x,t), \quad (x,t) \in \mathbb{R}_+ \times (0,T) \quad (4.4.3)$$

with the functions U and V given by the integrals with respect to t of A and B respectively, i.e.,

$$U(x,t) \;=\; \int_0^t \int_x^\infty G(x,u)\, b(u)\, F_x(u,\tau)\, du\, d\tau$$

$$V(x,t) \;=\; \int_0^t \mu(\tau) \int_0^x F_x(v,\tau) \int_{x-v}^\infty \frac{a(v,w)}{w}\, F_x(w,\tau)\, dw\, dv\, , d\tau \,,$$

$$(4.4.4)$$

The basis for our solution is the well–known Picard iteration scheme

$$F^{(n+1)}(x\,,t) = F_0(x) + U^{(n)}(x\,,t) - V^{(n)}(x\,,t), \quad n = 0,1,,\ldots,$$

$$(4.4.5)$$

where $U^{(n)}$ and $V^{(n)}$ are the same as in (4.3.4), except that each $F_x(\cdot,\cdot)$ in (4.3.4) has been replaced with $F^{(n)}(\cdot,\cdot)$, with $F^{(0)}(x\,,t) = F_0(x)$.

Our method of solution presented here illustrates the flexibility and power of Sinc methods for handling such a complicated problem. We proceed as follows:

1. Select $m_x = 2M + 1$ Sinc points $\{x_j\}_{j=-M}^{M}$ for variable $x \in (0, \infty)$ and $n_t = 2N + 1$ Sinc points $\{t_k\}_{k=-N}^{N}$ for variable $t \in (0, T)$.

2. Discretize $F_0(x)$ by forming an $m_x \times m_t$ matrix $\mathbf{F}_0 = [F_{j,k}]_0$, with $(F_{j,k})_0 = F_0(x_j)$.

3. For $n = 0, 1, \ldots$, let $\mathbf{F}^{(n)}$ denote an $m_x \times m_t$ matrix, where $\mathbf{F}^{(0)} = \mathbf{F}_0$, while for $n > 0$, $\mathbf{F}^{(n)}$ is gotten by evaluation of the right hand side of (4.3.5). This evaluation is carried out as follows:

   (i) Form a discretized approximation to $F_x(x, t)$ by application of the Sinc derivative matrix to the matrix $\mathbf{F}^{(n)}$, to from the matrix $\mathbf{F}_x^{(n)} = [(Fx)_{k,\ell}]$. Since differentiation of a Sinc approximation is unbounded at $x = 0$, we use the derivative polynomial derived in §1.7 to form $\mathbf{F}_x^{(n)}$ from $\mathbf{F}^{(n)}$.

   (ii) Discretize $a(x, y)/y$, using x–Sinc points for both x and y to form an $m_x \times m_x$ matrix $A = [a_{j,k}] = [a(x_j, x_k)/x_k]$. Then form a 3–d matrix $C = [c_{j,k,\ell}]$ with $c_{j,k,\ell}$ the product $c_{j,k,l} = a_{j\,k}(Fx)_{k,\ell}$.

   (iii) Letting $[\delta_{k',k}]$ denote the indefinite integration matrix for integrating from 0 to x with $x \in (0, \infty)$, we now form a new triple array, $[e_{j,k',\ell}]$ with $e_{j,k',\ell} = \sum_{k=-M}^{M} \delta_{k',k}\, c_{j,,k,\ell}$. Note that this triple array is just the discretization of the function

   $$e(v, z, \tau) = \int_z^\infty \frac{a(v, w)}{w}\, F_x^{(n)}(w, \tau)\, dw.$$

   (iv) Since we next wish to evaluate the convolution integral

   $$p(x, \tau) = \int_0^x F_x(v, \tau)\, e(v, x - v, \tau)\, dv,$$

   we need the Laplace transform (see §1.5.11)

   $$E(v, s, \tau) = \int_0^\infty \exp(-z/s)\, e(v, z, \tau)\, dz,$$

   for every eigenvalue s of the indefinite integration matrix used for integration from 0 to x. (See §1.5.9.) This

Laplace transform integral can be evaluated using using Sinc quadrature. (See §1.5.7).)

(v) We can now approximate $p(x, \tau)$ using the algorithm for $r(x)$ in §1.5.11.

(vi) The approximation of $\int_0^t p(x, \tau) \, d\tau$ can now be carried out applying Sinc indefinite integration from 0 to t over $(0, T)$, with T infinite in Example (i), and T finite in Example (ii).

In the program below we use the following functions:

$$F_0(x) \;=\; 1 - e^{-cx}$$

$$b(x) \;=\; \frac{1}{1 + x^2}$$

$$G(x, y) \;=\; \frac{x}{x^2 + y^2 + 1} \qquad\qquad (4.4.6)$$

$$a(x, y) \;=\; \frac{x}{x^2 + y^2 + 1}$$

$$\mu(t) \;=\; \frac{1}{1 + t^2}.$$

In function F_0 we selected $c = 2$ for program p_d_picard.m (Example (i)). In program p_d_T5.m (Example (ii)) we selected $c = 20$ - a normally computationally difficult boundary layer case, with boundary layer in the neighborhood of $x = 0$. Both programs use Picard iteration to compute a solution. However, Picard iteration does not converge over $\mathbb{R}_+ \times \mathbb{R}_+$ for Example (ii). Convergence is achieved by subdividing

$$\mathbb{R}_+ \times \mathbb{R}_+ = \mathbb{R}_+ \times \bigcup_{j=0}^{\infty} (T_j, T_{j+1}), \quad T_j = jT,$$

and performing Picard iteration anew over each subinterval (T_j, T_{j+1}), with initial condition, the approximate solution $F(x, T_j)$ obtained from the previous interval $(T_{j-1}.T_j)$.

```
%
% Program:  p_d_picard.m
```

```
%
% This program is copyrighted by SINC, LLC
% an belongs to SINC, LLC
%
% F(x,t) = F_0(x) + U(x,t) - V(x,t)
%
%   U(x,t) = int_0^t int_x^infinity G(x,u) b(u) "times"
%                   "times" F_x(u,tau) du dtau
%
%   V(x,t) = int_0^t mu(tau) int_0^x F_x(v,tau) "times"
%       "times" int_{x-v}^infinity w^{-1} a(v,w) "times"
%         "times" F_x(w,tau) dw dv dtau
%
% A successive approximation solution over
% (0,infinity) x (0,infinity) is computed.
% See Sec. 4.4
%
clear all
format long e
%
% Initial value inputs
%
% Note  M1 is slighly bigger than N1, to take care
% of the boundary layer.
%
% Begin timing
%
tic
M1 = 10;
N1 = 8;
m1 = M1 + N1 + 1;
M2 = 8;
N2 = 6;
m2 = M2 + N2 + 1;
ht = 0.4/sqrt(N2);
hx = 0.7/sqrt(N1);
% T = 1;
%
format long e
```

```
%
% Sinc Points, Weights, and Matrices
%
x = Sincpoints3(M1,N1,hx);
wx = weights3(M1,N1,hx);
%
% t = Sincpoints2(0,T,M2,N2,ht);
t = Sincpoints3(M2,N2,ht);
%
% Integration over (0,x) and (x,infinity)
%
Axp = Int3_p(M1,N1,hx);
Axm = Int3_m(M1,N1,hx);
%
% The derivative matrices
%
% The "x" derivative matrix on (0,infinity)
% Note the many possibilities of Sinc
% differentiation on (0, infinity).
%
Axd = Dpoly3_1(M1,N1,hx);
%
% Integration over (0,t)
%
% Atp = Int2_p(0,T,M2,N2,ht);
Atp = Int3_p(M2,N2,ht);
%
% First, some definitions of functions and
%
% some diagonal matrices.
%
for j=1:m1
%
% the initial F_0, b
%
F0(j) = 1 - exp(- x(j)/0.5);
b(j) = 1/(1+x(j)^2); % b(x) = 1/(1+x^2)
for k=1:m1
%
```

```
% The next two are values of G and a
%
  G(j,k) = x(j)/(x(j)^2 + x(k)^2+1); % x/(x^2 + y^2 + 1)
  a(j,k) = x(j)/(x(j)^2 + x(k)^2+1); % x/(x^2 + y^2 + 1)
end
end
%
% The diagonals of full matrices
%
for j=1:m1
    for k=1:m2
%
% The initial value of F
%
        FF0(j,k) = F0(j);
    end
end
%
% The approximation of "mu"
%
for k=1:m2
    mu(k) = 1/(1+t(k)^2);
end
%
Dmu = diag(mu);
%
% Start of iterative "contraction"  loop
%
F = FF0;   % FF0 = F(x,t)|_{t=0}
nn = 1
%
% The value of the "least squares"
% convergence norm, is here set to 1
%
while(nn > 0.00001)
%
% First compute the U(x,t) part -- see (4.3.5)
%
% First, compute the U = U(x,t) - Part
```

```
%
% F_x denoted by Fx
%
Fx = Axd * F;
%
% Next, G(x,u) b(u) F_x(u,tau)
%
for j=1:m1
    for k=1:m2
        for l=1:m1
        GbFx(j,k,l) = G(j,l) * b(l) * Fx(l,k);
        end
    end
end
%
% P2(x,tau) = int_x^infinity G(x,u) b(u) F_x(u,tau) du
%
for j=1:m1
    for k=1:m2
        P2(j,k) = 0;
        for l=1:m1
            P2(j,k) = P2(j,k) + Axm(j,l) * GbFx(j,k,l);
        end
    end
end
%
clear GbFx
%
% We now get U(x,t) at Sinc pts.:
% U(x,t) = int_0^t P2(x,tau) dtau
%
U = P2 * Atp.';
%
clear P2
%
%
%  V(x,t) is next:
%
for k=1:m2
```

```
        for l=1:m1
            Fxx(l) = F(l,k);
            for j=1:m1
                awFx(j,l) = a(j,l)/x(l)*Fxx(l);
            % awFx(v,w,tau) = a(v,w)/w*F_x(w,tau)
            end
        end
%
% Set ker_p(v,y,tau) = int_y^infinity awFx(v,w,tau) dw
%
        ker_p = Axm*awFx;
        p = real(conv_nl_discrete_p(x,x,wx,Axp,ker_p));
        for l=1:m1
            p(l) = Fxx(l)*p(l);
        end
        PP(k,l) = p(l);
end
%
% At this stage we have
% int_0^x F_x(v,tau) int_{x-v}^infty (a(v,w)/w) F(w,tau).
%
% We next multiply this by mu(tau) and integrate the result from
% 0 to t:
%
V = PP.'*Dmu*Atp.';
%
% We have thus completed the computation for V
%
% Next form the difference FF = F - FF0 - U + V:
%
FF = F - FF0 - U + V;
%
W = real(FF);
%
F = F - W;
nn = norm(W);
end
%
% end timing
```

```
%
toc
%
surf(t,x,F)
pause
print -dps2 p_d_picard.ps
```

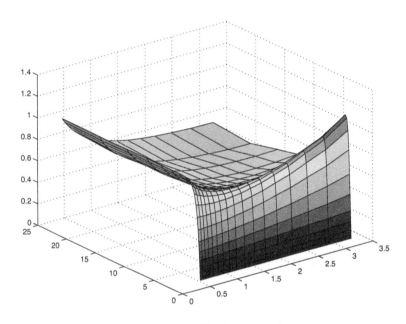

FIGURE 4.18. Density $F(x,t)$ on $(0,\infty) \times (0,\infty)$.

The following program, p_d_T5.m first solves the above integral equation problem over $(0,T)$, then over $(T,2T)$, and so on, up to $(4T,5T)$, with $T = 0.2$. The programs are self-contained except for the nonlinear convolution routines. The difficult to compute boundary layer in the initial value function F_0 can be effectively dealt with via a slight increase in the lower limit M for approximation with respect to the variable x.

```
%
%     Program: #2: p_d_T5.m
%
%
% This program is copyrighted by SINC, LLC
```

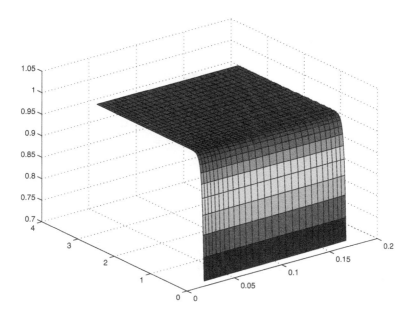

FIGURE 4.19. Density Boundary Layer $F(x,t)$ on $(0,\infty) \times (0,T)$.

```
% an belongs to SINC, LLC
%
% Sinc  successive approximation solution of
%
% F(x,t) = F_0(x) + U(x,t) - V(x,t)
%
% over
%
% (0,infinity) x {(0,T), (T,2T), (2T,3T), (3T,4T)
% and (4T,5T)}
%
% Here,
%
%   U(x,t) = int_0^t int_x^infinity G(x,u) b(u) "times"
%                    "times" F_x(u,tau) du dtau
%
%   V(x,t) = int_0^t mu(tau) int_0^x F_x(v,tau) "times"
%      "times" int_{x-v}^infinity w^{-1} a(v,w) "times"
%        "times" F_x(w,tau) dw dv dtau
%
```

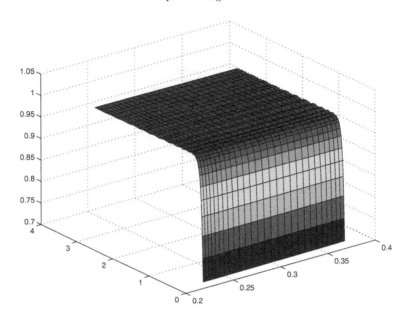

FIGURE 4.20. Density Boundary Layer $F(x,t)$ on $(0,\infty) \times (T, 2T)$.

```
% The initial condition has a bad boundary layer.  G(x,y)
% is also badly behaved near (x,y) = (0,0).
%
clear all
%
format long e
%
M1 = 12;
N1 = 9;
m1 = M1 + N1 + 1;
M2 = 7;
N2 = 7;
m2 = M2 + N2 + 1;
hx = .5/sqrt(N1);
ht = 0.3/sqrt(N2);
T = 0.05;
%
% Sinc Points, Weights, and Matrices
%
x = Sincpoints3(M1,N1,hx);
```

```
% x = Sincpoints4(M1,N1,hx);
wx = weights3(M1,N1,hx);
% wx = weights4(M1,N1,hx);
%
tau = Sincpoints2(0,T,M2,N2,ht);
Axp = Int3_p(M1,N1,hx);
Axm = Int3_m(M1,N1,hx);
% Axp = Int4_p(M1,N1,hx);
% Axm = Int4_m(M1,N1,hx);
%
% The derivative matrices
%
% The "x" derivative matrix on (0,infinity)
%
Axd = Dpoly3_1(M1,N1,hx);
% Axd = Dpoly4_1(M1,N1,hx);
%
% Indef integration wrt t over (0,T) matrix
%
Atp = Int2_p(0,T,M2,N2,ht);
%
% Now diagonalize the matrix Axp, for
% Sinc convolution
%
[Y S] = eig(Axp);
s = diag(S);
%
% First, some definitions of functions and
%
% some diagonal matrices.
%
for j=1:m1
% the initial F_0
      F0(j) = 1 - exp(- x(j)/0.05);
%
      b(j) = 1/(1+x(j)^2);
      for k=1:m1
% The next two are values of G and a
%
```

```
%              G(j,k) = 1/sqrt(1 + x(j)^4 + x(k)^4);
               G(j,k) = x(j)/(x(j)^2 + x(k)^2+1);
               a(j,k) = x(k)/(x(j)^2 + x(k)^2+1);
        end
end
%
% The diagonals of full matrices
%
for j=1:m1
    for k=1:m2
%
    % The initial value of F
    %
    FF0(j,k) = F0(j);
    end
end
%
% The approximation of "mu"
%
for k=1:m2
    mu(k) = 1/(1+tau(k)^2);
end
Dmu = diag(mu);
%
% This is a row vector of the diagonal
% entries of S
%
Yi = inv(Y);
%
% Beginning of "time-strip", ll - loop, to compute
% the solution F on the time interval I_l = ((ll-1)T,(ll)T)
%
%
for ll=1:5
%
% A bit more initialization for I_1, I_2, I_3, I_4 & I_5
%
    for j=1:m2
% Sinc points on  I_l
```

```
        t(j) = (11-1)*T+tau(j);
        Dmu(j,j) = 0.2/(1+t(j)^2);
    end
%
% Start of iterative "contraction"  loop
%
    F = FF0;   % FF0 = F(x,t)|_{t=0}
    nn = 1;
%
% The value of the "least squares"
% convergence norm, is here set to 1
    while(nn > 0.00001)
%
% First, compute the U = U(x,t) - Part
%
% F_x denoted by Fx
%
        Fx = Axd * F;
%
% Next, G(x,u) b(u) F_x(u,tau)
%
        for j=1:m1
            for k=1:m2
                for l=1:m1
                    GbFx(k,j,l) = G(j,l) * b(l) * Fx(l,k);
                end
            end
        end
%
% P2(x,tau) = int_x^infinity G(x,u) b(u) F_x(u,tau) du
%
        P2 = zeros(m2,m1);
        for k=1:m2
            for j=1:m1
                for l=1:m1
                    P2(k,j) = P2(k,j) + GbFx(k,j,l)*Axm(j,l);
                end
            end
        end
```

```
%
        clear GbFx PP2
%
% We now get U(x,t) at Sinc pts.:
% U(x,t) = int_0^t P2(x,tau) dtau
%
        U = (Atp*P2).';
%
        clear P2
%
%  V(x,t) is next:
%
        for k=1:m2
            for j=1:m1
                Fxx(1) = F(1,k);
                for l=1:m1
                    awFx(j,l) = a(j,l)/x(1)*Fxx(1);
                % awFx(v,w,tau) = a(v,w)/w*F_x(w,tau)
                end
            end


%
% Set ker_p(v,y,tau) = int_y^infinity awFx(v,w,tau) dw
%
            ker_p = Axm*awFx;
            p = real(conv_nl_discrete_p(x,x,wx,Axp,ker_p));
            for l=1:m1
                p(l) = Fxx(1)*p(l);
            end
            PP(k,l) = p(l);
        end
%
% At this stage we have
% int_0^x F_x(v,tau) int_{x-v}^infty (a(v,w)/w) F(w,tau).
%
% We next multiply this by mu(tau) and integrate the result fror
% 0 to t:
%
V = PP.'*Dmu*Atp.';
```

```
%
clear awFx
%

%
        UpV = real(U-V);
%
        FF = FF0 + UpV;
%
        clear UpV
        nn0 = norm(F - FF);
        F = FF;
        vector_iter_norm = nn0
        nn = nn0;
%
% End of "contraction iteration" loop
%
    end
%
% Printing a postscript file of the
% surface F
%
    if(ll<2)
        figure(1)
        surf(t,x,F)
        pause
        print -dps2 p_d_T5_1.ps
%
    elseif(ll>1) & (ll<3)
        figure(2)
        surf(t,x,F)
        pause
        print -dps2 p_d_T5_2.ps
%
    elseif(ll>2) & (ll<4)
        figure(3)
        surf(t,x,F)
        pause
```

```
            print -dps2 p_d_T5_3.ps
%
    elseif(11>3) & (11<5)
            figure(4)
            surf(t,x,F)
            pause
            print -dps2 p_d_T5_4.ps
%
    elseif(11>4)
            figure(5)
            surf(t,x,F)
            pause
            print -dps2 p_d_T5_5.ps
        end
  %
  % Initialization for next I_1
  %
  for j=1:m1
            for k=1:m2
                FF0(j,k) = F(j,m2);
            end
  end
            %
            % End of "time-strip" loop
            %
            %
  end
```

4.4.2 Navier–Stokes Example

We illustrate here the solution to the integral equation $(2.4.73)$,

$$\mathbf{u}(\bar{r}, t) = \int_{\mathbb{R}^3} \mathbf{u}^0(\bar{r})\, \mathcal{G}(\bar{r} - \bar{r}', t)\, d\bar{r}'$$

$$+ \int_0^t \int_{\mathbb{R}^3} \left\{ (\mathbf{u}(\bar{r}', t') \cdot \nabla' \mathcal{G})\, \mathbf{u}(\bar{r}', t') + p(\bar{r}', t')\, \nabla' \mathcal{G} \right\}\, d\bar{r}'\, dt',$$

$$(4.4.7)$$

in which we have written $\mathcal{G}$ for $\mathcal{G}(\bar{r} - \bar{r}', t - t')$.

The method of solution in `navier_stokes2.m` is based on the iteration scheme

$$\mathbf{u}_{n+1} = \mathbf{T}\,\mathbf{u}_n \qquad (4.4.8)$$

with $\mathbf{T}$ defined as in $(2.4.84)$, and where the operators $\mathbf{Q}$, $\mathbf{R}$, and $\mathbf{S}$ are defined as in $(2.4.85)$, and for which the Sinc approximation of these terms is described following $(2.4.85)$.

The program `navier_stokes2.m` uses the same spacial discretization in each dimension. This program uses the routines:

navier_stokes2.m	the main program
navier_stokes_ic.m	computes the initial condition vector $\mathbf{u}^0$
lap_tr_heat4.m	used for computing $\mathbf{Q}\mathbf{u}^0$
lap_tr_poi3.m	computes the $3-\mathrm{d}$ Laplace transform of $1/(4\pi r)$
navier_stokes_R.m	computes the integral $\mathbf{R}$ of $(2.4.85)$
navier_stokes_S.m	computes the integral $\mathbf{S}$ of $(2.4.85)$
conv4_vec.m	compute 4 dimensional convolution integrals
Sinpoints1.m	computes the spacial Sinc points
Sincpoints2.m	to compute the time Sinc points
D1_1.m	to compute derivatives
Int_both.m	spacial indefinite integration routines
Int2_p.m	time indefinite integration routine

Here the program `navier_stokes_ic.m` computes the divergence–free initial condition vector $\mathbf{u}^0$; this will need to be modified for a different initial condition vector. The present program computes a solution for the case of

$$\mathbf{u}^0(\bar{r}) = (x_2 - x_3, \ x_3 - x_1, \ x_1 - x_2) \exp\left(-r^2\right), \qquad (4.4.9)$$

for which the above divergence condition (2.4.66) is satisfied. The user may also wish to modify some of the parameters in the main program navier_stokes2.m.

Our program navier_stokes2.m converged for $T = 0.1$; this appears to be close to the largest value of T for which it converged. Indeed, the iterations took longer than we expected. A much faster rate of convergence can be achieved by first solving the equation over $\mathbb{R}^3 \times [0, T']$, and then solving the equation over $\mathbb{R}^3 \times [T', 2T']$, with e.g., $T' = 0.05$, where the initial conditions for the $\mathbb{R}^3 \times [T', 2T']$ solution is just the vector $\mathbf{u}(\bar{r}, T')$ obtained in the solution over $\mathbb{R}^3 \times [0, T']$. The modifications of the program navier_stokes2.m which would enable us to achieve this are similar to those used in the program p_d_T5.m of §4.3.1.

We compute two types of surface plots: one with coordinates at Sinc points followed by one with coordinates at a finer mesh, where the latter was obtained using the Sinc coordinate values in interpolation formulas of the Sinc–Pack manual. The plots with Sinc–point values (x, y, t) appear rougher, while the fine mesh plots are quite smooth. It is not difficult to show using results from Chapter 1 that our computations are uniformly accurate to at least 3 places on $\mathbb{R}^3 \times [0, T]$.

Remark: A Warning! Consider the two Green's functions, $\mathcal{G}$ and its derivative $\mathcal{G}_{x_1}$, where, $\mathcal{G}(x_1, x_2, x_3, t)$, is an even function in each of the variables x_1, x_2, x_3 whereas $\mathcal{G}_{x_1} = \mathcal{G}_{x_1}(x_1, x_2, x_3, t)$, is an even function of x_2 and x_3, but an odd function of x_1. Let $S(\bar{\Lambda}, \tau)$ and $S_{x_1}(\bar{\Lambda}, \tau)$ denote the four dimensional Laplace transforms of these functions. Set $\bar{r} = (x_1, x_2, x_3)$ and consider the approximation of the four dimensional convolution integrals

$$p(\bar{r}, t) = \int_0^t \int_{\mathbb{R}^3} \mathcal{G}(\bar{r} - \bar{r}', t - t') \, u(\bar{r}', t') \, d\bar{r}' \, dt'$$

$$\qquad (4.4.10)$$

$$q(\bar{r}, t) = \int_0^t \int_{\mathbb{R}^3} \mathcal{G}_{x_1}(\bar{r} - \bar{r}', t - t') \, u(\bar{r}', t') \, d\bar{r}' \, dt'$$

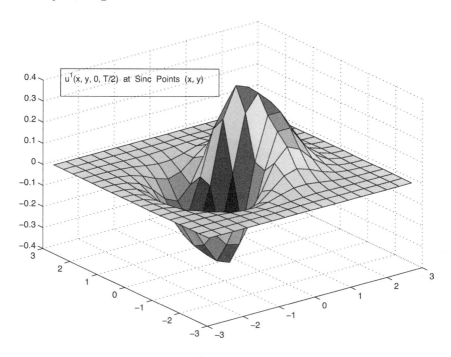

FIGURE 4.21. Plot of $u^1(x, y, 0, T/2)$ at Sinc Points (x, y)

through four dimensional Sinc convolution algorithms used in the program **navier_stokes2.m**. Letting U and V denote respectively, the four dimensional arrays gotten by evaluating u and v at the Sinc points $(x_1,\ x2,\ x_3, t)$ as in the above program, and in the notation for X, Y, Z used in **navier_stokes2.m**, the algorithm for approximating p at the Sinc points is given by

```
P = 0;
P = P + conv4_vec(X,m,X,m,X,m,Z,m,S,U);
P = P + conv4_vec(Y,m,X,m,X,m,Z,m,S,U);
P = P + conv4_vec(X,m,Y,m,X,m,Z,m,S,U);
P = P + conv4_vec(Y,m,Y,m,X,m,Z,m,S,U);
P = P + conv4_vec(X,m,X,m,Y,m,Z,m,S,U);
P = P + conv4_vec(Y,m,X,m,Y,m,Z,m,S,U);
P = P + conv4_vec(X,m,Y,m,Y,m,Z,m,S,U);
P = P + conv4_vec(Y,m,Y,m,Y,m,Z,m,S,U);
```

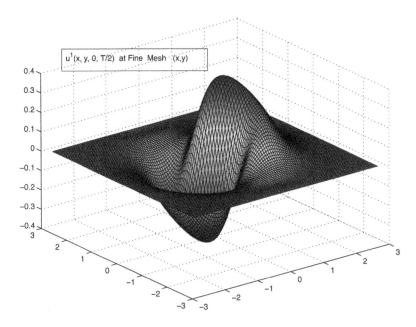

FIGURE 4.22. Plot of $u^1(x, y, 0, T/2)$ at Sinc Interpolated Points (x, t)

On the other hand, since $\mathcal{G}_{x_1}$ is an odd function of x_1, its one dimensional Laplace transform taken over the interval $-\mathbb{R}_+$ is negative, and consequently, so is its four dimensional Laplace transform, which is taken over $(x_1, x_2, x_3, t) \in (-\mathbb{R}_+) \times \pm\mathbb{R}_+ \times \pm\mathbb{R}_+ \times (0, T)$, where $\mathbb{R}_+ = (0, \infty)$. Hence the algorithm for approximating q at the Sinc points is given by

```
Q = 0;
Q = Q + conv4_vec(X,m,X,m,X,m,Z,m,Sx_1,V);
Q = Q - conv4_vec(Y,m,X,m,X,m,Z,m,Sx_1,V);
Q = Q + conv4_vec(X,m,Y,m,X,m,Z,m,Sx_1,V);
Q = Q - conv4_vec(Y,m,Y,m,X,m,Z,m,Sx_1,V);
Q = Q + conv4_vec(X,m,X,m,Y,m,Z,m,Sx_1,V);
Q = Q - conv4_vec(Y,m,X,m,Y,m,Z,m,Sx_1,V);
Q = Q + conv4_vec(X,m,Y,m,Y,m,Z,m,Sx_1,V);
Q = Q - conv4_vec(Y,m,Y,m,Y,m,Z,m,Sx_1,V);
```

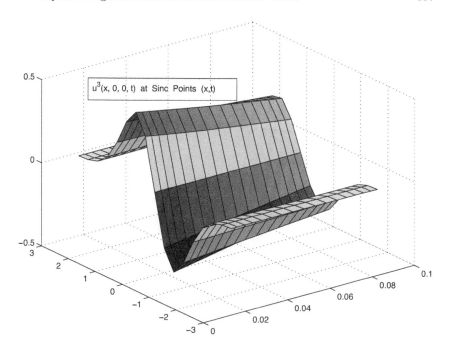

FIGURE 4.23. Plot of $u^3(x, 0, 0, t)$ at Sinc points (x, t)

```
%
% navier_stokes2.m
%
% This program is copyrighted by SINC, LLC
% an belongs to SINC, LLC
%
% Sinc Solution of (4.3.7)
% as outlined in Sec. 2.4.6.
%
% Initializations:
%
clear all
%
% Start time
%
tic
%
Eps = 0.5;
% T = 0.00008;
```

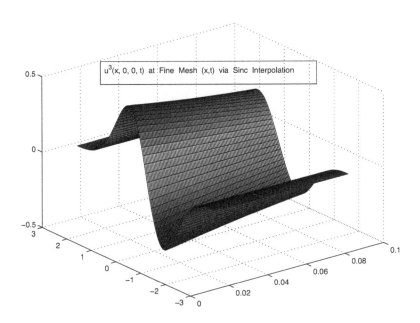

FIGURE 4.24. Plot of $u^3(x,0,0,t)$ at Sinc Interpolated Points (x,t)

```
T = 0.1;
%
% Sinc parameters
%
N = 8;
m = 2*N+1;
hs = 1/sqrt(N);
ht = 0.6*hs;
zs = Sincpoints1(N,N,hs); % Spacial map #1 Sinc pts
zt = Sincpoints2(0,T,N,N,ht); % Time Sinc points
Dr = D1_1(N,N,hs);
%
% Diagonalization of both spacial indefinite
% integration matrices:
%
[X,Y,SS] = Int_both1(N,N,hs);
Xi = inv(X);
Yi = inv(Y);
s = diag(SS);
```

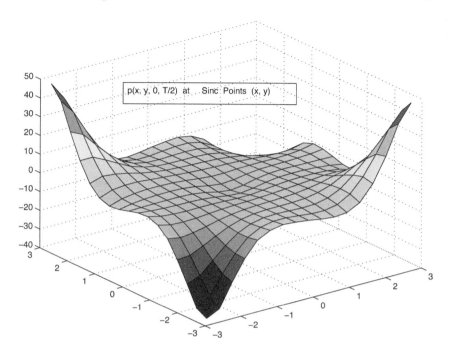

FIGURE 4.25. Plot of $p(x, y, 0, T/2)$ at Sinc points (x, y)

```
It = Int2_p(0,T,N,N,ht); % Time integration matrix on (0,T)
[Z,TT] = eig(It);
Zi = inv(Z);
tt = diag(TT);
clear SS TT It
%
% The initial velocity vector, u^0:
%
[U01,U02,U03] = navier_stokes_ic(zs);
%
% We first compute the Green's functions
% -- see the discussion following (4.3.11)
%
for ii = 1:m
    a = s(ii);
    for j = 1:m
        b = s(j);
        for k = 1:m
```

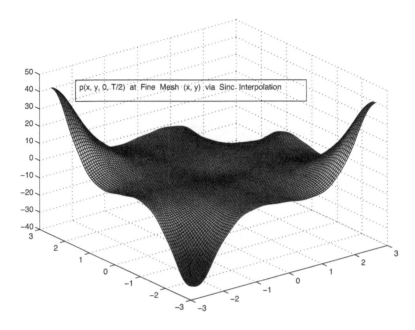

FIGURE 4.26. Plot of $p(x, y, 0, T/2)$ at Sinc Interpolated Points (x, y)

```
          c = s(k);
          G0(ii,j,k) = lap_tr_poi3(a,b,c);
          for l = 1:m
              t = tt(l);
Q(ii,j,k,l) = lap_tr_heat4(a,b,c,t,Eps)/t;
R = navier_stokes_R(a,b,c,t,Eps);
S = navier_stokes_S(a,b,c,t,Eps);
R2(j,k,ii,l) = R;
S2(j,k,ii,l) = S;
R3(k,ii,j,l) = R;
S3(k,ii,j,l) = S;
R1(ii,j,k,l) = R;
S1(ii,k,k,l) = S;
V1(ii,j,k,l) = U01(ii,j,k);
V2(ii,j,k,l) = U02(ii,j,k);
V3(ii,j,k,l) = U03(ii,j,k);
          end
     end
```

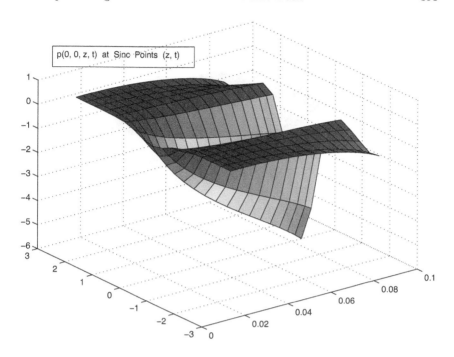

FIGURE 4.27. Plot of $p(0,0,z,t)$ at Sinc points (z,t)

```
      end
end
%
clear U01 U02 U03 R S
%
% Determination of the vector Q^0
%
U01 = zeros(m,m,m,m);
U02 = U01;
U03 = U01;
for n = 1:8
    if n == 1
      A = X; B = X; C = X;
      elseif n == 2
      A = X; B = X; C = Y;
      elseif n == 3
      A = X; B = Y; C = X;
      elseif n == 4
```

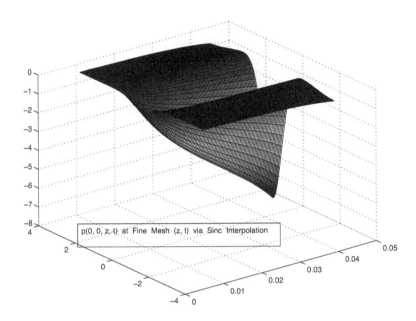

FIGURE 4.28. Plot of $p(0, 0, z, t)$ at Sinc Interpolated Points (z, t)

```
    A = X; B = Y; C = Y;
    elseif n == 5
     A = Y; B = X; C = X;
    elseif n == 6
     A = Y; B = X; C = Y;
    elseif n == 7
     A = Y; B = Y; C = X;
    else
     A = Y; B = Y; C = Y;
    end
    U01 = U01 + real(conv4_vec(A,m,B,m,C,m,Z,m,Q,V1));
    U02 = U02 + real(conv4_vec(A,m,B,m,C,m,Z,m,Q,V2));
    U03 = U03 + real(conv4_vec(A,m,B,m,C,m,Z,m,Q,V3));
end
%
clear Q  V1 V2 V3 A B C
%
% Start of Iteration Scheme
%
```

```
er = 1;
kkk = 0;
%
% Initial values of iterates
%
V1 = U01;
V2 = U02;
V3 = U03;
while(er > 10^(-4))
%
% First compute the "R-contributions"
%
U1c = U01;
U2c = U02;
U3c = U03;
%
B11 = V1.*V1;
%
U1c = U1c - conv4_vec(X,m,X,m,X,m,Z,m,R1,B11);
U1c = U1c + conv4_vec(Y,m,X,m,X,m,Z,m,R1,B11);
U1c = U1c - conv4_vec(X,m,X,m,Y,m,Z,m,R1,B11);
U1c = U1c + conv4_vec(Y,m,X,m,Y,m,Z,m,R1,B11);
U1c = U1c - conv4_vec(X,m,Y,m,X,m,Z,m,R1,B11);
U1c = U1c + conv4_vec(Y,m,Y,m,X,m,Z,m,R1,B11);
U1c = U1c - conv4_vec(X,m,Y,m,Y,m,Z,m,R1,B11);
U1c = U1c + conv4_vec(Y,m,Y,m,Y,m,Z,m,R1,B11);
%
clear B11
%
B22 = V2.*V2;
%
U2c = U2c - conv4_vec(X,m,X,m,X,m,Z,m,R2,B22);
U2c = U2c + conv4_vec(X,m,Y,m,X,m,Z,m,R2,B22);
U2c = U2c - conv4_vec(X,m,X,m,Y,m,Z,m,R2,B22);
U2c = U2c + conv4_vec(X,m,Y,m,Y,m,Z,m,R2,B22);
U2c = U2c - conv4_vec(Y,m,X,m,X,m,Z,m,R2,B22);
U2c = U2c + conv4_vec(Y,m,Y,m,X,m,Z,m,R2,B22);
U2c = U2c - conv4_vec(Y,m,X,m,Y,m,Z,m,R2,B22);
U2c = U2c + conv4_vec(Y,m,Y,m,Y,m,Z,m,R2,B22);
```

```
%
clear B22
%
B33 = V3.*V3;
%
U3c = U3c - conv4_vec(X,m,X,m,X,m,Z,m,R3,B33);
U3c = U3c + conv4_vec(X,m,X,m,Y,m,Z,m,R3,B33);
U3c = U3c - conv4_vec(X,m,Y,m,X,m,Z,m,R3,B33);
U3c = U3c + conv4_vec(X,m,Y,m,Y,m,Z,m,R3,B33);
U3c = U3c - conv4_vec(Y,m,X,m,X,m,Z,m,R3,B33);
U3c = U3c + conv4_vec(Y,m,X,m,Y,m,Z,m,R3,B33);
U3c = U3c - conv4_vec(Y,m,Y,m,X,m,Z,m,R3,B33);
U3c = U3c + conv4_vec(Y,m,Y,m,Y,m,Z,m,R3,B33);
%
clear B33
%
B12 = V1.*V2;
%
U1c = U1c - conv4_vec(X,m,X,m,X,m,Z,m,R2,B12);
U1c = U1c + conv4_vec(X,m,Y,m,X,m,Z,m,R2,B12);
U1c = U1c - conv4_vec(X,m,X,m,Y,m,Z,m,R2,B12);
U1c = U1c + conv4_vec(X,m,Y,m,Y,m,Z,m,R2,B12);
U1c = U1c - conv4_vec(Y,m,X,m,X,m,Z,m,R2,B12);
U1c = U1c + conv4_vec(Y,m,Y,m,X,m,Z,m,R2,B12);
U1c = U1c - conv4_vec(Y,m,X,m,Y,m,Z,m,R2,B12);
U1c = U1c + conv4_vec(Y,m,Y,m,Y,m,Z,m,R2,B12);
%
U2c = U2c - conv4_vec(X,m,X,m,X,m,Z,m,R1,B12);
U2c = U2c + conv4_vec(Y,m,X,m,X,m,Z,m,R1,B12);
U2c = U2c - conv4_vec(X,m,X,m,Y,m,Z,m,R1,B12);
U2c = U2c + conv4_vec(Y,m,X,m,Y,m,Z,m,R1,B12);
U2c = U2c - conv4_vec(X,m,Y,m,X,m,Z,m,R1,B12);
U2c = U2c + conv4_vec(Y,m,Y,m,X,m,Z,m,R1,B12);
U2c = U2c - conv4_vec(X,m,Y,m,Y,m,Z,m,R1,B12);
U2c = U2c + conv4_vec(Y,m,Y,m,Y,m,Z,m,R1,B12);
%
clear B12
%
B13 = V1.*V3;
```

```
%
U1c = U1c - conv4_vec(X,m,X,m,X,m,Z,m,R3,B13);
U1c = U1c + conv4_vec(X,m,X,m,Y,m,Z,m,R3,B13);
U1c = U1c - conv4_vec(X,m,Y,m,X,m,Z,m,R3,B13);
U1c = U1c + conv4_vec(X,m,Y,m,Y,m,Z,m,R3,B13);
U1c = U1c - conv4_vec(Y,m,X,m,X,m,Z,m,R3,B13);
U1c = U1c + conv4_vec(Y,m,X,m,Y,m,Z,m,R3,B13);
U1c = U1c - conv4_vec(Y,m,Y,m,X,m,Z,m,R3,B13);
U1c = U1c + conv4_vec(Y,m,Y,m,Y,m,Z,m,R3,B13);
%
U3c = U3c - conv4_vec(X,m,X,m,X,m,Z,m,R1,B13);
U3c = U3c + conv4_vec(Y,m,X,m,X,m,Z,m,R1,B13);
U3c = U3c - conv4_vec(X,m,X,m,Y,m,Z,m,R1,B13);
U3c = U3c + conv4_vec(Y,m,X,m,Y,m,Z,m,R1,B13);
U3c = U3c - conv4_vec(X,m,Y,m,X,m,Z,m,R1,B13);
U3c = U3c + conv4_vec(Y,m,Y,m,X,m,Z,m,R1,B13);
U3c = U3c - conv4_vec(X,m,Y,m,Y,m,Z,m,R1,B13);
U3c = U3c + conv4_vec(Y,m,Y,m,Y,m,Z,m,R1,B13);
%
clear B13
%
B23 = V2.*V3;
%
U2c = U2c - conv4_vec(X,m,X,m,X,m,Z,m,R3,B23);
U2c = U2c + conv4_vec(X,m,X,m,Y,m,Z,m,R3,B23);
U2c = U2c - conv4_vec(X,m,Y,m,X,m,Z,m,R3,B23);
U2c = U2c + conv4_vec(X,m,Y,m,Y,m,Z,m,R3,B23);
U2c = U2c - conv4_vec(Y,m,X,m,X,m,Z,m,R3,B23);
U2c = U2c + conv4_vec(Y,m,X,m,Y,m,Z,m,R3,B23);
U2c = U2c - conv4_vec(Y,m,Y,m,X,m,Z,m,R3,B23);
U2c = U2c + conv4_vec(Y,m,Y,m,Y,m,Z,m,R3,B23);
%
U3c = U3c - conv4_vec(X,m,X,m,X,m,Z,m,R2,B23);
U3c = U3c + conv4_vec(X,m,Y,m,X,m,Z,m,R2,B23);
U3c = U3c - conv4_vec(X,m,X,m,Y,m,Z,m,R2,B23);
U3c = U3c + conv4_vec(X,m,Y,m,Y,m,Z,m,R2,B23);
U3c = U3c - conv4_vec(Y,m,X,m,X,m,Z,m,R2,B23);
U3c = U3c + conv4_vec(Y,m,Y,m,X,m,Z,m,R2,B23);
U3c = U3c - conv4_vec(Y,m,X,m,Y,m,Z,m,R2,B23);
```

```
U3c = U3c + conv4_vec(Y,m,Y,m,Y,m,Z,m,R2,B23);
%
clear B23
%
% The function g of (2.4.73)
%
% First the derivatives
%
for l=1:m
    for j=1:m
        for k=1:m
            for ii=1:m
                W1(ii) = V1(ii,j,k,l);
                W2(ii) = V2(ii,j,k,l);
                W3(ii) = V3(ii,j,k,l);
            end
            DW1 = Dr*W1.';
            DW2 = Dr*W2.';
            DW3 = Dr*W3.';
            for ii=1:m
                U11(ii,j,k,l) = DW1(ii);
                U21(ii,j,k,l) = DW2(ii);
                U31(ii,j,k,l) = DW3(ii);
            end
        end
    end
    for ii=1:m
        for k=1:m
            for j=1:m
                W1(j) = V1(ii,j,k,l);
                W2(j) = V2(ii,j,k,l);
                W3(j) = V3(ii,j,k,l);
            end
            DW1 = Dr*W1.';
            DW2 = Dr*W2.';
            DW3 = Dr*W3.';
            for j=1:m
                U12(ii,j,k,l) = DW1(j);
                U22(ii,j,k,l) = DW2(j);
```

```
                        U32(ii,j,k,l) = DW3(j);
                end
            end
        end
        for ii=1:m
            for j=1:m
                for k=1:m
                    W1(k) = V1(ii,j,k,l);
                    W2(k) = V2(ii,j,k,l);
                    W3(k) = V3(ii,j,k,l);
                end
                DW1 = Dr*W1.';
                DW2 = Dr*W2.';
                DW3 = Dr*W3.';
                for k=1:m
                    U13(ii,j,k,l) = DW1(k);
                    U23(ii,j,k,l) = DW2(k);
                    U33(ii,j,k,l) = DW3(k);
                end
            end
        end
    end
%
% Now g --- see (2.4.73)
%
g = U11.*U11 + U22.*U22 + U33.*U33 + ...
    2*(U12.*U21 + U13.*U31 + U23.*U32);
%
clear U11 U22 U33 U12 U21 U13 U31 U23 U32
%
U1c = U1c - conv4_vec(X,m,X,m,X,m,Z,m,R1,g);
U1c = U1c + conv4_vec(Y,m,X,m,X,m,Z,m,R1,g);
U1c = U1c - conv4_vec(X,m,X,m,Y,m,Z,m,R1,g);
U1c = U1c + conv4_vec(Y,m,X,m,Y,m,Z,m,R1,g);
U1c = U1c - conv4_vec(X,m,Y,m,X,m,Z,m,R1,g);
U1c = U1c + conv4_vec(Y,m,Y,m,X,m,Z,m,R1,g);
U1c = U1c - conv4_vec(X,m,Y,m,Y,m,Z,m,R1,g);
U1c = U1c + conv4_vec(Y,m,Y,m,Y,m,Z,m,R1,g);
%
```

```
U2c = U2c - conv4_vec(X,m,X,m,X,m,Z,m,R2,g);
U2c = U2c + conv4_vec(X,m,Y,m,X,m,Z,m,R2,g);
U2c = U2c - conv4_vec(X,m,X,m,Y,m,Z,m,R2,g);
U2c = U2c + conv4_vec(X,m,Y,m,Y,m,Z,m,R2,g);
U2c = U2c - conv4_vec(Y,m,X,m,X,m,Z,m,R2,g);
U2c = U2c + conv4_vec(Y,m,Y,m,X,m,Z,m,R2,g);
U2c = U2c - conv4_vec(Y,m,X,m,Y,m,Z,m,R2,g);
U2c = U2c + conv4_vec(Y,m,Y,m,Y,m,Z,m,R2,g);
%
U3c = U3c - conv4_vec(X,m,X,m,X,m,Z,m,R3,g);
U3c = U3c + conv4_vec(X,m,X,m,Y,m,Z,m,R3,g);
U3c = U3c - conv4_vec(X,m,Y,m,X,m,Z,m,R3,g);
U3c = U3c + conv4_vec(X,m,Y,m,Y,m,Z,m,R3,g);
U3c = U3c - conv4_vec(Y,m,X,m,X,m,Z,m,R3,g);
U3c = U3c + conv4_vec(Y,m,X,m,Y,m,Z,m,R3,g);
U3c = U3c - conv4_vec(Y,m,Y,m,X,m,Z,m,R3,g);
U3c = U3c + conv4_vec(Y,m,Y,m,Y,m,Z,m,R3,g);
%
U1 = real(U1c);
U2 = real(U2c);
U3 = real(U3c);
%
clear U1c U2c U3c
%
% Error test
%
ERMAT = U1-V1+U2-V2+U3-V3;
ERVEC = reshape(ERMAT,m^4,1);
err = norm(ERVEC);
clear ERMAT ERVEC
%
V1 = U1;
V2 = U2;
V3 = U3;
%
kkk = kkk+1;
Iteration_no = kkk
Iteration_error = err
er = err;
```

```
end
%
% At this point we have computed the
% vector solution to Navier-Stokes
% equations.  Now, for the pressure:
%
qc = zeros(m,m,m);
%
for l=1:m
    AA = g(:,:,:,l);
    qc = qc + conv3_vec(X,m,X,m,X,m,G0,AA);
    qc = qc + conv3_vec(Y,m,X,m,X,m,G0,AA);
    qc = qc + conv3_vec(X,m,X,m,Y,m,G0,AA);
    qc = qc + conv3_vec(Y,m,X,m,Y,m,G0,AA);
    qc = qc + conv3_vec(X,m,Y,m,X,m,G0,AA);
    qc = qc + conv3_vec(Y,m,Y,m,X,m,G0,AA);
    qc = qc + conv3_vec(X,m,Y,m,Y,m,G0,AA);
    qc = qc + conv3_vec(Y,m,Y,m,Y,m,G0,AA);
    pc(:,:,:,l) = qc(:,:,:);
end
%
clear AA qc
%
p = real(pc);
%
toc
%
clear pc
%
% Printouts
%
% In what follows we produce several
% surface plots of solutions obtained
% at Sinc points. Some of these have
% the appearance of not being very
% accurate, whereas these solutions
% are, in fact quite accurate at the
% Sinc points.  We have thus also
% included some surface plots of
```

```
% solutions at finer meshes; these
% were obtained via use of Sinc
% interpolation formulas.
%
% Velocity fields at (x,y,0,T/2)
%
% These are
%
U1xy = U1(:,:,N+1,N+1);
U2xy = U2(:,:,N+1,N+1);
U3xy = U3(:,:,N+1,N+1);
%
% Velocity fields at (x,0,0,t) and
%
%  pressure at (x,0,0,t), (0,y,0,t),
%               (0,0,z,t)
%
for ii=1:m
    for j=1:m
        U1xt(ii,j) = U1(ii,N+1,N+1,j);
        U2xt(ii,j) = U2(ii,N+1,N+1,j);
        U3xt(ii,j) = U3(ii,N+1,N+1,j);
        pxt(ii,j) = p(ii,N+1,N+1,j);
        pyt(ii,j) = p(N+1,ii,N+1,j);
        pzt(ii,j) = p(N+1,N+1,ii,j);
    end
end
%
% Finer mesh values and Sinc inter-
% polation
%
steps = (zs(m) - zs(1))/100;
stept = (zt(m) - zt(1))/100;
%
bb = zeros(100,m);
% cc = zeros(100,m);
dd = zeros(100,m);
for j=1:100
    xx = zs(1) + (j-0.5)*steps;
```

```
      tt = zt(1) + (j-0.5)*stept;
      bb(j,:) = basis1(xx,N,N,hs);
%     cc(j,:) = basis2(0,T,tt,N,N,ht);
      dd(j,:) = basis_poly2(0,T,tt,N,N,ht);
      ys(j) = xx;
      yt(j) =tt;
end
%
% Plots
%
% 1. u_1 at Sinc points
%
figure(1)
surf(zs,zs,U1xy)
pause
print -dps2 n_s2_u1_x_y_0_halfT.ps
%
% 2. u1 at finer mesh
%
% The reader should compare the
% following figure(2) plot with
% the Sinc point plot of figure(1).
%
u1_fine = bb*U1xy*bb.';
figure(2)
surf(ys,ys,u1_fine)
pause
print -dps2 n_s2_u1_fine_x_y_0_halfT.ps
%
% 3. u2 at Sinc points
%
figure(3)
surf(zs,zs,U2xy)
pause
print -dps2 n_s2_u2_x_y_0_halfT.ps
%
% 4. u3 at Sinc points
%
figure(4)
```

```
surf(zs,zs,U3xy)
pause
print -dps2 n_s2_u3_x_y_0_halfT.ps
%
% 5. u1(x,0,0,t) at Sinc points
%
figure(5)
surf(zt,zs,U1xt)
pause
print -dps2 n_s2_u1x_0_0_t.ps
%
% 6. u2(x,0,0,t) at Sinc points
%
figure(6)
surf(zt,zs,U2xt)
pause
print -dps2 n_s2_u2x_0_0_t.ps
%
% 7. u3(x,0,0,t) at Sinc points
%
figure(7)
surf(zt,zs,U3xt)
pause
print -dps2 n_s2_u3x_0_0_t.ps
%
% 8. u3(x,0,0,t) on finer mesh
%
% The reader should compare the
% following plot with the Sinc point
% plot of figure(7)
%
u3xt_fine = bb*U3xt*dd.';
%
figure(8)
surf(yt,ys,u3xt_fine);
pause
print -dps2  n_s2_u3_fine_x_0_0_t.ps
%
% 9. Pressure at Sinc points (x,y,0,T/2)
```

```
%
Pxy = p(:,:,N+1,N+1);
%
figure(9)
surf(zs,zs,Pxy)
pause
print -dps2 n_s2_p_x_y_0_halfT.ps
%
% 10. Pressure p(x,y,0,T/2) on
% finer mesh.  Compare figs(9) and 10.
%
p_fine = bb*Pxy*bb.';
%
figure(10)
surf(ys,ys,p_fine)
pause
print -dps2 n_s2_p_fine_x_y_0_halfT.ps
%
% 11, 12,13. Pressures p(x,0,0,t),
% p(0,y,0,t), and p(0,0,z,t)   at Sinc
% points
%
figure(11)
surf(zt,zs,pxt)
pause
print -dps2 pxt_x_0_0_t.ps
%
figure(12)
surf(zt,zs,pyt)
pause
print -dps2 pyt_0_y_0_t.ps
%
figure(13)
surf(zt,zs,pzt)
pause
print -dps2 pzt_0_0_z_t.ps
%
% 14. p(0,0,z,t) on a finer
% mesh.  Compare figs(13) and (14).
```

```
%
pzt_fine = bb*pzt*dd.';
%
figure(14)
surf(yt,ys,pzt_fine)
pause
print -dps2 n_s2_pst_fine_0_0_z_t.ps
```

4.5 Performance Comparisons

The efficiency of Sinc allows it to take on important applied problems
not within the reach of classical methods. Here we present four simple
problems numerically accessible using one such classical method, the
finite element method (FEM), and make comparisons with Sinc. This
section will be brief and short on detail. We refer the reader to [SNJR]
for complete accounts of these and other examples.

4.5.1 THE PROBLEMS

1. **Poisson Problem**
   We consider

$$\Delta u(\bar{r}) = -(6 - 4r^2)e^{-r^2}, \quad \bar{r} \in \mathbb{R}^3, \tag{4.5.1}$$

with solution $u(\bar{r}) = e^{-r^2}$, where $\bar{r} = (x, y, z)$ and $r = \sqrt{x^2 + y^2 + z^2}$.

2. **Heat Problem**
   We consider

$$\frac{\partial u(\bar{r}, t)}{\partial t} - \Delta u(\bar{r}, t) = f(\bar{r}, t), \bar{r} \in \mathbb{R}^3 \tag{4.5.2}$$

with two heat sources

$$f_1(\bar{r}, t) = e^{-r^2 - 0.5 \cdot t}(1 + 5.5 \cdot t - 4 \cdot t \cdot r^2)$$

and

$$f_2(\bar{r}, t) = e^{-r^2 - 0.5 \cdot t}\left(\frac{1}{2\sqrt{t}} + 5.5\sqrt{t} - 4\sqrt{t} \cdot r^2\right)$$

having corresponding solutions

$$u_1(x, y, z, t) = t \cdot e^{-r^2 - 0.5 \cdot t}.$$

and

$$u_2(x, y, z, t) = \sqrt{t} \cdot e^{-r^2 - 0.5 \cdot t}.$$

3. Wave Problem

We consider the 2-d wave equation

$$\frac{\partial^2 u(\bar{r}, t)}{\partial t^2} - \nabla^2 u(\bar{r}, t) = f(\bar{r}, t), \quad \bar{r} \in \mathbb{R}^2, \ 0 < t \leq 1. \qquad (4.5.3)$$

with forcing function

$$f(\bar{r}, t) = \left(\frac{3}{4 \cdot \sqrt{t}} - \frac{3}{2}\sqrt{t} + \sqrt{t}^3 \cdot \left(\frac{17}{4} - 4 \cdot r^2 \right) \right) e^{(-r^2 - 0.5 \cdot t)},$$

and with solution

$$u(\bar{r}, t) = t^{3/2} \cdot e^{(-r^2 - 0.5 \cdot t)}.$$

Here, $\bar{r} = (x, y) \in \mathbb{R}^2$ and $r = \sqrt{x^2 + y^2}$.

4.5.2   THE COMPARISONS

1. The Green's function solution u to (4.5.1) is the sum of eight indefinite convolution integrals of the form

$$u^{(1)}(x, y, z) = \int_{a_1}^{x} \int_{y}^{b_2} \int_{a_3}^{z} \frac{g(\xi, \eta, \zeta) \, d\zeta \, d\eta \, d\xi}{4\pi \sqrt{(x - \xi)^2 + (y - \eta)^2 + (z - \zeta)^2}},$$
$$(4.5.4)$$

evaluated for all points (x, y, z) equal to Sinc points $\{(i\,h, j\,h, k\,h) : i, j, k = -N, \cdots, N\}$.

Let $U = U_{ijk}$ be the solution array whereby at Sinc points ih, jh and kh,

$$U_{ijk} = u(ih, jh, kh) \approx \sum_{\ell=1}^{8} u^{(\ell)}(ih, jh, kh). \qquad (4.5.5)$$

Numerically computing the elements of U, these three dimensional convolutions $u^{(\ell)}$, then proceeds by taking the Laplace transform $\hat{G}$ of the Green's function $1/(4\pi r)$ - as performed in §2.4.1. Algorithm 2.5.2 for computing the three dimensional indefinite convolution may now be applied. Accordingly, we define

$$A_i = A \;\; = \;\; h\,I^{(-1)} = X\,S\,X^{-1}, \;\; i = 1,3$$
$$A_2 = A^T \;\; = \;\; h\,(I^{(-1)})^T = Y\,S\,Y^{-1},$$

$$(4.5.6)$$

where $S = \mathrm{diag}[s_{-N}, \ldots, s_N]$ is a diagonal matrix of eigenvalues of the matrix A, and X and Y are the corresponding matrices of eigenvectors. We then evaluate the array $[g_{ijk}] = [g(ih, jh, kh)]$, and form a vector $\mathbf{g}$ from the array, where the subscripts appear in a linear order dictated by the order of appearance of the subscripts in the do loops, do k = -N, N, followed by do j = -N, N, followed by do i = -N, N. We then also form the diagonal matrix $\hat{\mathbf{G}}$ in which the entries are the values $\hat{G}_{ijk} = \hat{G}(s_i, s_j, s_k)$, where we also list the values $\hat{G}_{ijk}$ in the same linear order as for $\mathbf{g}$.

Similarly from the arrays $U_{ijk}^{(\ell)}$, we can define vector $\mathbf{u}_\ell$ by listing the elements $U_{ijk}^{(\ell)}$ in the linear order. Algorithm 2.5.2 may then symbolically be expressed using the Kronecker matrix product, which in the case $\ell = 1$ takes the form

$$\mathbf{u}_1 \;\; = \;\; \mathbf{U}^{(1)}\,\mathbf{g}$$
$$\mathbf{U}^{(1)} \;\; = \;\; [X \otimes Y \otimes Z]\,\hat{\mathbf{G}}\,[X^{-1} \otimes Y^{-1} \otimes Z^{-1}],$$

$$(4.5.7)$$

Similar formulas hold for the seven other $\mathbf{U}^{(\ell)}$ matrices. The vector $\mathbf{u}$ approximating the solution u of PDE (4.5.1) at the Sinc points is then given by

$$\mathbf{u} = \left(\sum_{\ell=1}^{8} \mathbf{U}^{(\ell)} \right) \mathbf{g}. \qquad (4.5.8)$$

We emphasize that the matrices $\mathbf{U}^{(\ell)}$ need never be computed since our algorithm involves performing a sequence of one–dimensional matrix multiplications. For example, with $N = 20$ we get at least 3 places of accuracy, and the size of the corresponding matrix $\mathbf{U}^\ell$ is

$41^3 \times 41^3$ or size $68,921 \times 68,921$. This matrix, which is full, thus contains more than 4.75×10^9 elements. If such a matrix were to be obtained by a Galerkin scheme, with each entry requiring the evaluation of a three dimensional integral, and with each integral requiring 41^3 evaluation points, then more than 3.27×10^{14} function evaluations would be required, an enormous task indeed! Yet Sinc accurately gives us *all* of these values with relatively little work.

A FEM solution was computed (4.5.1) on a cubic area $[-6,6]^3$ with zero boundary conditions. This restriction caused no problems because u and g are rapidly decreasing functions.

We computed the Sinc based solution with $2N + 1$ Sinc points. The corresponding values of h were computed as $h = \pi/\sqrt{N}$. The FEM solution was computed afterward using more and more nodes until the accuracy of the Sinc solutions was achieved. Differences are dramatic. For example, with N as small as 20, Sinc achieved an absolute uniform error of 0.0040 in a time 14 times faster than the FEM, which had about 10,000 nodes. Throughout, Sinc maintained an error bounded by $e^{-1.5\sqrt{N}}$, whereas FEM required 53706 nodes to achieve .0062 error, and 143497 nodes to achieve error of .0026.

2. Sinc Green's function solutions to (4.5.2) are obtained following much the same procedure as with the Poisson problem. Here, we rely on finding the Laplace transform of the Green's function for parabolic problems in $\mathbb{R}^3$, derived in §2.4.5.

The total number of Sinc points used in each direction as well as in time t was $2N + 1$. The FEM solution was computed on a cubic area with center the origin, side length 12 and zero boundary conditions, and the time interval was chosen as $[0,1]$ with constant step size.

For heat source $f_1(\bar{r},t)$, a uniform error of about .01 was achieved with $N = 12$ for Sinc, and achieved with time step size .01 for FEM, taking about 7 times longer than Sinc computation. When both methods reached an error of about .006, and N equal to 14, Sinc again ran about 7 times faster than FEM.

For the second heat source $f_2(\bar{r},t)$, memory constraints early on became an issue for FEM so that with main memory of .5GB, a lower limit of .02 accuracy was achieved with time step of .005, which took about 4 times as long as Sinc to run, this with $N = 11$. A ten fold increase resulted Sinc achieving an error bounded by .0006, this with

$N = 30$.

The Sinc error was approximately $e^{-1.3\sqrt{N}}$ in this second case. Early failure in the second case for FEM can be attributed to a singularity of f_2 at $t = 0$, thus requiring a finer mesh.

3. The lower limit of accuracy due to memory constraints for FEM stopped at .015 error, and took about 100 times longer than the Sinc solution for the same error, this with $N = 22$. For $N = 60$, Sinc reached an error 2e-07 in a time increased by a factor of 38. The Sinc error converged at a rate about $e^{-1.6\sqrt{N}}$.

5

Directory of Programs

ABSTRACT This chapter provides cross references between theorems and equations of this handbook and the programs of Sinc–Pack. In these tables, programs followed by a star (*) are programs of this package which are not illustrated in this handbook.

5.1 Wavelet Formulas

These formulas were derived in §1.4. For these, we list the following programs corresponding to the equations in §1.4:

```
    Where Used                      Program

Formula (1.4.3-4)             basis_period.m (*)
Formula (1.4.13)              wavelet_basis_even_int.m
Formula (1.4.16)              wavelet_basis_even_half.m
Formula (1.4.19)              wavelet_basis_odd_int.m
Formula (1.4.22)              wavelet_basis_odd_half.m
Formula (1.4.17)              wavelet_basis_even_half_cos.m
Formula (1.4.18)              wavelet_basis_even_half_sin.m
Formula (1.4.95)             waveletintmat_e_int.m
formula (1.4.94)             waveletintmat_o_int.m
Example 3.10.1               wavelet_main_odd_half.m
Example 3.10.1               wavelet_main_even_int.m (*)
Example 3.10.1               wavelet_main_odd_int.m (*)
Deriv. similar to (1.4.94)   waveletintmat_e_half.m (*)
                             waveletintmat_o_half.m (*)
                  -- uses    wavelet_basis_odd_half.m (*)
DFT: Formula (1.4.110)        ftransmat.m
DFT inv. Formula (1.4.116)    ftransmatinv. m
Example 3.10.2               wavelet_main_even_half.m
          -- uses            wavelet_basis_even_half_cos.m
                             wavelet_basis_even_half_sin.m
Sec. 3.10.2                  wavelet_ode_ivp_nl1.m
```

	-- uses	`waveletintmat_e_int.m`
		`wavelet_basis_even_int.m`
Equation 1.4.93		`waveletint_e.m`
Equation 1.4.93		`waveletint_o.m (*)`

5.2   One Dimensional Sinc Programs

Formulas of Sinc–Pack are one dimensional. But because of *separation of variables*, these one dimensional formulas can be used to solve multidimensional PDE problems, i.e., without use of large matrices. These one–dimensional matrix–vector operations are discussed in §3.1.

5.2.1   STANDARD SINC TRANSFORMATIONS

We list here Sinc–Pack programs of standard Sinc transformations, and also, programs for first and second derivatives of these transformations. The number following the word "phi", e.g., phi3, refers to the transformation number as described in §1.5.3.

Equation	Prog.	1st Deriv. Prog.	2nd Deriv. Prog.
Identity	phi1.m	phi1_1.m	phi1_2.m
$\log\left(\dfrac{x-a}{b-x}\right)$	phi2.m	phi2_1.m	phi2_2.m
$\log(x)$	phi3.m	phi3_1.m	phi3_2.m
$\log(\sinh(x))$	phi4.m	phi4_1.m	phi4_2.m
$\log\left(x+\sqrt{1+x^2}\right)$	phi5.m	phi5_1.m	phi5_2.m
$\log(\sinh(x+\sqrt{1+x^2}))$	phi6.m	phi6_1.m	phi6_2.m
$i*(v-u)/2$ $+\log\left(\frac{x-\exp(i*u)}{\exp(i*v)-x}\right)$	phi7.m	phi7_1.m	phi7_2.m

$$(5.2.1)$$

We also list the functions ψ, where $x = \psi(w)$ corresponds to the inverse of the transformation $w = \varphi(x)$. These programs `psi*.m` are used to generate the Sinc points, based on the formula, $z_k = \psi(k\,h)$; these Sinc point programs are listed in the section which follows.

$$\psi(w) \qquad\qquad\qquad \text{Prog., psi} * .m$$

$$
\begin{array}{cc}
\text{Identity} , \text{w} & \text{psi1}.m \\[4pt]
\dfrac{a + b\,e^w}{1 + e^w} & \text{psi2}.m \\[4pt]
\exp(w) & \text{psi3}.m \\[4pt]
\log\left(e^w + \sqrt{1 + e^{2\,w}}\right) & \text{psi4}.m \\[4pt]
\sinh(w) & \text{psi5}.m \\[4pt]
\dfrac{1}{2}\left(t(w) + \dfrac{1}{t(w)}\right) & \\[4pt]
t(w) = \log\left(e^w + \sqrt{1 + e^{2\,w}}\right) & \text{psi6}.m \\[4pt]
\dfrac{e^{w+i\,v} + e^{i\,(u+v)/2}}{e^w + e^{i\,(v-u)/2}} & \text{psi7}.m
\end{array}
\qquad (5.2.2)
$$

5.2.2 SINC POINTS AND WEIGHTS

The Sinc points, $z_k = \psi(k\,h)$ and weights, $h\,\psi'(k\,h) = h/\varphi'(z_k)$ are ubiquitous throughout this handbook, in formulas of interpolation, in derivatives, in quadrature, indefinite integration, Hilbert transforms, analytic continuation, Laplace transform inversion, and in the solution of PDE and IE (partial differential and integral equations). The programs corresponding to these formulas are `Sincpoints*.m` and `weights*.m` respectively. Each of these programs generates a row vector of m numbers, with $m = M + N + 1$, and with M and N corresponding to the limits in the sum (1.5.27), and with the integer "*" referring to the transformation number of the transformation in §1.5.3.

Sincpoint Programs	Weights Programs
Sincpoints1.m	weights1.m
Sincpoints2.m	weights2.m
Sincpoints3.m	weights3.m
Sincpoints4.m	weights4.m
Sincpoints5.m	weights5.m
Sincpoints6.m	weights6.m
Sincpoints7.m	weights7.m

5.2.3 INTERPOLATION AT SINC POINTS

There are two classes of algorithms.

The first class is based on the results of Sec. 1.5.4, and more specifically, Equation (1.5.24). Each of the formulas basis$k$.m, with $k = 1, 2, 3, 4, 5, 6$ and 7 uses the formula basis.m, which corresponds to the case of $k = 1$ for approximation on the real line R. The other integers from 2 to 7 correspond to the transformation number in §1.5.3. For example, the call statement basis3(M,N,h,x), with $x \in (0, \infty)$ returns a row vector of $m = M + N + 1$ basis values $(\omega_{-M}(x), \ldots, \omega_N(x))$, enabling the evaluation of the formula (1.5.27). For the case of a finite interval (a, b) we also need to specify a and b in the call of the program basis2.m, e.g., with any $x \in (a, b)$, one would write "basis2(a,b,M,N,x)".

The second class of programs is based on the methods of §1.7. These are denoted by basis_poly*.m, with $k = 1, 2, 3, 4, 5, 6$ and 7 corresponding to the transformation $(*)$ in §1.5.3. As for the first class above, these programs also return a row vector of $m = M+N+1$ numerical values $b_{-M}(y), \ldots, b_N(y)$, which enables the evaluation

$$p(y) = \sum_{k=-M}^{N} b_k(y) F(z_k),$$

and which is exact at the Sinc points z_k. Indeed, $p(y)$ is a polynomial of degree $m-1$ in the variable $y = \rho(x)/(1+\rho(x))$, with $\rho(x)$ defined as in Definition 1.5.2. This approximation is frequently more accurate than than the Sinc interpolation formula (1.5.27) in cases of when Γ has a finite end–point.

Formula	Program
(1.5.24) & (1.5.27)	basis.m
	basis1.m
	basis2.m
	basis3.m
	basis4.m
	basis5.m
	basis6.m
	basis7.m

(1.7.1), (1.7.16)   basis_poly1.m
                           basis_poly2.m
                           basis_poly3.m
                           basis_poly4.m
                           basis_poly5.m
                           basis_poly6.m
                           basis_poly7.m

5.2.4 DERIVATIVE MATRICES

There are four classes of programs for one dimensional differentiation. The first two classes are based on differentiation of Sinc interpolation formulas. The first class, D*_j.m is based on differentiating the interpolation formula

$$F_h(x) = \sum_{k=-M}^{N} F_k \, S(k,h) \circ \varphi(x);$$

the second class, Domega*_1.m is based on differentiating the formula (1.5.27) (see Theorem 1.5.14), while the third class uses the inverses of the indefinite integration matrices obtained via the programs Int*_p.m, or the negative of the inverses of the matrices obtained via the programs Int_m.m (see below). While these classes produce results that converge when $\Gamma = \mathbb{R}$, and more generally, on closed subsets of all open arcs Γ, they do not give accurate results in a neighborhood of a finite end–point of an arc Γ, even when the function F does have bounded derivatives at such points. The reason for this is that the derivative of either the above formula for F_h or the formula (1.5.27) contains a factor $\varphi'(x)$ which is unbounded at a finite end–point of an arc Γ.

The fourth class of programs, Dpoly*_1.m, is derived in §1.7, and is based on differentiating a composite polynomial that interpolates a function at the same Sinc points. As proved in §1.7, these formulas are able to produce uniformly accurate derivative results when such results exist, even for arcs that have finite end–points.

Other Illustrations of Derivative Programs:

Handbook Location Method   Program   Interval

Example 3.3.1   # 1 above   der_test33.m   (0,1)
                 # 1 above   der_test01.m*   (0,1)
                 # 2 above   der_test3.m*   (0,1)
                 # 1 above   der_test4.m*   (0,infinity)
                 # 3 above   der_test_Int2.m*   (0,1)

5.2.5 QUADRATURE

I. Let us first list the programs Quad $*$ .m , where the "$*$" corresponds to the $*^{th}$ transformation in §1.5.3, i.e., for use of evaluation of integrals over Γ_* based on the transformation

$$\int_{\Gamma_*} F(z)dz = \int_{\mathbb{R}} F(\psi_*(w))\, \psi'_*(w)\, dw , \qquad (5.2.3)$$

and then evaluation of the integral via use of the Trapezoidal rule, i.e., via use of Quad1.m. There are seven integrals, Quad$*$.m , for $* = 1, 2, \ldots, 7$. The user must write the routine FF$*$.m for use in Quad$k$.m. The routines themselves use the transformation from the left of (5.2.8) to the right, and they then evaluate the integral. The user is cautioned however: *Often – particularly for the cases of $* = 2, 6$ and 7 it is preferable to first make the transformation $z = \psi_*(w)$ in the integral on the left of (5.2.8), to get the integral on the right of (5.2.8), and then to apply the rule* Quad1.m. Calls take the form, e.g., "$c = Quad3(Eps)$, where Eps is the error to which the integral is to be evaluated. We must, however, be more specific for the cases of $* = 2$ and $* = 7$, in which case the calls take the form $c = Quad2(a, b, Eps)$ and $c = Quad7(u, v, Eps)$ respectively.

*	int_{Gamma_k} F(x) dx	Program
1		Quad1.m
2		Quad2.m
3		Quad3.m
4		Quad4.m
5		Quad5.m
6		Quad6.m
7		Quad7.m

II. The quadrature routines in the following table are also contained in Sinc–Pack:

Handbook Location   Program

Algorithm 1.5.18   quad1_FF.m*
                                                     quadd1.m
                                                     quad1.m*
                                                     quad1_FF.m*
                                                     ex_1_4_quad.m*

All of these programs are essentially the same, except that quadd1.m is most informative re printouts.

III. The following programs could be combined with Algorithm 1.5.18, or the program **quadd1.m** but they have not – see (1.5.46) – (1.5.47), and the corresponding entries in Tables 1.1 and 1.2. Doing so could be an interesting and worthwhile project. These programs return four numbers corresponding to the entries $q(x)$ and $q'(x)$ in Tables 1.1 and 1.2, for x, and then also, for $1/x$ (see **quadd1.m**), corresponding to the transformation number k in §1.5.3.

$$y = sqrt(1 + x^2)$$

*	q(x)	q'(x)	program
1	log(x)	1/x	q_both1.m*
2	(x-a)/(b-x)	(b-a)x/(1+x)^2	q_both2.m*
3	x	1	No Program
4	log(x + y)	1/y	q_both4.m*
5	(1/2)(x - 1/x)	(1/2) y^2/x^2	q_both5.m*
6	(1/2)(w - 1/w)	(1/2)(1 + 1/w^2)/y	q_both6.m*

$$w = log(x + y)$$

7	V^2(x+U/V)/(x+V/Y)	V^2x*V/U-U/V)/(x_U/V)^2	q_both7.m*

$$U = exp(iu/2), \quad V = exp(iv/2)$$

5.2.6   INDEFINITE INTEGRATION

The matrices for indefinite integration are all a product of an intrinsic matrix – Im.m, for indefinite integration over the real line, $\mathbb{R}$, and a diagonal matrix of weights. There are two matrices for each Γ_*, with $*$ corresponding to the transformation # * in §1.5.3, one for integration from the left end–point a of Γ_* to a point $x \in \Gamma_*$, and the other, for integration from $x \in \Gamma_*$ to the right end–point b of Γ_*. Recall, if V is the operator that transforms a function f defined on Γ to a column vector $V f = (f_{-M}, \ldots, f_N)^T$, with $f_k = f(z_k)$, and where z_k denotes a Sinc point, then $V \int_a^x f(t)\, dt \approx A\,(V\,f)$ where A denotes the indefinite integration matrix for integration from a to x. Similarly, $V \int_x^b f(t)\, dt \approx B\,(V\,f)$, where B denotes the indefinite integration matrix for integration from x to b. In the table which follows, the integer $*$ refers to the transformation # * in §1.5.3.

$*$	A--Matrix Program	B--Matrix Program
1	Int1_p.m	Int1_m.m
2	Int2_p.m	Int2_m.m
3	Int3_p.m	Int3_m.m
4	Int4_p.m	Int4_m.m
5	Int5_p.m	Int5_m.m
6	Int6_p.m	Int6_m.m
7	Int7_p.m	Int7_m.m

The program indef_ex_1.m gives an illustrative example of the approximation of an indefinite integral on $(0, \infty)$.

5.2.7   INDEFINITE CONVOLUTION

Linear Convolutions. "Linear convolution" is based on the formula (1.5.55) of §1.5.9. There are two linear one dimensional convolution programs:

$$\text{conv1\_mat}.m \quad \text{and} \quad \text{conv1\_vec}.m \qquad (5.2.4)$$

Each of these is written for for purposes of approximating indefinite convolution integrals over an interval $(a, b) \subseteq \mathbb{R}$ of the form of p and q in §1.5.9. For each, there is an implied indefinite integration matrix – call it $A : A = XSX^{-1}$ for indefinite integration either from a to x, or from x to b, and where $S = \text{diag}[s_{-M}, \ldots, s_N]$ is a diagonal matrix. Then, with F the Laplace transform of f, one needs to externally form the diagonal matrix FF with j^{th} diagonal entry s_j.

The call

$$W = \text{conv1\_mat}(X, FF)$$

then returns $F(A)$. An application of this routine is the approximation of a solution to a Wiener–Hopf integral equation.

The routine `conv1_vec.m` approximates either p or q of (1.5.51), and for this purpose we also require an externally defined routine to determine the vector $gg = Vg$, for the approximation $V \int_a^x f(x - t)g(t)dt \approx X\,FF\,X^{-1}\,gg$, and similarly, $\int_x^b f(t - x)g(t)dt \approx X\,FF\,X^{-1}\,gg$.

A direct illustration is given in the example program `ex_abel.m` in §3.6. But this procedure is also applied several examples in §3.9, e.g., in the program example `ode_airy.m` in §3.9.4, and in a more sophisticated manner in the program examples of §4.3.

Nonlinear Convolutions. These are programs resulting from the approximations over $\mathbb{R}$ or a subinterval (a, b) of $\mathbb{R}$ of the functions p, q, and r of §1.5.11. The derivations of these approximations is based on the linear convolution procedure of §1.5.9. These formulas are basic for enabling Sinc solutions to PDE over curvilinear regions. There are two sets of programs, each requiring its own inputs:

(i) An array of values (x, t, A, ker_p) for computing p, and array of values (t, x, A, ker_q) for computing q, or an array of (t, x, A, ker_r) values for computing r; and

(ii) Arrays of values (X, KK) as in (i) above, but with ker_p replaced with KK, where X is the matrix of eigenvectors of the indefinite integration matrix (call it A) (for integration form either a to x or from x to b with $x \in (a, b)$) and where the entries of KK are:

(s_j, z_k) for the case of p;

(z_j, x_k) for the case of q; and

(z_j, s_k, z_l) for the case of r,

and where, with the Laplace transforms are defined as in (1.5.62) for p, as in (1.5.66) for q, and as in (1.5.70) for r, where x_j denote the Sinc points of (a, b), and where s_j denoting the eigenvalues of A. The programs are called **lap_transf_ker_*.m**, where "*" denotes either p, q, or r, where the Laplace transforms are defined as in (1.5.62) for p, as in (1.5.66) for q, and as in (1.5.70) for r. The variable s in these Laplace transforms becomes the discretized s_j, with the s_j denoting the eigenvalues of A.

The programs are named as follows:

Function	Program (i)	Program (ii)
p(x)	conv_nl_discrete_p.m	convnl1_p.m
q(x)	conv_nl_discrete_q.m	convnl1_q.m
r(x)	conv_nl_discrete_r.m	convnl1_r.m

5.2.8 LAPLACE TRANSFORM INVERSION

There is only one program, **lap_inv_g.m**, for all intervals. The application of this procedure to curvilinear arcs requires more research. There is a one dimensional example program, **ex_lap_inv4.m**, in §3.7, illustrating Laplace transform inversion for reconstructing the function $t^{-1/2} \sin(t)$ on $(0, \infty)$. The method is also used for programming of the operator $\mathbf{Q}$ in our illustration of the the solution of a Navier–Stokes equation on $\mathbb{R}^3 \times (0, \infty)$.

5.2.9 HILBERT TRANSFORM PROGRAMS

The Hilbert transform programs of the following table are included in Sinc–Pack. The first group, **Hilb_yam*.m** are formulas essentially for intervals, given in Theorem 1.5.24, which were first discovered by Yamamoto ([Ya2]), via a different procedure. In these, the "*" refers the number of the transformation in §1.5.3. We then list a program based on (1.5.106), which gives results that are, in fact, uniformly accurate, even for functions that do not vanish at end points of the interval. Finally, we also list a program of Hunter's method, which is accurate for functions that one can accurately approximate via polynomials on an interval, and finally, programs of some examples.

Equation, or Method	Program
(1.5.100)	Hilb_yam1.m
	Hilb_yam2.m
	Hilb_yam3.m
	Hilb_yam4.m
	Hilb_yam5.m
	Hilb_yam6.m
	Hilb_yam7.m
Example of Hilb_yam	ex_hilb_yam.m
Example of Hilb_yam	ex1_hilb_yam.m
(1.5.84)	Hilbert_t2.m
Hunter's	Hilb_hunter.m
Example of applics of Hunter's	Hilb_hunter_ex.m
Other Simple Hilbert transforms	Hilb2_ex.m

5.2.10 CAUCHY TRANSFORM PROGRAMS

These programs compute several (n) vectors of the Cauchy transformed sinc functions c_j defined in Sec. 1.5.12 – see Eq. (1.5.78) – for performing the Cauchy transform over an interval or contour.

They are labeled Cauchy*.m, where the '*' corresponds to the transformation #(*) in §1.5.3, and program selection should correspond to
functional behavior as described in the particular example of §1.5.3..
Function calls take the form, e.g., Cauchy3(M,N,h,u), with u a complex number. For example, for the case of $n = 1$, m function values
$c_j(u)$ are returned, and one then gets a uniformly accurate approximation of the form

$$\frac{1}{2\pi i} \int_0^\infty \frac{f(t)}{t-u}\, dt \approx \sum_{j=-M}^{N} f(z_j)\, c_j(u), \qquad (5.2.5)$$

even in the case when f (although bounded at 0) does not vanish
at the origin. Although these formulas work for any $u \in \mathbb{C}$, the user
must ensure that phi$k$(u) is properly defined.

Contour	Program
Gamma1	Cauchy1.m
Gamma2	Cauchy2.m
Gamma3	Cauchy3.m
Gamma4	Cauchy4.m
Gamma5	Cauchy5.m
Gamma6	Cauchy6.m
Gamma7	Cauchy7.m

5.2.11 ANALYTIC CONTINUATION

These algorithms are based on the results of Sec. 1.5.13,
and more specifically, Equation (1.5.107). Each of the formulas
basis_harm*.m, with $* = 1, 2, 3, 4, 5, 6$ and 7 uses the program basis_harm.m, which corresponds to the case of $k = 1$ for
approximation using data on the real line $\mathbb{R}$. The other integers "*"
from 2 to 7 correspond to the transformation number in §1.5.3. For
example, the call statement basis_harm3(M,N,h,x), returns a row
vector of the $m = M + N + 1$ basis functions as defined in (1.5.107),
i.e., $(\delta_{-M}(x), \ldots, \delta_N(x))$, where these functions are defined for
$\varphi(x) = \log(x)$ enabling the evaluation of the formula (1.5.109) at
$x \in \mathbb{C}$. For the case of a finite interval (a, b) we also need to specify
a and b in the call of the program basis_harm.m, e.g., with data
in (a, b), and $x \in \mathbb{C}$, one would write basis_harm2(a,b,M,N,x). Al-

though these formulas work for any $x \in \mathbb{C}$, the user must ensure that $\varphi(x)$ is well defined.

Formula	Program
(1.5.107)	basis_harm.m
	basis_harm1.m
	basis_harm2.m
	basis_harm3.m
	basis_harm4.m
	basis_harm5.m
	basis_harm6.m
	basis_harm7.m

5.2.12 CAUCHY TRANSFORMS

Only two programs appear, Cauchy1.m, and Cauchy2.m. The first is used for approximating a Cauchy program over the real line, at several values in the upper half plane, while the second is used for approximating a Cauchy transform over a finite interval (a, b), at several values not on (a, b). The basis_harm*.m programs of the previous subsection can also be used for approximation of Cauchy transforms.

5.2.13 INITIAL VALUE PROBLEMS

These are based, mainly, on the indefinite integration routines. The examples in the following table are contained in Sinc–Pack.

Program Location	Program
Section 3.9.1	ode_ivp_nl1.m
Section 3.9.2	ode_ivp_li1.m
Section 3.9.3	ode_ivp_li2.m
Section 3.9.4	ode_airy.m

5.2.14 WIENER–HOPF EQUATIONS

These are discussed in §1.5.15, and an example is discussed in §3.9.5. The program example, `wiener_hopf.m` is given in §3.9.5.

5.3 Multi–Dimensional Laplace Transform Programs

We list here, Sinc programs of Sinc–Pack, for multidimensional Laplace transforms of Green's functions that are used for approximating multidimensional convolution integrals. We list Laplace transforms of Green's functions for Poisson problems, for wave equation (i.e., hyperbolic) problems, for heat equation (i.e., parabolic) problems, for Navier–Stokes problems, and for two dimensional biharmonic problems.

We shall use the notation "G–F" to denote the words "Green's Function(s)".

5.3.1 Q – FUNCTION PROGRAM

The function $Q(a)$ is defined as an in (2.4.1), over the arc $\mathcal{C}$ defined in (2.4.2). This function is required for evaluation of many multidimensional Laplace transforms. The program `Qfunc.m` evaluates $Q(a)$ for any complex number a, based on the results of Theorem 2.4.1.

5.3.2 TRANSF. FOR POISSON GREEN'S FUNCTIONS

The Greens function of the one dimensional Laplacian,

$$\left(\frac{d}{dx}\right)^2$$

is just $-(1/2)|x|$. That is, the function

$$w(x) = -\frac{1}{2}\int_a^b |x-t|\,g(t)dt \qquad (5.3.1)$$

is a particular solution on (a,b) to the problem $w_{xx} = -g(x)$. The Laplace transform of the G–F $-(1/2)|x|$ is, trivially,

$$G_0^{(1)}(s) = \int_0^\infty -\frac{x}{2}\exp(-x/s)\,dx = -\frac{s^2}{2}. \qquad (5.3.2)$$

The G–F of the Laplacian in two and three dimensions are

$$-(2\pi)^{-1} \log \sqrt{x_1^2 + x_2^2} \quad \text{and} \quad (4\pi \sqrt{x_1^2 + x_2^2 + x_3^2})^{-1}$$

respectively. These G–F have Laplace transforms as given in Theorems 2.4.3 and 2.4.4 respectively. The programs for evaluating these Laplace transforms are called `lap_tr_poi2.m` and `lap_tr_poi3.m` respectively.

5.3.3 TRANSFORMS OF HELMHOLTZ GREEN'S FUNCTION

The G–F for the Helmholtz equation in one, two and three dimensions are given in Equation (2.4.29). These G–F satisfy the differential equation(s) (2.4.30). The Laplace transforms of these G–F are given in Theorems 2.4.6, 2.4.7 and 2.4.8.

Formulas and corresponding computer programs of Laplace transforms of the G–F of Helmholtz equations are intimately related to formulas and computer programs of the Laplace transforms of the G–F of wave and heat equations. The computer programs of these Laplace transforms are given in the following table:

```
         G--F                          Laplace transform program

(2 k)^(-1) exp(- k|x_1|)                     lap_tr_helm1.m

(2 pi)^(-1) K_0(k r)                         lap_tr_helm2.m

r = ( (x_1)^2 + (x_2)^2 )^(1/2)

   (4 pi R)^(-1) exp(- k R)                  lap_tr_helm3.m

R = ((x_1)^2 + (x_2)^2 + (x_3)^2 )^(1/2)
```

5.3.4 TRANSFORMS OF HYPERBOLIC GREEN'S FUNCTIONS

Green's Functions for hyperbolic equations satisfy the differential equation and initial conditions (2.4.52). Performing a Laplace transform with respect to the variable t on the differential equation for $\mathcal{G}^{(d)}$ for $d = 2,\ 3$, transforms this differential equation into a differential equation very similar to that of the differential equation (2.4.30) for the d dimensional Helmholtz equation. The Laplace transforms of

Helmholtz equations can thus be used to express the Laplace transforms of hyperbolic (wave) equations, as spelled out in Theorem 2.4.10. The following programs, which, then obviously use the programs for Helmholtz equations, are then used to compute the Laplace transforms of the Green's Functions for wave equations. The reader is instructed to examine (2.4.51) and (2.4.52) regarding the role of d in these results.

n	Green's Functions	Laplace transform program
2	See (2.4.52)	lap_tr_wave2.m
3		lap_tr_wave3.m
4		lap_tr_wave4.m

5.3.5   TRANSFORMS OF PARABOLIC GREEN'S FUNCTIONS

Green's functions for parabolic equations satisfy the differential equation and initial conditions (2.3.1). Performing a Laplace transform with respect to the variable t on the differential equation (2.3.2) for the Green's Functions $\mathcal{G}^{(d)}$ for $d = 2,\ 3,\ 4$, transforms this differential equation into a differential equation very similar to that of the differential equation (2.4.30) for the $n-1$ dimensional Helmholtz equation. The Laplace transforms of Helmholtz equations can thus be used to express the Laplace transforms of parabolic (heat) equations, as spelled out in Theorem 2.4.11. The following programs, which, then obviously use the programs for Helmholtz equations, are then used to compute the Laplace transforms of the Green's Functions for wave equations.

d	Green's Functions	Laplace transform program
1	See (2.4.61)	lap_tr_heat2.m
2		lap_tr_heat3.m
3		lap_tr_heat4.m

5.3.6 TRANSF. OF NAVIER–STOKES GREEN'S FUNCTIONS

The Green's Functions for which we will list Laplace transform programs are based on the following formulation of the Navier–Stokes equations:

$$\mathbf{u}(\bar{r}, t) = \int_{\mathbb{R}^3} \mathcal{G}(\bar{r} - \bar{r}', t)\, \mathbf{u}^0(\bar{r})\, d\bar{r}'$$

$$+ \int_0^t \int_{\mathbb{R}^3} \{((\nabla' \mathcal{G}) \cdot \mathbf{u}(\bar{r}', t'))\, \mathbf{u}(\bar{r}', t') + p(\bar{r}', t')\, (\nabla' \mathcal{G})\}\, d\bar{r}'\, dt',$$

$$(5.3.3)$$

in which we have written $\mathcal{G}$ for $\mathcal{G}(\bar{r} - \bar{r}', t - t')$.

The the four dimensional Laplace transform of $-\mathcal{G}_{x_1}(\bar{r}, t)$ is expressed in (2.4.97), and this can be evaluated using navier_stokes_R.m. The program for the four dimensional Laplace transform of $-\mathcal{K}_{x_1}(\bar{r}, t)$, with $\mathcal{K}(\bar{r}, t)$ defined in (2.4.91) is given in (2.4.98) and is called navier_stokes_S.m.

5.3.7 TRANSFORMS OF BIHARMONIC GREEN'S FUNCTIONS

The biharmonic Green's Functions which satisfies the differential equation

$$\Delta^2 \mathcal{G}(x, y) = \delta(x)\, \delta(y) \tag{5.3.4}$$

is just the function $\mathcal{G}(x, y) = (16\pi)^{-1} R \log(\mathcal{R})$, as is given in (2.4.102), where $R = x_1^2 + x_2^2$. The Laplace transform of this function, as well as of the functions $\mathcal{G}_x(x, y)$ and $\mathcal{G}_{xx}(x, y)$ (see 2.4.103)) are given in the statement of Theorem 2.4.12.

Laplace Transform	Program
G--F Eq. (2.4.105)	lap_tr_bihar2.m
G_x -- Eq. (2.4.107)	lap_tr_bihar2x.m
G_xx -- Eq. (2.4.107)	lap_tr_bihar2xx.m
G_y -- See (2.4.107)	lap_tr_bihar2y.m

```
G_yy -- See (2.4.107)                    lap_tr_bihar2yy.m
```

There is, of course an obvious matrix transpose relationship between the outputs of the programs for G_x and G_y, as well as between the outputs of the programs for G_{xx} and G_{yy}, in the case when the defining indefinite integration matrices of these functions are the same.

5.3.8   Example Programs for PDE Solutions

Included in Sinc–Pack are programs for solution of the following problems:

1. *Harmonic Sinc Approximation Program.* The program `continue.m` enables analytic continuation.

2. *Programs for Solving Poisson and Dirichlet Problems.*

   (a) The program `pois_R_2.m` in §4.1.2 solves a Poisson problem over $\mathbb{R}^2$;

   (b) The program, `lap_square.m` in §4.1.3 solves a Poisson–Dirichlet problem on a square;

   (c) The program, `lemniscate.m` in §4.1.4 converts Dirichlet to Neumann data on a lemniscate;

   (d) The programs `lap_poi_disc.m` in §4.1.5 gets a particular solution to a Poisson problem on the unit disc. The program `lap_harm_disc.m` in §4.1.5 solves a Dirichlet problem with discontinuous data on the boundary of the unit disc; and

   (e) The program `pois_R_3.m` solves a Poisson problem on $\mathbb{R}^3$.

3. *Biharmonic Problems.* The program `biharm.m` computes three approximate solution tasks for the PDE problem over the square $B = (-1, 1) \times (-1, 1)$

$$\Delta^2 w(x, y) = f(x, y), \quad \text{in } B,$$

$$w = g \text{ and } \quad \frac{\partial w}{\partial n} = h \quad \text{on } \partial B, \tag{5.3.5}$$

where ∂B denotes the boundary of B, and where g and h are given functions:

(a) Particular solutions, w, w_x and w_y, based on evaluating the integral (4.1.48) via Sinc convolution, using the Laplace transform programs, `lap_tr_bihar2.m`, `lap_tr_bihar2x.m` and also, `lap_tr_bihar2y.m`;

(b) It solves a boundary value problem

$$\Delta^2 V(x,y) = 0 \quad \text{in } B;$$

$$V = g_1, \quad \frac{\partial U}{\partial n} = h_1 \quad \text{on } \partial B,$$

(5.3.6)

where g_1 and h_1 are given functions. To this end, it determines values of functions u, v and ϕ on the boundary of B, where these functions are harmonic in B, such that $V = (x-a)u + (y-b)v + \phi$ in B, and with a and b any constants.

(c) Finally, it evaluates the solution V in the interior B via use of the analytic continuation procedure that was used in the above program, `lap_square.m`.

4. The program `wave_3p1.m` in §4.2.1 solves a *wave equation problem* over $\mathbb{R}^2 \times (0,T)$.

5. The programs `p_d_picard.m`, and `p_d_T5.m` in §4.4.1 solve various forms of the nonlinear integro–differential equation (4.4.1) for a probability density. The second of these computes a solution for the case of a bad boundary layer.

6. The program, `navier_stokes2.m` of §4.4.2 computes a solution to a *Navier–Stokes problem* discussed in §2.4.6, but over the region $(\bar{r}, t) \in \mathbb{R}^3 \times (0,T)$.

Bibliography

[Ab] N.H. Abel, *Solution de quelques problèmes à l'aide d'integrales dèfinies*, Œuvres complètes D'Abel, V. 1, (1823) 11–27.

[AiQTr1] P.N.D. Aalin, P.H. Quan, & D.D. Tron, *Sinc Approximation of Heat Flux on the Boundary of a Two–Dimensional Finite Slab*, Numerical Functional Analysis and Optimization, V. 27 (2006) 685–695.

[AjSo] I. Abu-Jeib & T. Shores, *On Properties of Matrix $I^{(-1)}$ of Sinc Methods*, New Zealand Journal of Mathematics, V. 32 (2003), 1–10.

[AlCo] J. Albrecht & L. Collatz, editors *Numerical Treatment of Integral Equations*, Birkhäuser–Verlag, Stuttgart, Germany (1980).

[Al1] K. Al–Khaled, *Theory and Computation in Hyperbolic Model Problems*, Ph.D. Thesis, University of Nebraska (1996).

[Am] W.F. Ames. *Numerical Methods for Partial Differential Equations*. Academic Press, New York, 2nd edition (1977).

[AnjBob] J.–E. Andersson & B.D. Bojanov, *A Note on the Optimal Quadrature in $\mathbf{H}^p$*, Numer. Math., V. 44 (1984) 301–308.

[AnrHoLs] R. Anderssen, F. de Hoog, & M. Lukas, eds., *The Application and Numerical Solution of Integral Equations*, Sijthof & Noordhoff, Alphen aan den Rijn, The Netherlands (1980).

[ATP] D.A. Anderson, J.C. Tannenhill, & R.H. Pletcher, *Computational Fluid Mechanics and Heat Transfer*. McGraw-Hill Book Company, Washington (1984).

[AgGoVT] D.D. Ang, R. Gorenflo, L.K. Vy & D.D. Trong, *Moment Theory and some Inverse Problems in Potential Theory*. Lecture Notes in Mathematics, Springer (2002).

[ALuS] D. D. Ang, J. Lund, & F. Stenger, *Complex Variable and Regularization Methods for Inversion of the Laplace Transform*, Math. Comp., V. 83 (1989) 589–608.

[An] M.H. Annaby, *On Sampling Theory Associated with the Reolvents of Singular Sturm–Liouville Problems*, Proc. Amer. Math. Soc. V. 131 (2003) 1803-1812.

[AnTh] M.H. Annaby, & M.M. Tharwat, *Sinc–Based Computations of Eigenvalues of Dirac Systems*, BIT Numerical Mathamatics, V. 47 (2007) 690–713.

[AnAsTh] M.H. Annaby, R.M. Asharabi, & M.M. Tharwat, *Computing Eigenvalues of Boundary Value Problems Using Sinc–Gaussian Method*, Sampling Theory in Signal and Image Processing V. 1 (2002) 1–50.

[Ans] P.M. Anselone, *Collectively Compact Operator Approximation Theory and Application to Integral Equations*, Prentice–Hall, Englewood Cliffs, N.J. (1971).

[AsMjRl] U.M. Ascher, R.M. Mattheij & R.D. Russell, *Numerical Solution of Boundary Value Problems for Ordinary Differential Equations*, Classics Appl. Math. V. 13 SIAM, (1995).

[Asr] R. Askey, *Orthogonal Polynomials and Special Functions*, SIAM (1975).

[At] K. Atkinson, *A Survey of Numerical Methods for the Solution of Fredholm Integral Equations of the Second Kind*, Society for Industrial and Applied Mathematics, Philadelphia, PA (1976).

[BaBojBopPl] D.H. Bailey, J.M. Borwein, P.B. Borwein, & S. Plouffe, *The Quest for Pi*, Mathematical Intelligencer, V. 19 (1997) 50–57.

[BaXiLiJe] David H. Bailey, Xiaoye S. Li, & Karthik Jeyabalan, *A Comparison of Three High-Precision Quadrature Schemes*, Experimental Mathematics, V. 14 (2005) 317-329.

[BaBojKaWe] David H. Bailey, Jonathan M. Borwein, Vishal Kapoor, & Eric Weisstein, *Ten Problems in Experimental Mathematics*, American Mathematical Monthly, V. 113 (2006) 481–409.

[BaBoj1] David H. Bailey, & Jonathan M. Borwein, *Highly Parallel, High-Precision Numerical Integration*, manuscript, Jun 2006, http://crd.lbl.gov/d̃hbailey/dhbpapers/quadparallel.pdf.

[BaBoj12] David H. Bailey & Jonathan M. Borwein, *Effective Error Bounds in Euler-Maclaurin-Based Quadrature Schemes*, Proceedings of the 2006 High-Performance Computing Conference (HPCS), http://crd.lbl.gov/d̃hbailey/dhbpapers/hpcs06.pdf.

[BaBojCr1] David H. Bailey, Jonathan M. Borwein, & Richard E. Crandall, *Box Integrals*, Journal of Computational and Applied Mathematics, to appear, Jun 2006, http://crd.lbl.gov/d̃hbailey/dhbpapers/boxintegrals.pdf.

[BaBojCr12] David H. Bailey, Jonathan M. Borwein, & Richard E. Crandall, *Integrals of the Ising Class*, "Journal of Physics A: Mathematical and General", to appear, Jun 2006, http://crd.lbl.gov/d̃hbailey/dhbpapers/ising.pdf.

[BakMi] C. Baker & G. Miller, editors *Treatment of Integral Equations by Numerical Methods*, Academic Press, New York (1982).

[BawFoWa] R.W. Barnard, W.T. Ford, & Hsing Y. Wang, *On Zeros of Interpolating Polynomials*, SIAM J. Math. Anal., V. 17 (1986) 734–744.

[BesNob] S. Beignton & B. Noble, *An Error Estimate for Stenger's Quadrature Formula*, Math. Comp., V. 38 (1982) 539–545.

[BemJoWiS] M.J. Berggren, S. Johnson, C. Wilcox, & F. Stenger, *Rational Function Frequency Extrapolation in Ultrasonic Tomography*, pp. 19–34 of "Wave Phenomena, Modern Theory and Applications," C. Rogers and T.B. Moodie, eds. Elsevier Science Publishers (B.B), North–Holland (1984).

[Ber1] J.-P. Berrut, *Barycentric Formulae for Cardinal (SINC)–Interpolants*, Numer. Math. V. 54 (1989) 703–718.

[Ber2] J.-P. Berrut, *A Formula for the Error of Finite Sinc–Interpolation over a Finite Interval*, Numerical Algorithms V. 45 (2007) 369–374.

[Ber3] J.-P. Berrut, *A Formula for the Error of Finite Sinc–Interpolation, with an Even Number of Nodes*, to appear in Numerical Algorithms (2010).

[Ber4] J.–P. Berrut, *Integralgleichungen und Fourier–Methoden zur Numerischen Konformen Abbildung*, Ph.D. Thesis, Eidgenössische TechniShe Hochschule Zürich (1985).

[Bia1] B. Bialecki, *Sinc Method of Solution of CSIE and Inverse Problems*, Ph.D. Thesis, University of Utah (1987).

[Bia2] B. Bialecki, *Sinc–Type Approximation in $\mathbf{H}^1$–Norm, with Application to Boundary Value Problems*, Journ. Comp. and Appl. Math., V. 25 (1989) 289–303.

[Bia3] B. Bialecki, *A Modified Sinc Quadrature Rule for Functions with Poles Near the Arc of Integration*, BIT, V. 29 (1989) 464–476.

[Bia4] B. Bialecki, *A Sinc Quadrature Rule for Hadamard Finite Part Integrals*, Numer. Math., V. 57 (1990) 263–269.

[Bia5] B. Bialecki, *Sinc–Hunter Quadrature Rule for Cauchy Principal Value Integrals*, Math. Comp., V. 55 (1990) 665–681.

[Bia5] B. Bialecki, *Sinc–Nyström Method for Numerical Solution of a Dominant System of Cauchy Singular Integral Equations Given on a Piecewise Smooth Contour*, Siam J. Numer. Anal. V. 26 (1989) 1194–1211.

[BiaS] B. Bialecki, & F. Stenger, *Sinc–Nyström Method for Numerical Solution of One–Dimensional Cauchy Singular Integral Equations Given on a Smooth Arc in the Complex Plane*, Math. Comp., V. 51 (1988) 133–165.

[BL] G. Birkhoff & R.E. Lynch. *Numerical Solution of Elliptic Problems*, SIAM, Philadelphia (1984).

[BlH] N. Bleistein, & R.A. Handelsman. *Asymptotic Expansions of Integrals*, Dover Publications, Inc., New York (1986).

[Bar] R.P. Boas, Jr., *Summation Formulas and Band Limited Signals*, Tôhuko Math. J., V. 24 (1972) 121–125.

[Bob] B.D. Bojanov, *Best Quadrature Formula for a Certain Class of Analytic Functions*, Zastosowania Matematyki Appl. Mat. V. XIV (1974) 441–447.

[dBPi] C. De Boor & A. Pinkus, *The B-spline Recurrence Relations of Chakalov and of Popoviciu*, J. Approx. Theory V. 124 (2003) 115–123.

[Bor1] E. Borel, *Sur l'interpolation*, C.R. Acad. Sci. Paris, V. 124 (1897) 673–676.

[Bor2] E. Borel, *Mèmoire sur les Sèries Divergente*, Ann. École Norm. Sup., V. 16 (1899) 9–131.

[BPn] J.F. Botha, & G.F. Pinder. *Fundamental Concepts in the Numerical Solution of Differential Equations*. John Wiley and Sons, New York (1987).

[BoLu] K.L. Bowers, & J. Lund, *Numerical Simulations: Boundary Feedback Stabilization*, IEEE Control Systems Society (1992) 175–2002.

[BoCaLu] K.L. Bowers, T.S. Carlson, & John Lund, *Advection–Diffusion Equations: Temporal Sinc Methods*, Numerical Methods for Partial Differential Equations, V. 11 (1995) 399–422.

[BoKo1] K.L. Bowers, & S. Koonprasert, *Block Matrix Sinc-Galerkin Solution of the Wind–Driven Current Problem*, Applied Mathematics and Computation, V. 155 (2004) 607–635.

[BoKo2] K.L. Bowers, & S. Koonprasert, *The Fully Sinc–Galerkin Method for Time–Dependent Boundary Conditions*, Numerical Methods for Partial Differential Equations, V. 20 (2004) 494–526.

[BoLu] K.L. Bowers, & J. Lund, *Numerical Solution of Singular Poisson Solvers, via the Sinc–Galerkin Method*, SIAM J. Numer. Anal., V. 24 (1987) 36–51.

[ByDiPi] W.E. Boyce, & R. C. DiPrima. *Elementary Differential Equations and Boundary Value Problems*, John Wiley and Sons, New York (1965).

[Bre] Claude Brezinski, *Convergence Acceleration Methods: The Past Decade*, J. Comput. Appl. Math., V. 12/13 (1985) 19–36.

[Bro] Vincent Broman, *Initial Value Methods for Integral Equations*, M.Sc. Thesis, University of Utah, (1979).

[BuRi] H. Brunner, & H. de Riele, *The Numerical Solution of Volterra Equations*, Elsevier, The Netherlands (1986).

[dB] N.G. de Bruijn, *Asymptotic Methods in Analysis*. Dover Publications, Inc., New York (1981).

[BrHo] H.G. Burchard, & K. Höllig, *N–Width and Entropy of $\mathbf{H}^p$ Classes in $\mathbf{L}^q(-1, 1)$*, SIAM J. Math. Anal., V. 16 (1985) 405–421.

[Bu] R. Burke, *Application of the Sinc–Galerkin Method to the Solution of Boundary Value Problems*, M.Sc. Thesis, University of Utah (1977).

[Br] J.C. Butcher, *The Numerical Analysis of Ordinary Differential Equations: Runge–Kutta and General Linear Methods*, Wiley–Intersci. Publ., Wiley, Chichester (2008)

[Bk] A.G. Butkovski, *Green's Functions and Transfer Functions Handbook*, Ellis Horwood Limited (1982).

[CHQ] C. Canuto, M.Y. Hussaini, A. Quarteroni, & T.A. Zang. *Spectral Methods in Fluid Dynamics*, Springer–Verlag, New York (1987).

[CaLuBo] T.S. Carlson, J. Lund, & K.L. Bowers, *A Sinc–Galerkin Method for Convection Dominated Transport*, Computation and Control III, (Proceedings of the Third Bozeman Conference on Computation and Control, August 5–11, 1992, Bozeman, Montana), K. L. Bowers & J. Lund, eds., V. 15 "Progress in Systems and Control Theory", Birkhäuser Boston, Inc. (1993) 121–139.

[Cu] W. Cauer, *Bemerkung Über eine Extremalaufgabe, von E. Zolotareff*, Z. Angew. Math. Mech., V. 20 (1940) 358.

[CwKa] M.M. Chawla, & V. Kaul, *Optimal Rules for Numerical Integration Round the Unit Circle*, BIT, V. 13 (1973) 145–152.

[Ch] E.W. Cheney. *Approximation Theory*. Chelsea Publishing Company, New York, (1982).

[Ci] Y.H. Chiu, *Integral Equation Solution of $\nabla^2 u + k^2 u = 0$ in the Plane*, Ph.D. Thesis, University of Utah (1977).

[CrScV] P.G. Ciarlet, M.H. Schultz, & R.S. Varga, *Numerical Methods of High Order Accuracy for Nonlinear Boundary Value Problems, I. One Dimensional Problem*, Numer. Math., V. 2 (1967) 394–430.

[ClCu] C.W. Clenshaw, & A.R. Curtis, *A Method for Numerical Integration on an Automatic Computer*, Numer. Math. V. 2 (1960) 197–205.

[Dq1] G. Dahlquist, *Convergence and Stability in the Numerical Solution of Differential Equations*, Math. Scand. V. 4 (1956) 33–53.

[Dq2] G. Dahlquist, *A Special Stability Criterion of Linear Multistep Methods*, BIT V. 3 (1963) 22–43.

[DqS] G. Dahlquist, *On the Summation Formulas of Plana, Abel and Lindelöf, and Related Gauss–Christoffel Rules*, in manuscript.

[Da1] P.J. Davis, *Errors of Numerical Approximation for Analytic Functions*, J. Rational Mech. Anal., V. 2 (1953) 303–313.

[Da2] P.J. Davis, *On the Numerical Integration of Periodic Analytic Functions*, pp. 45–59 of "On Numerical Approximation," R. Langer, ed., University of Wisconsin Press, Madison. (1959).

[Da3] P.J. Davis, *Interpolation and Approximation*. Dover Publications, Inc., New York (1975).

[Da4] P.J. Davis, *The Schwarz Function and Its Applications*, Carus Mathematical Monographs # 17 MAA (1974).

[DRz] P.J. Davis, & P. Rabinowitz, *Methods of Numerical Integration*, Academinc Press, N.Y. (1975).

[dB] C. de Boor, *A Practical Guide to Splines*, Springer–Verlag, New York (1978).

[dDPi] E. de Donker & R. Piessens, *Automatic Computation of Integrals with Singular Integrand over a Finite or Infinite Range*, Report TW22, Applied Mathematics and Programming Division, Katholike Universiteit, Leuven (1975).

[dH] F. de Hoog, *A New Algorithm for Solving Toeplitz Systems of Equations*, Lin. Algebra Appl., Vols. 88/89 (1987) 122–133.

[Do] De-A. Dorman, *Methods for Accelerating Convergence*, M.Sc. Thesis, University of Utah (1983).

[DvWs] L. Delves, & J. Walsh, editors, *Numerical Solution of Integral Equations*, Clarendon Press, Oxford, England, UK (1974).

[dVSc] R.A. deVore & K. Scherer, *Variable Knot, Variable Degree Spline Approximation to x^β*, in "Quantitative Approximation", (R.A. deVore and K. Scherer, Eds.), Academic Press (1980) 121–131.

[DrRaSr] D.P. Dryanov, Q.I. Rahman, & G. Schmeisser, *Converse Theorems in the Theory of Approximate Integration*, Constr. Approx., V. 6 (1990) 321–334.

[Ec1] U. Eckhardt, *Einige Eigenschaften Wilfscher Quadraturformeln*, Numer. Math., V. 12 (1968) 1–7.

[EdSaV] A. Edrei, E.B. Saff, & R.S. Varga, *Zeros of Sections of Power Series*, Springer Lecture Notes in Mathematics # 1002 Springer–Verlag, New York (1983).

[EgJaLu] N. Eggert, M. Jarratt, & J. Lund, *Sinc Function Computation of the Eigenvalues of Sturm–Liouville Problems*, J. Comput. Phys., V. 69 (1987) 209–229.

[EfLu] N. Eggert & J. Lund, *The Trapezoidal Rule for Analytic Functions of Rapid Decrease*, J. Computational and Appl. Math., V. 27 (1989) 389–406.

[EiLiV] M. Eiermann, X. Li, & R.S. Varga, *On Hybrid Semi–Iterative Methods*, SIAM J. Numer. Anal. V. 26 (1989) 152–168.

[EiVNi] M. Eiermann, R.S. Varga, & W. Niethammer, *Iterationsverfahren für Nichtsymmetrische Gleichungssysteme und Approximationsmethoden in Komplexen*, Jber. d. Df. Math.-Verein., V. 89 (1987) 1–32.

[E1] M. El–Gamel, *A Numerical Scheme for Solving Nonhomogeneous Time–Dependent Problems*, Z.Angew. Math. Phys., V. 57 (2006) 369–383.

[E2] M. El–Gamel, *Sinc and the Numerical Solution of Fifth–Order Boundary Value Problems*, AMC , V. 187 (2007) 1417–1433.

[E3] M. El–Gamel, *A Comparison Between the Sinc–Galerkin and the Modified Decomposition Methods for Solving Two–Point Boundary–value Problems*, J. of Computational Physics V. 223 (2007) 369–383.

[EZ1] M. El–Gamel, & A.I. Zayed, *A Comparison Between the Wavelet–Galerkin and the Sinc–Galerkin Methods in Solving Mon–Homogeneous Heat Equations*, Contemporary Mathematics of the American Mathematical Society, Series, Inverse Problem, Image Analysis and Medical Imaging Edited by Zuhair Nashed and Otmar Scherzer, Vol. 313, AMS, Providence (2002) 97–116.

[EZ2] M. El–Gamel, & A.I. Zayed, *Sinc–Galerkin Method for Solving Nonlinear Boundary-Value Problems*, Computers and Math. Appl. V. 48 (2004) 1285–1298.

[ECa] M. El-Gamel, & J.R. Cannon, *On the Solution of Second Order Singularly–Perturbed Boundary Value Problem by the Sinc–Galerkin Method*, Z. Angew. Math. Phys. V. 56 (2005) 45–58.

[ECaZa] M. El-Gamel, J. R. Cannon, & A. Zayed, *Sinc–Galerkin Method for Solving Linear Sixth Order Boundary–Value Problems*, Math. Comp. V. 73 (2004) 1325–1343.

[E1] D. Elliott, *Truncation Errors in Two Chebyshev Series Approximations*, Math. Comp. V. 19 (1965) 234–248.

[E2] D. Elliott, *Some Observations on the Convergence Theories of Anselone, Noble, and Linz*, University of Tasmania Mathematics Department, Technical Report No. 122 (1979).

[E3] D. Elliott, *A Note on the Convergence of Operator Equations*, University of Tasmania Mathematics Department, Technical Report No. 124 (1979).

[ES] D. Elliott, & F. Stenger, *Sinc Method of Solution of Singular Integral Equations*, "IMACS Conference on CSIE", Philadelphia, P.A. (1984) 155–166.

[Eg] H. Engels, *Numerical Quadrature and Cubature*, Academic Press, London (1980).

[EnGr] H.W. Engl, & C.W. Groetsch, *Inverse and Ill–Posed Problems*, Academic Press, Inc. (1987).

[Er] A. Erdélyi. *Asymptotic Expansions*, Dover Publications, Inc., New York (1956).

[Ff] C. Fefferman, *Existence and Smoothness of the Navier–Stokes Equation*, Clay Math. Institute,
http://www.claymath.org/millennium/ .

[Fi] H.E. Fettis, *Numerical Calculation of Certain Definite Integrals by Poisson's Summation Formula*, M.T.A.C., V. 9 (1955) 85–92.

[Fr] G. Fichera, *Asymptotic Behavior of the Electric Field Near the Singular Points of the Conductor Surface*, Academi Nazionale dei Lincei, V. 60 (1976) 1–8.

[FyW] G. Forsythe, & W. Wasow, *Finite Difference Methods in Partial Differential Equations*. Wiley, N.Y. (1960).

[Fr] K.O. Friedrichs, *Special Topics in Analysis*. N.Y. Univ. Inst. of Math. Science 1953–54.

[Gb1] B.G. Gabdulhaev, *A General Quadrature Process and Its Application to the Approximate Solution of Singular Integral Equations*, Soviet Math., Dokl., V. 9 (1968) 386–389.

[Gb2] B.G. Gabdulhaev, *Approximate Solution of Singular Integral Equations by the Method of Mechanical Quadratures*, Soviet Math., Dokl., V. 9 (1968) 329–332.

[Ga] D. Gaier, *Konstruktive Methoden der konformen Abbildung*, Springer–Verlag (1964).

[Gv] F.D. Gakhov, *Boundary Value Problems*, translated from Russian by I.N. Sneddon, Pergamon Press, Oxford (1966).

[Gu2] W. Gautschi, *Barycentric Formulae for Cardinal (SINC–) Interpolants, by Jean–Paul Berrut (Remark)* Numer. Math., V. 87 (2001) 791–792.

[GuV] W. Gautschi, & R.S. Varga, *Error Bounds for Gaussian Quadrature of Analytic Functions*, SIAM J. Numer. Anal. V. 20 (1983) 1170–1186.

[Gr] G.W. Gear, *Numerical Initial Value Problems in Ordinary Differential Equations*, Prentice-Hall, Inc., Englewood Cliffs, New Jersey, 1971.

[GeS] W.B. Gearhart, & F. Stenger, *An Approximate Convolution Equation of a Given Response*, pp. 168–196 of "Optimal Control Theory and Its Applications" Springer–Verlag Lecture Notes on Economics and Mathematical Systems, v. 106, Springer-Verlag, New York (1974).

[Ge] N.H. Getz, *Discrete Periodic Wavelet Transform Toolbox* (1992).
http://www.inversioninc.com/wavelet.html

[GiTu] D. Gilbarg, & N.S. Trudinger, *Elliptic Partial Differential Equations of Second Order.* (Second Edition) Springer–Verlag, New York (1983).

[Go] M. Golberg, editor, *Solution Methods for Integral Equations*, Plenum Press, New York (1978).

[Gl] G.M. Golusin, *Geometric Theory of Functions of a Complex Variable*, Translations of Mathematical Monographs, AMS (1969).

[GcS] R.F. Goodrich, & F. Stenger, *Movable Singularities and Quadrature*, Math. Comp., V. 24 (1970) 283–300.

[Goo] E.T. Goodwin, *The Evaluation of Integrals of the Form $\int_{-\infty}^{\infty} a(x)\, e^{-x^2}\, dx$*, Proc. Camb. Philos. Soc., V. 45 (1949) 241–245.

[GvBd] A.D. Gorbunov, & B.B. Budak, *On the Convergence of Certain Finite Difference Processes for the Equations $y' = f(x, y)$ and $y'(x) = f(x, y(x), y(x - \tau(x)))$*, Dokl. Akad. Nauk SSSR V.119 (1958) 644–647.

[GoOr] D. Gottlieb, & S.Z. Orszag, *Numerical Analysis of Spectral Methods: Theory and Applications*, SIAM, Philadelphia (1977).

[GdRy] I. S. Gradshteyn, & I. M. Ryzhik, *Tables of Integrals, Series and Products*, 4th Edition, Academic Press, New York (1965).

[GrSz] Grenander, & G. Szegö, *Toeplitz Forms and Their Applications*, University of California Press, Berkeley (1958).

[Gs1] S.-Å. Gustafson, *Convergence Acceleration on a General Class of Power Series*, Computing, V. 21 (1978) 53–69.

[Gs2] S.-Å. Gustafson, *Two Computer Codes for Convergence Acceleration*, Computing, V. 21 (1978) 87–91.

[Ha1] S. Haber, *The Error in the Numerical Integration of Analytic Functions*, Quart. Appl. Math., V. 29 (1971) 411–420.

[Ha2] S. Haber, *The Tanh Rule for Numerical Integration*, SIAM J. Numer. Anal., V. 14 (1977) 668–685.

[Ha3] S. Haber, *Two Formulas for Numerical Indefinite Integration*, Math. Comp., V. 60 (1993) 279–296.

[Hc] W. Hackbusch, *Integral Equations: Theory and Numerical Treatment*. Birhäuser Verlag, Basel (1994).

[HgS] M. Hagmann, & F. Stenger, *Unique Advantages of Sinc Function Basis in the Method of Moments*, in Proceedings of the Conference on Electromagnetic Insonification, IEEE, V. 54 (1980) 35–37.

[H] R. V. L. Hartley, *The Transmission of Information*. Bell System Tech. J. V. 7 (1928) 535–560.

[Hn] P. Henrici, *Applied Computational and Complex Analysis*, V. 1 (1974); V. 2 (1977); V. 3 (1986) John Wiley and Sons, New York.

[Ho] M.H. Hohn, *Solution of Singular Elliptic PDEs on a Union of Rectangles Using Sinc Methods*, ETNA, Vol. 23 (2006) 88–104.

[Hu] T.E. Hull, *Numerical Integration of Ordinary Differential Equations*, MAA (1966).

[Ik] Y. Ikebe, *The Galerkin Method for Numerical Solution of Fredholm Integral Equations of the Second Kind*, SIAM Rev., V. 14 (1972) 465–491.

[IKoS] Y. Ikebe, M. Kowalski, & F. Stenger, *Rational Approximation of the Step, Filter and Impulse Functions*, pp. 441–454 of "Asymptotic and Computational Analysis," Edited by R. Wong, Marcel Dekker, New York (1990).

[ILS] Y. Ikebe, T.Y. Li, & F. Stenger, *Numerical Solution of Hilbert's Problem*, in "Theory of Approximation with Applications", Acad. Press, Inc. (1976) eds. A.G. Law and B. Sahney, 338–358.

[Im] J.P. Imhof, *On the Method of Numerical Integration of Clenshaw and Curtis*, Numer. Math. V. 5 (1963) 138–141.

[In] E.L. Ince, *Ordinary Differential Equations*. Dover Publications, Inc., New York (1956).

[Is] M.E.H. Ismail, *Contiguous Relations, Basic Hypergeometric Functions, and Orthogonal Polynomials*, Inst. for Constructive Mathematics, ICM # 88–103 University of South Florida (1988).

[Ja1] M. Jarratt, *Approximation of Eigenvalues of Sturm–Liouville Differential Equations by the Sinc–Collocation Method*, Ph.D. Thesis, Montana State University (1987).

[Ja2] M. Jarratt, *Eigenvalue Approximations on the Entire Real Line*, pp. 133–144 of "Computation and Control," Edited by K.L. Bowers and J. Lund, Birkhäuser, Basel (1989).

[JaLBo] M. Jarratt, J. Lund, & K.L. Bowers, *Galerkin Schemes and the Sinc–Galerkin Method for Singular Sturm–Liouville Problems*, J. Comput. Phys., V. 89 (1990) 41–62.

[Je] A. Jerri, *The Shannon Sampling Theorem—Its Various Extensions and Applications: a Tutorial Review*, Proc. IEEE, V. 65 (1977) 1565–1596.

[JoZhTrBrS] S.A. Johnson, Y. Zhou, M.L. Tracy, M.J. Berggren, & F. Stenger, *Inverse Scattering Solutions by Sinc Basis, Multiple Source Moment Method,—Part III: Fast Algorithms*, Ultrasonic Imaging, V. 6 (1984) 103–116.

[Ke] R.B. Kearfott, *A Sinc Approximation for the Indefinite Integral*, Math. Comp., V. 41 (1983) 559–572.

[KeSiS] R.B. Kearfott, K. Sikorski, & F. Stenger, *A Sinc Adaptive Algorithm for Solving Elliptic Partial Differential Equations*, in manuscript.

[Ke] J.P. Keener. *Principles of Applied Mathematics: Transformation and Approximation.* Addison-Wesley Publishing Co., Redwood City, California (1988).

[Kn] F. Keinert, *Uniform Approximation to $|x|^\beta$ by Sinc Functions*, J. Approx. Th., V. 66 (1991) 44–52.

[Ku] W. Knauff, *Fehlernormen zur Quadratur Analytischer Funktionen*, Computing, v. 17 (1977) 309–322.

[KuKr] W. Knauff, & R. Kreß, *Optimale Approximation Linearer Funktionale auf Periodischen Funktionen*, Numer. Math., V. 22 (1974) 187–205.

[Ko] H. Kober, *A Dictionary of Conformal Representation.* Dover, New York (1957).

[Ko] S. Koonprasert, *The Sinc–Galerkin Method for Problems in Oceanography*, Ph.D. Thesis, Montana State University (2003).

[KoBo1] S. Koonprasert, & K.L. Bowers, *The Fully Sinc–Galerkin Method for Time–Dependent Boundary Conditions*, Numercical Methods for Partial Differential Equations, V. 20 (2004) 494–526.

[KoBo2] S. Koonprasert, & K.L. Bowers, *Block Matrix Sinc–Galerkin Solution of the Wind–Driven Current Problem*, Applied Math. and Computation V. 20 (2004) 607–735.

[Kt] V.A. Kotel'nikov, *On the Carrying Capacity of the 'Ether' and Wire in Telecommunications*, "Material for the First All–Union Conference on Questions of Communication," Izd. Red. Upr. Svyazi RKKA, Moscow (1933).

[KoSiS] M. Kowalski, K. Sikorski, & F. Stenger, *Selected Topics of Approximation and Computation.* Oxford University Press (1995).

[Kr] H.P. Kramer, *A Generalized Sampling Theorem*, J. Math. Phys. V. 38 (1959) 68–72.

[Kß1] R. Kreß, *Linear Integral Equations.* Springer–Verlag, Berlin (1989).

[Kß2] R. Kreß, *On Gerneral Hermite Trigonometric Interpolation*, Numer. Math. V. 20 (1972) 125–138.

[Kß3] R. Kreß, *Interpolation auf einem Unendlichen Intervall*, Computing, V. 6 (1970) 274–288.

[Kß4] R. Kreß, *Über die numerische Berechnung Konjugierter Funktionen*, Computing, V. 10 (1972) 177–187.

[Kß5] R. Kreß, *Ein Ableitungsfrei Restglied für die trigonometrische Interpolation Periodischer Analytischer Funktionen*, Numer. Math. V. 16 (1971) 389–396.

[Kß6] R. Kreß, *Zur Numerischen Integration Periodischer Funktionen nach der Rechteckregel*, Numer. Math., V. 20 (1972) 87–92.

[Kß7] R. Kreß, *On Error Norms of the Trapezoidal Rule*, SIAM J. Numer. Anal., V. 15 (1978) 433–443.

[Kß8] R. Kreß, *Zur Quadratur Unendlicher Integrale bei Analytischen Funktionen*, Computing, V. 13 (1974) 267–277.

[La] L. Lapidus, & G.F. Pinder. *Numerical Solution of Partial Differential Equations in Science and Engineering*, John Wiley and Sons, New York (1982).

[Lx] P. Lax, *Lax-Milgram Theorem*. From MathWorld–A Wolfram Web Resource.
http://mathworld.wolfram.com/Lax-MilgramTheorem.html

[Le] D.L. Lewis, *A Fully Galerkin Method for Parabolic Problems*, Ph.D. Thesis, Montana State University (1989).

[LeLuBo] D.L. Lewis, J. Lund, & K.L. Bowers, *The Space–Time Sinc–Galerkin Method for Parabolic Problems*, Int. J. Numer. Methods Eng., V. 24 (1987) 1629–1644.

[LpS] P. Lipow, & F. Stenger, *How Slowly Can Quadrature Formulas Converge?*, Math. Comp., V. 26 (1972) 917–922.

Lp1] A. Lippke, *Analytic Solution and Sinc Function Approximation in Thermal Conduction with Nonlinear Heat Generation*, J. Heat Transfer (Transactions of the ASME), V. 113 (1991) 5–11.

[LoWe] H.L. Loeb, & H. Werner, *Optimal Numerical Quadratures in $\mathbf{H}^p$ Spaces*, Math. Z., V. 138 (1974) 111–117.

[Lu1] J. Lund, *Numerical Evaluation of Integral Transforms*, Ph.D. Thesis, University of Utah, (1978).

[Lu2] J. Lund, *Sinc Function Quadrature Rule for the Fourier Integral*, Mathematics of Computation, V. 41 (1983) 103–113.

[Lu3] J. Lund, *Bessel Transforms and Rational Extrapolation*, Numerische Mathematik, V. 47 (1985) 1–14.

[Lu4] J. Lund, *Symmetrization of the Sinc–Galerkin Method for Boundary Value Problems*, Math. Comp., V. 47 (1986) 571–588.

[Lu5] J. Lund, *Accuracy and Conditioning in the Inversion of the Heat Equation*, Progress in Systems and Control Theory, Vol. I (1989) 179–196.

[Lu6] J. Lund, *Sinc Approximation Method for Coefficient Identification in Parabolic Systems*, Progress in Systems and Control Theory, Vol. 4 (1990) 507–514.

[LuBo] J. Lund, & K.L. Bowers, *Sinc Methods for Quadrature and Differential Equations*, SIAM (1992).

[LuBoCa] J. Lund, K.L. Bowers, & T.S. Carlson, *Fully Sinc–Galerkin Computation for Boundary Feedback Stabilization*, Jour. Math. Syst. Est. and Control, V. 2 (1991) 165–182.

[LuBoMca] J. Lund, K.L. Bowers, & K. McArthur, *Symmetrization of the Sinc-Galerkin Method with Block Techniques for Elliptic Equations*, IMA Jour. Num. Anal., V 9 (1989) 29–46.

[Ly] N.J. Lybeck, *Sinc Domain Decomposition Methods for Elliptic Problems*, Ph.D. Thesis, Montana State University (1994).

[LyBo1] N.J. Lybeck, & K.L. Bowers, *Domain Decomposition in Conjunction with Sinc Methods for Poisson's Equation*, Numerical Methods for Partial Differential Equations, V. 14 (1996) 461–487.

[LyBo2] N.J. Lybeck, & K.L. Bowers, *Sinc Methods for Domain Decomposition*, Applied Mathematics and Computation, V. 75 (1996) 13–41.

[LyBo3] N.J. Lybeck, & K.L. Bowers, *Sinc–Galerkin Schwarz Alternating Method for Poisson's Equation*, Computation and Control IV, (Proceedings of the Fourth Bozeman Conference on Computation and Control, August 3–9, 1994, Bozeman, Montana), K. L. Bowers & J. Lund, eds., V. 20 "Progress in Systems and Control Theory", Birkhäuser Boston, Inc., (1997) 247–258.

[LyBo4] N.J. Lybeck, & K.L. Bowers, *Domain Decomposition via the Sinc–Galerkin Method for Second Order Differential Equations*, Domain Decomposition Methods in Scientific and Engineering Computing, (Proceedings of the Seventh International Conference on Domain Decomposition, October 27–30, 1993, The Pennsylvania State University), D. E. Keyes & J. Xu, eds., V. 180 "Contemporary Mathematics", American Mathematical Society, Providence, (1994) 271–276.

[LyBo5] N.J. Lybeck, & K.L. Bowers, *The Sinc–Galerkin Patching Method for Poisson's Equation*, Proceedings of the 14th IMACS World Congress on Computational and Applied Mathematics, July 11-15, 1994, Georgia Institute of Technology, W. F. Ames, ed., "International Association for Mathematics and Computers in Simulation" (IMACS), V. 1 (1994) 325–328.

[LyBo6] N.J. Lybeck, & K.L. Bowers, *The Sinc–Galerkin Patching Method for Poisson's Equation*, Proceedings of the 14th IMACS World Congress on Computational and Applied Mathematics, July 11-15, 1994, Georgia Institute of Technology, W. F. Ames, ed., International Association for Mathematics and Computers in Simulation (IMACS), V. 1 (1994) 325–328.

[Mca] K.M. McArthur, *Sinc–Galerkin Solution of Second-Order Hyperbolic Problems in Multiple Space Dimensions*, PhD. Thesis, Montana State Univ. (1987).

[McaBoLu] K.M. McArthur, K.L. Bowers, & J. Lund, *Numerical Implementation of the Sinc–Galerkin Method for Second-Order Hyperbolic Equations*, Numerical Methods for Partial Differential Equations, V. 3 (1987) 169–185.

[McaSrLuBo] K.M. McArthur, R.C. Smith, J. Lund, & K.L. Bowers, *The Sinc-Galerkin Method for Parameter Dependent Self-Adjoint Problems*, Applied Mathematics and Computation, V. 50 (1992) 175–202.

[MhE] A. Mohsen, & M. El-Gamel, *A Sinc–Collocation Method for the Linear Fredholm Integro–Differential Equations*, Z. Angew. Math. Phys. V. 58 (2007) 380–390.

[MSu] P. Monk, & E. Süli, *Convergence Analysis of Yee's Scheme on Nonuniform Grids*, SIAM J. Numer. Anal., V. 31 (1994) 393–412.

[Mo1] M. Mori, *On the Superiority of the Trapezoidal Rule for the Integration of Periodic Analytic Functions*, Memoirs of Numerical Mathematics, V. 1 (1974) 11–19.

[Mo2] M. Mori, *An IMT–Type Double Exponential Formula for Numerical Integration*, Publ. RIMS, Kyoto Univ., V. 14 (1978) 713–729.

[Mo3] M. Mori, *Analytic Representations Suitable for Numerical Computation of Some Special Functions*, Numer. Math. V. 35 (1980) 163–174.

[Mo4] M. Mori, *Quadrature Formulas Obtained by Variable Transformation and the DE–Rule*, J. Comput. Appl. Math. V. 12 & 13 (1985) 119–130.

[Mo5] M. Mori, *An Error Analysis of Quadrature Formulas Obtained by Variable Transformatiom*, Algebraic Analysis, V. 1 ed., M. Kashiwara & T. Kawai (1988) 423–437, Academic Press.

[Mo6] M. Mori, *Developments in the Double Exponential Formulas for Numerical Integration*, Proceedings of the International Congress of Mathematicians Kyoto (1009), Springer–Verlag (1991) 1585–1594.

[Mo7] M. Mori, *Optimality of the Double Exponential Transformation in Numerical Analysis*, Sugaku Expositions, V. 14 (2001) 103–123.

[Mo8] M. Mori, *Discovery of the Double Exponential Transformation and its Developments*, Publ. RIMS Kyoto Univ., V. 41 (2005) 897—935.

[MoMh1] M. Mori, & M. Muhammad, *Numerical Indefinite Integration by the Double Exponential Transformation* (in Japanese), Transactions of the Japan Society for Industrial and Applied Mathematics, V. 13 (2003) 361–366.

[MoMh2] M. Mori, & M. Muhammad, *Numerical Iterated Integraion by the Double Exponential Transformation*, (in Japanese), Transactions of the Japan Society for Industrial and Applied Mathematics, V. 13 (2003), 485–493.

[MoNt] M. Mori, & M. Natori, *Error Estimation in the Linear Approximation of Analytic Functions*, Report of the Computer Centre, Univ. of Tokyo, V. 4 (1971/72) 1–17.

[MoOo] M. Mori, & T. Ooura, *Double Exponential Formulas for Fourier Type Integrals with a Divergent Integrand*, Contributions in Numerical Mathematics, ed., R.P.Agarwal, World Scientific Series in Applicable Analysis Vol.2 (1993) 301–308.

[MoSu] M. Mori & M. Sugihara, *The Double–Exponential Trans-formation in Numerical Analysis*, J. Comput. Appl. Math., V. 127 (2001) 287–296.

[MlLyBo1] A.C. Morlet, N.J. Lybeck, & K. L. Bowers, *Convergence of the Sinc Overlapping Domain Decomposition Method*, Applied Math. and Computations, V. 98 #2–3 (1999) 209–227.

[MlLyBo2] A.C. Morlet, N.J. Lybeck, & K. L. Bowers, *The Schwarz Alternating Sinc Domain Decomposition Method*, Applied Math. and Computations, V. 25 #4 (1997) 461–483.

[MlLyBo1] A.C. Morlet, N.J. Lybeck, & K. L. Bowers, *Convergence of the Sinc Overlapping Domain Decomposition Method*, Applied Math. and Computations, V. 98 (1999) 209–227.

[MlLyBo2] A.C. Morlet, N.J. Lybeck. & K. L. Bowers, *he Schwarz Alternating Sinc Domain Decomposition Method*, Applied Math. and Computations, V. 25 (1997) 461–483.

[MrFe] P.M. Morse & H. Feshbach, *Methods of Theoretical Physics*, §5.1, Vol. 1, 464–523 & 655–666 (1953).

[Mu1] J. Mueller, *Inverse Problems for Singular Differential Equations*, Ph.D. Thesis, University of Nabraska (1997).

[MuSo1] J. Mueller, & T. Shores, *Uniqueness and Numerical Recovery of a Potential on the Real Line*, Inverse Problems, V. 13 (1997) 781–800.

[MuSo2] J. Mueller, & T. Shores, *A New Sinc–Galerkin Method for Convection–Diffusion Equations with Mixed Boundary Conditions*, Comp. and Math. Apps. V. 47 (2004) 803–822.

[MhMo1] M. Muhammad, & M. Mori, *Double Exponential Formulas for Numerical Indefinite Integration*, J. Comput. Appl. Math., V. 161 (2003) 431–448.

[MhMo2] M. Muhammad, & M. Mori, *Numerical Iterated Integration Based on the Double Exponential Transformation*, Japan J. Indust. Appl. Math., V. 22 (2005) 77–86.

[MhNuMoSu] M. Muhammad, A. Nurmuhammad, M. Mori, & M. Sugihara, *Numerical Solution of Integral Equations by Means of the Sinc Collocation Method Based on the Double Exponential Transformation*, J. Comput. Appl. Math., V. 177 (2005) 269–286.

[N] A. Naghsh-Nilchi, *Iterative Sinc–Convolution Method for Solving Three–Dimensional Electromagnetic Models*, Ph.D. Thesis, University of Utah (1997).

[NS] A. Naghsh–Nilchi, & F. Stenger, *Sinc–Convolution Method of Solving the Electric Field Integral Equations over* $\mathbb{R}^2 \times \mathbb{R}^+ \times [0, T]$, in 2001 Marakesh Conference on Optimal Algorithms (2002) 155–256.

[NrMjS] S. Narashimhan, J. Majdalani, & F. Stenger, *A First Step in Applying the Sinc Collocation Method to the Nonlinear Navier–Stokes Equations*, Numerical Heat Transfer, V. 41, B (2002) 447–462.

[Na] National Bureau of Standards, *Handbook of Mathematical Functions*. Natioanl Bureau of Standards Applied Math. Series (1964).

[Nr] S. Narasimhan, *Solving Mechanincal Engineering Problems via Sinc Methods*, Ph.D. thesis, Univ. of Utah (1998).

[NuMhMoSu] A. Nurmuhammad, M. Muhammad, M. Mori, & M. Sugihara, *Double Exponential Tranformation in the Sinc–Collocation Method for a Boundary Value Problem with Fourth–Order Ordinary Differential Equation*, J. Comput. Appl. Math. V. 182 (2005) 32–50.

[NuMhMoSu] A. Nurmuhammad, M. Muhammad, & M. Mori, *Numerical Solution of Initial Value Problems*, Publ. RIMS Kyoto Univ., V. 41 (2005) 937–948.

[Ny] H. Nyquist, *Certain Topics in Telegraph Transmission Theory*, Trans. Amer. Inst. Elect. Engrg. V. 47 (1928) 617–644.

[O] F.W.J. Olver, *Asymptotics and Special Functions*, Academic Press, New York (1974).

[OS] F.W.J. Olver, & F. Stenger, *Error Bounds for Asymptotic Solution of Second Order Differential Equations Having an Irregular Singularity of Arbitrary Rank*, SIAM J. Numer. Anal. V. 12 (1965) 244–249.

[OoMo1] T. Oura, & M. Mori, *The Double Exponential Formula for Oscillatory Functions Over the Half Infinite Interval*, J. Comput. Appl. Math. V. 38 (1991) 353–360.

[OoMo2] T. Oura, & M. Mori, *A Robust Rouble Rxponential Formula for Fourier–Type Integrals*, J. Comput. Appl. Math. V. 112 (1999) 229–241.

[Pa] K. Parker, *PTOLEMY, A Sinc Collocation Sub–System*, Ph.D. thesis, University of Utah (1997).

[PeT] R. Peyret, & T.D. Taylor. *Computational Methods for Fluid Flow*. Springer-Verlag, New York (1983).

[Pf] K. Pfabe, *A Problem in Nonlinear Ion Transport*, Ph.D. Thesis, University of Newbraska (1995).

[PfSo] K. Pfabe, & T. Shores, *Numerical Methods for an Ion Transport Problem*, Applied Numerical Mathematics 32 (2000) 175–193.

[Pw] M.J.D. Powell, *Approximation Theory and Methods*, Cambridge University Press, Cambridge (1981).

[Ps] S. Prößdorf & B. Silbermann *Projectionsverfahren und die näherungsweise Lösungen von Integralgleichungen*, Teubner, Leipzig, Germany (1977).

[RaSe] Q.I. Rahman, & G. Schmeisser, *The summation formulæof Poisson, Plana, Euler–Maclaurin and their relationship*, Journal of Mathematical Sciences, V. 28 (1994) 151–174.

[Rm] D. Ramkrishna, *Population Balances, Chapter II*, Academic Press, NY (2000).

[RsZa1] J. Rashidinia, & M. Zarebnia, *Solution of a Volterra Integral Equation by the Sinc–Collocation Method*, J. Comput. Appl. Math. V. 206 (2007) 801–813.

[RsZa2] J. Rashidinia, & M. Zarebnia, *Convergence of Approximate Solution of System of Fredholm Integral Equations*, J. Math. Anal. Appl., V. 333 (2007) 1216–1227.

[RsZa3] J. Rashidinia, & M. Zarebnia, *The Numerical Solution of Integro–Differential Equation by Means of the Sinc Method*, Applied Mathematics and Computation, V. 188 (2007) 1124–1130.

[RsZa4] J. Rashidinia, & M. Zarebnia, *New Approach for Numerical Solution of Hammerstein Integral Equations*, Applied Mathematics and Computation, V. 185 (2007) 147–154.

[RsZa5] J. Rashidinia, & M. Zarebnia, *Numerical Solution of Linear Integral Equations by using Sinc–Collocation Method*, Applied Mathematics and Computation, V. 168 (2005) 806–822.

[Re] E.Y. Remez, *Sur un procédé d'approximation successives pour déterminer les polynômes d'approximation*, C.R. V. 198 (1934) 2063–2065.

[Ri] T.J. Rivlin, *An Introduction to the Approximation of Functions*, Dover Publications, Inc., New York (1969).

[Schc] N.A. Schclar, *Anisotropic Analysis Using Boundary Elements*, Topics in Engineering, V. 20, Computational Mechanics Publications, Southampton, UK and Boston, USA (1994).

[Sg1] G. Schmeisser, *Sampling Theorems for Entire Harmonic Functions of Exponential Type*, In: F. Marvasti (ed.), 1995 Workshop on Sampling Theory & Applications, (ISBN 9984-9109-0-3), pp. 140–145, Institute of Electronics and Computer Science, Riga (1995).

[Sg2] G. Schmeisser, *Uniform Sampling of Complex–Valued Harmonic Functions*, I In: A.I. Zayed (editor), "Proceedings of the 2001 International Conference on Sampling Theory and Applications", (ISBN 0-7803-9813-0), pp. 39–43, The University of Central Florida, Orlando (2001).

[Sg3] G. Schmeisser, *Nonuniform Sampling of Complex–Valued Harmonic Functions*, Sampling Theory in Signal and Image Processing V. 2 (2004) 217–233.

[Sg4] G. Schmeisser, *Approximation of Entire Functions of Exponential Type by Trigonometric Polynomials*, Sampling Theory in Signal and Image Processing, Report (2007).

[SgBuSs] J. Schmeisser, P.L. Butzer, & L. Stens, *An Introduction to Sampling Analysis*, In: F. Marvasti (ed.), "Nonuniform Sampling: Theory and Practice" Chap. 2, pp. 17–121, Kluwer Academic–Plenum Publishers, New York (2001).

[SgRh] G. Schmeisser, R. Gervais & Q.I. Rahman, *A Bandlimited Function Simulating a Duration–Limited One*, In: P.L. Butzer, R.L. Stens, B. Sz.–Nagy (eds.) Anniversary Volume on Approximation Theory and Functional Analysis. ISNM V.65, (1984) 355–362, Birkh"auser, Basel.

[SGHiVo] G. Schmeisser, J.R, Higgins & J.J. Voss, *The Sampling Theorem and Several Equivalent Results in Analysis*, Journal of Computational Analysis and Applications V. 2 (2000), 333–371.

[SgRh] G. Schmeisser, & Q.I. Rahman, *Representation of Entire Earmonic Functions by Given Values*, J. Math. Anal. Appl., V. 115 (1986) 461–469.

[SgSr] G. Schmeisser, & H. Schirmeier, *Practische Mathematik*, Walter de Gruyter, Berlin (1976).

[SeS] G. Schmeisser, & F. Stenger, *Sinc Approximation with a Gaussian Multiplier*, Sampling Theory in Signal and Image Processing V. 6 (2007) 199–221.

[SgVo] G. Schmeisser, & J.J. Voss, *A New Interpretation of the Sampling Theorem and its Extensions*, In: G. Nürnberger et al. (eds.), "Multivariate Approximation and Splines, Proceedings of the Mannheim Conference 1996" ISNM V. 125 275–288, Birkhäuser, Basel (1997).

[Scw] C. Schwab, *p and hp Finite Element Methods: Theory and Applications in Solid and Fluid Mechanics*, Numerical Methods and Scientific Computation, Clarendon Press, Oxford (1998).

[ShSiS] J. Schwing, K. Sikorski, & F. Stenger, *ALGORITHM 614. A FORTRAN Subroutine for Numerical Integration in $\mathbf{H}^p$*, ACM TOMS V. 10 (1984) 152–160.

[Sh] C. E. Shannon, *A Mathematical Theory of Communication*, Bell System Tech. J. V. 27 (1948) 379–423, 623–656.

[So] T. Shores, *Numerical Methods for Parameter Identification in a Convection–Diffusion Equation*, ANZIAM J. (E) (2004) 660–675.

[Si] A. Sidi, *Practical Extrapolation Methods: Theory and Applications*, #10 of Cambridge Monographs on Applied and Computational Mathematics, Cambridge University Press (2003).

[Sm] G. D. Smith. *Numerical Solution of Partial Differential Equations: Finite Difference Methods*. Clarendon Press, Oxford, 2nd edition (1978).

[Srm] R.C. Smith, *Numerical Solution of Fourth–Order Time–Dependent Problems with Applications to Parameter Identification*, Ph.D. Thesis, Montana State University (1990).

[SrmBgBoLu] R.C. Smith, G.A. Bogar, K.L. Bowers, & J. Lund, *The Sinc–Galerkin Method for Fourth–Order Differential Equations*, SIAM J. Num. Anal. V. 28 (1991) 760–788.

[SrmBo1] R.C. Smith, & K.L. Bowers, *A Fully Galerkin Method for the Recovery of Stiffness and Damping Parameters in Euler–Bernoulli Beam Models*, "Computation and Control II", Proceedings of the Second Bozeman Conference on Computation and Control, August 1-7, 1990, Bozeman, Montana), "Progress in Systems and Control Theory", Birkhäuser Boston, Inc. (1991) 289–306.

[SrmBo2] R.C. Smith, & K.L. Bowers, *Sinc-Galerkin Estimation of Diffusivity in Parabolic Problems*, Inverse Problems, V. 9 (1993) 113–135.

[SrmBoLu] R.C. Smith, K.L. Bowers & J. Lund, *A Fully Sinc–Galerkin Method for Euler-Bernoulli Beam Models*, Numer. Methods for Partial Diffrential Equations, V. 8 (1992) 172–202.

[SrmBoLu] R.C. Smith, K.L. Bowers, & J. Lund, *Efficient Numerical Solution of Fourth–Order Problems in the Modeling of Flexible Structures*, "Computation and Control" (Proceedings of the Bozeman Conference on Computation and Control, August 1–11, 1988, Bozeman, Montana), in "Progress in Systems and Control Theory", Birkhäuser Boston, Inc. (1989) 183–297.

[SrmBoVo] R.C. Smith, K.L. Bowers, & C.R. Vogel, *Numerical Recovery of Material Parameters in Euler–Bernoulli Beam Models*, Jour. Math. Systems, Estimation, and Control, V. 7 (1997) 157–195.

[Sol] J. Soller, *The Automated Detection of Delirium*, Ph.D. Dissertation, Salt Lake City: University of Utah Department of Computer Science.

[SolS1] J. Soller & F. Stenger, *The Sinc Tensor Product Network and Nonlinear Signal Processing. Neural Information Processing Systems*, Workshop on Nonlinear Signal Processing with Time Delay Connections. Denver: NIPS (1994).

[SolS2] J. Soller, & F. Stenger, *Regularized Sinc Tensor Product Networks*, "Machines That Learn Workshop", Abstracts (1995).

[S1] F. Stenger, *Numerical Methods Based on Sinc and Analytic Functions*, Computational Math. Series, Vol. 20, Springer–Verlag (1993).

[S2] F. Stenger, *Explicit, Nearly Optimal, Linear Rational Approximations with Preassigned Poles*, Math. Comp. V. 47 (1986) 225–252.

[S3] F. Stenger, *Numerical Methods Based on Whittaker Cardinal, or Sinc Functions*, SIAM Review, V. 23 (1981) 165–224.

[S4] F. Stenger, *Explicit, Nearly Optimal, Linear Rational Approximations with Preassigned Poles*, Math. Comp. V. 47 (1986) 225–252.

[S5] F. Stenger, *The Asymptotic Approximation of Certain Integrals*, SIAM J. Math. Anal. V. 1 (1970) 392–404.

[S6] F. Stenger, *Transform Methods of Obtaining Asymptotic Expansions of Definite Integrals*, SIAM J. Math. Anal. V. 3 (1972) 20–30.

[S7] F. Stenger, *Collocating Convolutions*, Math. Comp., V. 64 (1995) 211–235.

[S8] F. Stenger, *Sinc Approximation for Cauchy–Type Singular Integrals over Arcs*, Aus. Math. Soc. V. 42 (2000) 87–97.

[S9] F. Stenger, *A Sinc Galerkin Method of Solution of Boundary Value Problems*, Math. Comp. V. 33 (1979) 85–109.

[S10] F. Stenger, *Kronecker Product Extension of Linear Operators*, SIAM J. Numer. Anal. 5 (1968) 422–435.

[S11] F. Stenger, *Integration Formulas Based on the Trapezoidal Formula*, J. Inst. Maths Applics, V. 12 (1973) 103–114.

[S12] F. Stenger, *The Approximate Solution of Wiener–Hopf Integral Equations*, J. Math. Anal. Appl., V. 37 (1972) 687–724.

[S13] F. Stenger, *Polynomial Function and Derivative Approximation of Sinc Data*, J. Complexity, Vol. 25 (2009) 292–302.

[S14] F. Stenger, *An Analytic Function which is an Approximate Characteristic Function*, SIAM J. Numer. Anal. Vol. 12 (1975) 239–254.

[S15] F. Stenger, *Polynomial function and derivative approximation of Sinc data*, Jour. of Complexity, Vol. 25 (2009) 292–302.

[SRA] F. Stenger, R. Ramlau & R. Anderssen, *Sinc Solution of the Wave Equation in* $\mathbb{R}^3 \times (0, T)$, to appear.

[SBaVa] F. Stenger, B. Barkey & R. Vakili, *Sinc Convolution Method of Solution of Burgers' Equation*, pp. 341–345 of "Proceedings of Computation and Control III", edited by K. Bowers and J. Lund, Birkhäuser (1993).

[SCK] F. Stenger, T. Cook & M. Kirby, *Sinc Methods and Biharmonic Problems*, Canad. Appl. Math. Quarterly, V. 12 (2006) 391-414.

[SGKOrPa] F. Stenger, SÅ Gustafson, B. Keyes, M. O'Reilly & K. Parker, *ODE – IVP – PACK via Sinc Indefinite Integration and Newton's Method*, Numerical Algorithms V. 20 (1999) 241–268. A FORTRAN program of this package can be downloaded from *Netlib*®.

[SNJR] F. Stenger, A. Naghsh–Nilchi, J. Niebsch, & R. Ramlau, *Sampling Methods for Solution of PDE*, with A. Naghsh–Nilchi, J. Niebsch, and R. Ramlau, pp. 199-250 of "Inverse Problems and Image Analysis", AMS, edited by Z. Nashed and O. Scherzer, 2002.

[SBCHK] F. Stenger, B. Baker, C. Brewer, G. Hunter, & S. Kaputerko, *Periodic Approximation Based on Sinc*, International Journal of Pure and Applied Math., Vol. 49 (2008) 63–72.

[SRi] F. Stenger, & T. Ring, *Aggregation Modeling Using Sinc Methods of Solution to the Stieltjes Formulation of the Population Balance Equation*, to appear.

[SSc] F. Stenger, & R. Schmidtlein, *Conformal Maps via Sinc Methods*, in Computational Methods in Function Theory (CMFT '97), N. Papamichael, St. Ruscheweyh and E.B. Saff (Eds.), World Scientific Publishing Co. Pte. Ltd., (1999) 505–549.

[ST] F. Stenger & D. Tucker, *Sinc Solution of Navier–Stokes Equations*, to appear.

[Sh] H.J. Stetter, *Analysis of Discretization Methods for Ordinary Differential Equations*, Tracts Nat. Philos., V. 23 Springer, N.Y. (1973).

[StBu] J. Stoer, & R. Bulirsch, *Introduction to Numerical Analysis*, Springer–Verlag (1991).

[Sr1] M. Stromberg, *Solution of Shock Problems by Methods Using Sinc Functions*, Ph.D. Thesis, University of Utah (1988).

[Sr2] M. Stromberg, *Sinc Approximation of Derivatives in Multiple Dimensions*, in manuscript.

[Sr3] M. Stromberg, *Sinc Approximation and Quadrature on Polyhedra*, in manuscript.

[Sr4] M. Stromberg, *Sinc Approximate Solution of Quasilinear Equations of Conservation Law Type*, Computation and Control, Progress in Systems and Control Theory, V. 1, Birkhäuser (1989) 317–331.

[SrGi] M. Stromberg, & X. Gilliam, *Sinc–Galerkin Collocation Method for Parabolic Equations in Finite Space–Time*, Computation and Control III, Progress in Systems and Control Theory, V. 15, Birkhäuser (1992) 355–366.

[StSs] A.H. Stroud, & D.H. Secrest, *Gaussian Quadrature Formulas*, Prentice–Hall, Englewood Cliffs, N.J. (1966).

[Su1] M. Sugihara, *Near Optimality of Sinc Approximation*, Math. Comp. V. 72 (2003) 767–786.

[Su2] M. Sugihara, *Double Exponential Transformation in the Sinc– Collocation Method for Two–Point Boundary Value Problems*, Scientific and Engineering Computations for the 21st Century – Methodologies and Applications (Shizuoka, 2001). J. Comput. Appl. Math. V. 149 (2002) 239–250.

[Su3] M. Sugihara, *Optimality of the Double Exponential Formula – Functional Analysis Approach*, Numer. Math. V. 75 (1997) 379–395.

[SuMa] M. Sugihara, & T. Matsuo, *Recent Developments of the Sinc Numerical Methods*, Proceedings of the 10th International Congress on Computational and Applied Mathematics (ICCAM-2002). J. Comput. Appl. Math. V. 164/165 (2004) 673–689.

[Sy1] G.T. Symm, *An Integral Equation Method in Conformal Mapping*, Numer. Math. V. 9 (1966) 250–258.

[Sy2] G.T. Symm, *Numerical Mapping of Exterior Domains*, Numer. Math. V. 10 (1967) 437–445.

[Ta] H. Takahasi, *Complex Function Theory and Numerical Analysis*, Publ. RIMS, Kyoto Univ. V. 41 (2005) 979-988.

[TaMo1] H. Takahasi, & M. Mori, *Error Estimation in the Numerical Integration of Analytic Functions*, Report of the Computer Centre, Univ. of Tokyo, V. 3 (1970) 41–108.

[TaMo2] H. Takahasi, & M. Mori, *Estimation of Errors in the Numerical Quadrature of Analytic Functions*, Applicable Analysis, V. 1 (1971) 201–229.

[TaMo3] H. Takahasi, & M. Mori, *Quadrature Formulas Obtained by Variable Transformation*, Numer. Math., V. 21 (1973) 206–219.

[TnSuMu] K. Tanaka, M. Sugihara, & K Murota, *Numerical Indefinite Integration by Double Exponential Sinc Method*, Math. Comp. V. 74 (2005) 655–679.

[Te] V.K. Tewary, *Computationally Efficient Representation for Elastic and Elastodynamic Green's Function for Anisotropic Solids*, Physical Review B, V. 51 (1995) 15695–15702.

[Ti] A.F. Timan, *Theory of Approximation of Functions of a Real Variable.* Pergamon Press (1963).

[TM1] H. Takahasi, & M. Mori, *Quadrature Formulas Obtained by Variable Transformation*, Numer. Math., V. 21 (1973) 206–219.

[TM2] H. Takahasi, & M. Mori, *Double Exponential Formulas for Numerical Integration*, Publ. RIMS, Kyoto Univ., V. 9 (1974) 721–741.

[To] J. Todd, *Survey of Numerical Analysis.* McGraw–Hill Book Company, Inc., (1962).

[Tr1] L.N. Trefethen, *Spectral Methods in Matlab.* SIAM (2000).

[Tr1] L.N. Trefethen, *Is Gauss Quadrature Better than Clenshaw–Curtis ?*, Report # 06/07, Oxford University Computing Laboratory (2006) 1–19.

[W] J.H. Wilkinson, *Rounding Errors in Algebraic Processes*, Prentice–Hall, Englewood Cliffs, N.J. (1963).

[Va1] R.S. Varga, *Matrix Iterative Analysis*, Prentice–Hall, Englewood Cliffs, N.J. (1962).

[Va2] R.S. Varga, *Functional Analysis and Approximation Theory in Numerical Analysis.* SIAM (1987).

[Va3], R.S. Varga, *Scientific Computations on Mathematical Problems and Conjectures.* SIAM (1990).

[Va4] R.S. Varga, *Gersgorin and His Circles*, Springer–Verlag (2004).

[Ve1] E. Venturino, *On Solving Singular Integral Equations via a Hyperbolic Tangent Quadrature Rule*, Mathematics of Computation, V. 47 (1985) 159–167.

[Ve2] E. Venturino, *Recent Developments in the Numerical Solution of Singular Integral Equations*, Journal of Mathematical Analysis and Applications, V. 115, (1986) 239-277.

[Ve3] E. Venturino, *Stability and Convergence of a Hyperbolic Tangent Method for Singular Integral Equations*, Mathematische Nachrichten, V. 164 (1993) 167-186.

[VrPMS] D. Verkoeijen, G.A. Pouw, G.M.H Meesters, & B. Scarlett, *Population Balances for Particulate Processes – A Volume Approach*, Chem. Eng. Sci. V 57, #12 (2002) 2287–2302.

[WeRe] J.A.C. Weideman, & S.C. Reddy, *A MATLAB Differentiation Matrix Suite*, "ACM Transactions of Mathematical Software", V. 26 (2000) 465–519.

[WiBoLu] D.F. Winter, Kenneth, L. Bowers, & John Lund, *Wind–Driven Currents in a Sea with a Variable Eddy Viscosity Calculated via a Sinc–Galerkin Technique*, International Journal for Numerical Methods in Fluids, V. 33 (2000) 1041–1073.

[Wo] R. Wong, *Asymptotic Approximation of Integrals*. Academic Press, Boston (1989).

[Ya1] T. Yamamoto, *Toward the Sinc–Galerkin Method for the Poisson Problem in one Type of Curvilinear Coordinate Domain*, ETNA, V. 23 (2006) 63–75.

[Ya2] T. Yamamoto, *Approximation of the Hilbert Transform via use of Sinc Convolution*, ETNA V. 23 (2006) 320–328.

[Ye] L. Ye, *Numerical Quadrature, Theory and Computation*, MSC Thesis, Computer Science, Dalhousie Univ. (2006).

[Y] K. Yee, *Numerical Solution of Boundary Value Problems Involving Maxwell's Equations in Isotropic Media*, IEEE Trans., Antennas and Propagation, V. AP–16 (1966) 302–307.

[Yi] G. Yin, *Sinc Collocation Method with Orthonormalization for Singular Poisson–Like Problems*, Math. Comp. V. 62 (1994) 21–40.

[Yo] D. Young, *Iterative Solution of Large Linear Systems*, Academic Press, N.Y. (1971).

[Za] P.P. Zabreyko etal., *Integral Equations – A Reference Text*. Nordhoff International Publishing (1975).

[ZcTh] E.C. Zachmanoglou & D.W. Thoe, *Introduction to Partial Differential Equations*. Dover Publications, Inc., New York (1986).

[Zy] A.I. Zayed, *Advances in Shannon Sampling Theory*, CRC Press, Boca Raton, Florida (1993).

Index

analytic arc: Def. 2.6.2

analytic continuation
§1.5.13, S4.2.1, §5.2.11

Approximation:
one-dimensional Sinc:
§1.3, S 1.5
rational §1.6
wavelet §1.4, §3.11

arc Ex. 1.5.1, Eq. (1.5.101)
§2.2.2, Def. 2.6.2

band limited §1.2

Burgers' equation §4.4.6

Cardinal function §1.2

Cauchy-Kowalevsky theorem
Thm 2.6.9, Thm 2.6.10

Cauchy-Riemann Equations §2.6

Cauchy transform §1.5.12, 2.6,
5.2

Chebyshev polynomials §1.4.4

continued fraction §1.6.2

contour Def. 2.6.4

conformal map: A one–one
analytic function
mapping a domain
into another.

curve Def. 2.6.3

Dirichlet kernel §1.4

Dirichlet problem
§2.5.2, §4.1.3–§4.1.4

Dunford integral §1.5

FFT — Fast Fourier
transform §1.2,
Introduction

Fourier series §1.2, §1.4.3

Fourier transform §1.2

Fredholm alternative Example
2.2.6
Problem 2.2.2, §2.6

Green's function §2.1.4, §2.2–§2.3

Hadamard product §2.4, §2.5.

harmonic §1.5.13, §2.5

heat equation §2.2.1, §4.4.4

Hilbert transform §1.5.12, §2.5.2,
§3.8

hyperbolic PDE §2.22, §4.2

indefinite integral §1.5.8, §3.5

integral equation §1.5.15, §2.5

interpolation via Sinc §1.2,
§1.5.4, §1.4

Laplace's PDE §2.1.4, §2.5.2,
§4.1, §2.5.4

Laplace transform of
Green's functions §2.3

Laplace transform inversion
§1.5.10, §3.7

Laurent expansion §1.4.4

461

Midordinate rule §1.4.4

Nonlinear convolutions
        §1.5.11
ODE via Sinc §1.5.14, §2.5.3,
        §3.9
PDE Chapters 2 & 4

Poisson Summation Formula
        §1.3.
Poisson's equation §2.1.4, §2.3.1,
        §2.4, §2.5.2,
        §4,1,2–§4.1.5
principal value integral §1.5.12

rational functions §1.6
Sinc methods: Chapters 1 & 2
Sinc approximation §1.3,
        §1.5

Sinc approximation of
        derivatives §1.2, §1.5.5
Sinc basis §1.1.2
Sinc collocation, §1.5.6

Sinc indefinite convolution
        §1.5.9
Sinc indefinite integration
        §1.5.8
Sinc interpolation §1.2, §1.3,
        §1.4.1, §1.4.9, §1.5.1,
        §1.5.4, §1.7.1, §1.7.4,
        §3.2, §4.4.1, §5.2.4

Sinc Laplace transform
        inversion §1.5.10, §3.7
Sinc points Preface, §1.2, §1.5.1,
        §1.5.4, §3.2.1
Sinc quadrature §1.2, §1.3,
        §1.5.7, §3.4
Sinc spaces §1.5.1, §1.5.2
Sinc transformations §1.5.3
Stirling's formula §1.5

Thiele extrapolation §1.6.2
trapezoidal §1.1, §2.
Trapezoidal rule §1.3, §1.4.3

Sinc weights Ex 1.1.1, §3.1,
        §3.6, §5.2.2

Sinc–wavelets §1.4
Thiele interpolation §1.6.2
wave equation §2.2.2, §4.2.1, §4.4.5

Wiener space §1.2

SYMBOLS

$\mathfrak{C}$ §1.2
$\mathbf{C}^1$ §2.2.1
$C(f, h)$ §1.2

$D(u)$ §1.5.4
$\delta_{i,k}$ §1.2
$\delta_k^{(m)}$ §1.2

Γ §1.5.2, §1.5.4
$[x]$ greatest integer $\leq x$
$\Im a$ imaginary part of a

$\mathcal{J}$ §1.5.8
$\mathbf{Lip}_\alpha$ §1.5.2
$\mathbf{L}^2(a, b)$ §1.2

LF §1.5.1, §1.5.2
$\mathbf{L}_{\alpha,d}(\mathrm{id})$ §1.3
$\mathbf{L}_{\alpha,d}(\varphi)$ §1.5.2

$\mathbf{M}$, $\mathbf{M}'$ Def. 2.6.1
$\mathbf{M}_{\alpha,d}(\mathrm{id})$ §1.3
$\mathbf{M}_{\alpha,d}(\varphi)$ §1.5.2

$\| \cdot \|$ §1.3
ω_j §1.5.4
φ §1.2, §1.5

ψ §1.5.2
PV §1.2
$\mathbb{R}$ Preface, §1.2

$\rho(z)$ §1.5.2
$\Re a$ = real part of a
$\mathrm{sinc}(x)$ *Preface*, §1.1

$S(k, h)(x)$ §1.1
$\mathcal{S}$ §1.2, §1.6.3
$s_k(x)$ §1.5.12

σ_k §1.2
$t_k(x)$ §1.5.12
$T(k, h)(x)$ S1.2

ε_N §1.3
$V(u)$ §1.5.4
$\mathbf{w}$ §1.5.4

$\mathbf{W}(\cdot)$ §1.2,
z_k §1.5